DATE DUE

12-14-9	Ii: 2648698	
1/31/08 ill	3776768	
JUN 0 3 2011		

Demco, Inc. 38-293

Indoor
Air
Research
Series

Max
Eisenberg
Series
Editor

Center
for
Indoor
Air
Research

CIAR

Second Edition

The Chemistry of Environmental Tobacco Smoke:
Composition and Measurement

R. A. Jenkins

M. R. Guerin

B. A. Tomkins

Lewis Publishers
Boca Raton   London   New York   Washington, D.C.

Indoor
Air
Research
Series

Max
Eisenberg
Series
Editor

Center
for
Indoor
Air
Research

Titles in the Series

**The Chemistry of Environmental Tobacco Smoke:
Composition and Measurement**
M.R. Guerin, R.A. Jenkins, and B.A. Tomkins

Bioaerosols
Edited by *Harriet A. Burge*

Risk Assessment and Indoor Air Quality
Edited by *Elizabeth L. Anderson and Roy E. Albert*

**Asthma: Causes and Mechanisms of an Epidemic
Inflammatory Disease**
Edited by *Thomas A.E. Platts-Mills*

**The Chemistry of Environmental Tobacco Smoke:
Composition and Measurement, Second Edition**
R.A. Jenkins, M.R. Guerin, and B.A. Tomkins

Series Preface

The field of indoor air science is of growing interest and concern given that modern society spends the better part of each day indoors. Since the indoor air environment is a major, continual exposure medium for occupants, it is important to study what is present and if and how it affects the health and comfort of occupants.

Volumes in this Indoor Air Research Series are intended to provide state-of-the-art information on many areas germane to indoor air science including chemical and biological sources, exposure assessment, dosimetry, engineering controls, and perception of indoor air quality. In each volume, authors known for their expertise on the topic will present comprehensive and critical accounts of our current understanding in the area.

It is hoped that the series will advance knowledge and broaden interest among the scientific community at large in the indoor air science field.

Max Eisenberg, Ph.D.
Series Editor

Library of Congress Cataloging-in-Publication Data

Guerin, M. R. (Michael Richard), 1941–
   Chemistry of environmental tobacco smoke : composition and measurement / Michael
R. Guerin, Roger A. Jenkins, Bruce Tomkins.—2nd ed.
      p.   cm. — (Indoor air research series)
   Includes bibliographical references and index.
   ISBN 1-56670-509-6 (alk. paper)
      1. Tobacco smoke pollution. 2. Tobacco smoke—Composition. I. Jenkins, R. A. (Roger
   A.) II. Tomkins, B. A. (Bruce A.) III. Series.
RA577.5 .G84 2000
615.9′02—dc21   00-021747
                                                                              CIP

© 2000 by CRC Press LLC
Lewis Publishers is an imprint of CRC Press LLC

No claim to original U.S. Government works
International Standard Book Number 1-56670-509-6
Library of Congress Card Number 00-021747
Printed in the United States of America 1 2 3 4 5 6 7 8 9 0
Printed on acid-free paper

Preface

In the eight years since the authors completed the first edition of this monograph, there have been several important advances in the state of knowledge regarding environmental tobacco smoke (ETS). These advances include (a) the use of multiple markers of ETS to describe indoor air concentrations more accurately; (b) a more complete understanding by the scientific community that most components of indoor air, except those that are specific to tobacco, exist in both the presence and absence of smoking, and that it is not a trivial matter to sort out the contribution from ETS; and (c) a shift in the research from studies in which area sampling is used to measure a few selected indoor environments to larger studies in which personal ETS exposure of randomly chosen subjects is measured.

These changes are largely reflected in the degree to which individual chapters in this second edition have been updated. The emphasis of the overall monograph continues to be on actual concentrations encountered under field conditions because that is where humans are exposed to ETS. While all of the chapters have been updated to reflect the current state of scientific knowledge, Chapters 4, 6, and 7 have undergone the greatest degree of change. With the exception of Appendix 7, for which there was an absence of new relevant studies, all of the appendices have been revised and updated. This is reflected in their increased length. However, we recommend that the original manuscripts cited be studied, as it is impossible to capture the wealth of information reported in a few tables or paragraphs.

It is the authors' hope that the updating will extend the useful life of this monograph and will provide the reader a sense of both the breadth of human exposures to ETS and the magnitude of "typical" exposures.

Roger Jenkins
Mike Guerin
Bruce Tomkins

Oak Ridge, Tennessee
August 1999

Credit

The preparation of both editions of this monograph was sponsored by the Center for Indoor Air Research, Linthicum, Maryland, under contract no. ERD-88-812 with Oak Ridge National Laboratory, managed by Lockheed Martin Energy Research Corp., Oak Ridge, Tennessee, for the U.S. Department of Energy under contract no. DE-AC05-84OR9622464.

Acknowledgment

This is to especially acknowledge the contribution of Patricia M. Trentham to both editions of this volume. Pat saw to it that the authors never had "input delay" as an excuse for slowed progress. She was also largely responsible for finding, cataloging, and re-finding the many references reviewed in this work. During the preparation of the second edition, in addition to all of her other duties, she patiently scanned all of the original text into electronic format, the authors having inadvertently discarded most of the original disks.

Author Information

Roger A. Jenkins is a group leader in the Chemical and Analytical Sciences Division of Oak Ridge National Laboratory (ORNL). He received a baccalaureate degree from Michigan State University in 1969 and a Ph.D. from the University of Wisconsin–Madison in 1975. He has worked in the area of tobacco smoke chemistry for more than 23 years, has published a number of papers and proceedings related to the development of analytical methods and instrumentation for the characterization of mainstream and environmental tobacco smoke, and has acted as principal investigator for several large studies of personal exposure to environmental tobacco smoke. His research interests include the development of analytical methods for the physical and chemical characterization of complex organic mixtures and aerosols related to inhalation toxicology and environmental matrices. He has acted as a consultant and reviewer to the National Cancer Institute, the National Institute on Drug Abuse, the Federal Trade Commission, and Health Canada in the areas of tobacco smoke chemistry and the aerosol inhalation exposure of experimental animals.

Michael R. Guerin is the associate director of the Chemical and Analytical Sciences Division of ORNL. He received a baccalaureate degree from Northern Illinois University in 1963 and a Ph.D. in analytical chemistry from Iowa State University in 1967. Dr. Guerin was appointed principal investigator of a project designed to provide analytical chemical support to the National Cancer Institute Less Hazardous Cigarette Program in 1968. Subsequent work included the development of methods for the determination of trace constituents of cigarette smoke, the development of inhalation bioassay instrumentation for studies of cigarette smoke, and the development of methods for the characterization of environmental tobacco smoke. Dr. Guerin has served as a member of the National Cancer Institute Tobacco Working Group, as a consultant to the Office on Smoking and Health and the Federal Trade Commission, and as principal investigator of projects dealing with cigarette smoke for federal agencies and for tobacco industry research consortia. He is frequently called upon to review proposals, journal articles, and those portions of reports by the Surgeon General of the United States dealing with cigarette smoke chemistry. Current research emphasizes the development of analytical technologies for the determination of trace organic chemicals in the environment and in physiological media.

Bruce A. Tomkins is a research staff member in the Analytical Chemistry Division of ORNL. He received a baccalaureate degree in 1973 from the University of Connecticut (Storrs), and his M.S. (1975) and Ph.D. (1978) degrees from the University of Illinois at Urbana-Champaign. He has worked in the area of trace and ultratrace determination of organic constituents for more than 20 years and has published a number of papers and proceedings related to the development of analytical methods for these compounds in complex mixtures and environmental matrices. His research interests include the development of gas and liquid chromatographic methods for the chemical characterization of tobacco smoke, fossil fuels, drinking water and groundwater, airborne particulate matter, and both volatile and semivolatile organic constituents of liquid radioactive process waste.

Contents

1

Introduction

Environmental tobacco smoke (ETS) is the term used to describe the material in indoor air that results from tobacco smoking. Pipe, cigar, and cigarette smoke all contribute to ETS, but cigarette smoke is of principal interest because it is by far the most common form of tobacco smoking.

ETS from cigarettes is produced primarily by the release of smoke from the burning tip of the product between puffs (i.e., sidestream smoke) and the smoke exhaled by the smoker (i.e., exhaled mainstream smoke). A small additional contribution issues from the tip of the cigarette and through the cigarette paper during a puff and through the paper and from the mouth-end of the cigarette between puffs. The smoke is rapidly diluted and dispersed in the indoor environment with attendant changes in its physicochemical properties and decreases in its concentrations of individual constituents. The principal physical change is a decrease in the proportion of smoke constituents found in the particulate phase as opposed to the vapor phase of the smoke. Mean ETS particle size is subsequently somewhat smaller than that of mainstream or sidestream smoke. The principal chemical change is the change in composition (the relative quantities of the individual constituents present) brought about by differences in ways individual constituents respond to ventilation and to contact with indoor surfaces. There is some indication that chemical transformation of reactive species also occurs but studies with regard to the nature and magnitude of these changes are in their infancy.

There continues to be considerable ambiguity as to the definition of "environmental tobacco smoke." ETS has been treated as if it were a single entity and compared with materials such as asbestos and radon-222. ETS has also been treated as a collection of independent individual constituents with the presence of those constituents being ascribed to ETS because they are known constituents of tobacco smoke or because tobacco smoke is observed in the area of study. In addition, ETS has been treated as if its properties (chemical, physical, and biological) were identical with those of cigarette sidestream smoke. This is clearly incorrect. The current and seemingly most promising approach is to address individual constituents of concern and to determine the degree to which tobacco smoking contributes to their presence. The possibly more rigorous approach of treating ETS as a complex mixture with its attendant composite properties is hampered by a poor understanding of the toxicology of complex mixtures and by the fact that the composition of ETS changes with time and environmental conditions.

Human exposure to ETS is variously termed "passive smoking," "involuntary smoking," or exposure to "second-hand smoke." Exposure to ETS is probably of popular concern in part because it is popularly equated with active smoking and in part because it is readily detected by the human senses. Exposure to ETS is of scientific interest because findings of several epidemiological and risk assessment studies indicate that exposure to ETS contributes to increased risks of respiratory and cardiopulmonary disease and lung cancer. Interest is heightened by knowledge that ETS contains chemicals identified as carcinogenic or otherwise hazardous by various federal regulatory agencies and scientific review groups. A further concern is that populations exposed to ETS (or any indoor air contaminant) include some who may be particularly sensitive and thus at higher risk to exposure than the general population. The ill, aged, allergic, and very young might constitute sensitive populations.

Controlled experimental studies involving the exposure of human subjects to ETS clearly demonstrate that exposure results in some degree of uptake. Experimentally exposed non-smokers exhibit elevated levels of carboxyhemoglobin in their blood and nicotine (and its metabolite, cotinine) in their blood and urine. Nicotine and cotinine are also detected at elevated levels in the saliva, blood, and urine of non-smokers exposed to sufficient quantities of ETS under natural conditions. There are also data available suggesting that increased levels of other species (e.g., thiocyanate in blood, tobacco-specific nitrosamine and 4-aminobiphenyl adducts with hemoglobin, hydrocarbons in breath) are detectable in individuals exposed to ETS. The relationship between the uptake of any one constituent of ETS to the uptake of other constituents remains a subject of research. Competing sources (e.g., diet) of many of the constituents of ETS are also a subject of research.

The degree to which ETS exposure represents a health hazard remains a point of contention. This is partly because, except for a few tobacco-specific species, it is difficult to quantify the excess exposure due solely to tobacco smoking in environments most frequently encountered by the general population. Commonly encountered environments contain ETS constituents at low concentrations and other sources of many of these constituents contribute to their concentrations. The assessment of public health impact is further complicated by the uncertain relative importance of a chronic exposure to slightly elevated concentrations of ETS constituents versus periodic exposure to clearly elevated concentrations of those constituents.

The scientific and technical literature presents a wide range of concentrations at which individual ETS constituents may occur in indoor air. However, these reports also point to the fact that the extreme (high-concentration end of observed ranges) concentrations rarely occur. The contribution of ETS to the concentration of indoor air contaminants in commonly encountered environments is much less than is implied by the extreme values included in many tabulations of ranges observed.

Important progress has been made toward determining the quantities of ETS constituents and establishing the contribution of ETS to common indoor air contaminants in environments associated with chronic exposure. The development of reliable, highly sensitive methods for determining nicotine in indoor air has resulted in extensive surveys of nicotine concentrations in a wide variety of environments. Expanding the list of other, perhaps more reliable, markers of ETS seems to have spurred numerous large ETS personal exposure studies. Excess indoor air concentrations of carbon monoxide, benzene, and several alkylbenzenes due to ETS have been determined for a few environments.

The kinds and magnitudes of health effects from passive smoking are likely to be the subjects of research for some time. Such research, whether epidemiological, biochemical, or risk assessment modeling, is highly dependent on measures of exposure. Until technology is available to determine and confidently interpret measures of actual human dose, indoor air concentrations are likely to remain the primary measure of exposure. The purpose of this volume is to compile and describe observations to date on the properties of ETS and on concentrations of its constituents in indoor air. Common natural indoor environments and those associated with chronic exposure are focused on, as well as measurement methods and competing sources of the constituents. The chemistry of mainstream and sidestream smoke is briefly treated for reference.

The reader is referred to the 1986 National Research Council (NRC 1986) and Surgeon General (Surgeon General 1986) Reports for descriptions of the general issue of ETS exposure. Weiss (1989) provides a perspective on the carcinogenicity of ETS. O'Neill et al. (1987) and Gold et al. (1990) provide representative descriptions of measurement methods and indoor air concentrations of ETS-related constituents.

REFERENCES

Gold, K. W., Naugle, D. F. & Berry, M. A. (1990) *Indoor Concentrations of Environmental Carcinogens.* RTI Report No. 4479/07-F, Research Triangle Institute, P.O. Box 12194, Research Triangle Park, NC 27709, 40 pp.

(NRC) National Research Council (1986) *Environmental Tobacco Smoke. Measuring Exposures and Assessing Health Effects.* National Academy Press, 2101 Constitution Avenue, NW, Washington, D.C. 20418, 337 pp.

O'Neill, I. K., Brunnemann, K. D., Dodet, B. & Hoffmann, D. (1987) *Environmental Carcinogens. Methods of Analysis and Exposure Assessment. Vol. 9- Passive Smoking (IARC Scientific Publication No. 81)* International Agency for Research on Cancer, Lyon France, 372 pp.

Surgeon General (1986) *The Health Consequences of Involuntary Smoking. A Report of the Surgeon General.* DHHS, PHS, CDC, Office on Smoking and Health, Rockville, MD 20857, 359 pp.

Weiss, W. (1989) *Selected Abstracts on Environmental Tobacco Smoke and Cancer. Oncology Overview.* DHHS, PHS, NCI U.S. Government Printing Office, Washington, D.C. 20404, 37 pp.

INTRODUCTION

Indoor air contains a great variety of physical, biological, and chemical contaminants. They arise from multiple sources, differ greatly in quantity, and usually vary both temporally and spatially in concentration. Many of the chemical constituents from these multiple sources are the same as those in tobacco smoke and thus complicate experimental designs and data interpretation for studies intended to address ETS. Knowledge of confounding sources is especially important for those studies that seek to address general population exposure in natural environments associated with chronic (e.g., residential) exposure to ETS. It is important to be aware of common background concentrations and their sources. The term "background" is used here to mean "ETS-free."

Identifying ETS-free concentrations of contaminants from published studies is more difficult than might be expected. Most indoor air studies not addressing ETS as a variable do not specify whether environments are ETS-free. Other studies (Akland et al. 1985) specify the use of only a nonsmoking population and principally address a non-ETS issue but include ETS exposure as a side issue. Turk et al. (1987) define a nonsmoking condition as one where smoking is not observed within 30 ft of the sampling point. In a study (Klus et al. 1987) of background levels of constituents in office air, the investigators report a mean nicotine concentration of 5.7 $\mu g/m^3$; a concentration well within the range (Chap. 7) of known ETS-containing environments. It is likely that the investigators chose the absence of observed smoking in the sampled environment during sampling as being a background condition. Differing definitions of "background" or lack of attention to ETS as a factor in any given study can provide over-estimates or under-estimates of the importance of ETS exposure. More recent studies of personal exposure to ETS (Heavner et al. 1996, Jenkins and Counts 1999, Chaps. 6 and 7) include conditions of nonsmoking workplaces and nonsmoking homes but provide for normal activities that may include unintentional exposure to ETS. A special effort has been made to select data from environments that appear to be truly ETS-free (unless otherwise noted) for inclusion in the tables accompanying this chapter.

Tobacco smoke contains several thousand chemical constituents (Chap. 3), and ETS can thus contribute to the indoor air concentrations of these constituents. ETS can also be a major source of respirable suspended particulate matter (Chap. 6). Other important indoor air contaminants, in particular, bacteria, molds, fungi, asbestos, and pesticides are clearly not ETS-related. Some important constituents (halogenated hydrocarbons and radon-222) are essentially unrelated to ETS because of their massive contribution from other sources and trace (if any) presence in tobacco smoke. Constituents such as carbon monoxide, nitrogen oxides, formaldehyde, and petroleum hydrocarbons can be measurably contributed to indoor air by ETS but are frequently the principal result of other sources.

Few ETS-unique constituents have been detected, making it difficult to identify constituents that can be used as a measure of ETS-specific contamination (Chap. 4). Nicotine and its related alkaloids are unique to ETS but they appear to behave differently in the ambient environment than particulate matter or constituents of concern. Tobacco leaf constituents such as solanesol and very high molecular weight alkanes (e.g., hentriacontane, n-$C_{31}H_{64}$; dotriacontane, n-$C_{32}H_{66}$) may be ETS-specific and serve as useful ETS-particulate matter markers, but their relationship to the expected hazardous constituents of ETS has not been clearly established as yet. Pyrolytic nitrogenous breakdown products of nicotine and tobacco proteins such as pyridines, pyrrolidines, and nitriles might be ETS-specific and serve as markers of ETS-constituents. Currently favored markers for ETS exposure are solanesol and scopoletin for a measure of particulate matter, 3-ethenyl pyridine for volatile organics, and nicotine (Daisey 1999, Rickert 1995, Chaps. 6 and 7). Nicotine and 3-ethenyl pyridine also serve as markers for the exposure to ETS as a whole. Iso-alkanes and ante-iso-alkanes from tobacco leaves are being used (Kavouras et al. 1998, Rogge et al. 1994) as markers of particulate matter in outdoor environments. Background (non-ETS) sources of common, high-molecular-weight tobacco leaf constituents and of breakdown products of tobacco leaf nitrogenous constituents have generally not yet been sought in indoor air.

Background sources and characteristics are discussed here and in several chapters dealing with individual contaminants. Background data are also included in the appendices. The reader is encouraged to study the tables accompanying this chapter. They have been structured to be stand-alone and are generally referred to only peripherally in the narrative.

GENERAL CONSIDERATIONS

Indoor air contaminants of current concern can be divided into two general categories: (a) those most commonly associated with occupant discomfort or acute illness and (b) those thought to be carcinogenic or to contribute to other chronic illnesses. Microorganisms, carbon dioxide, formaldehyde, petroleum-derived solvents, halogenated solvents, and a variety of other VOCs are most frequently the cause of discomfort and acute illness ("Sick Building

Syndrome"). Microorganisms, volatile organic compounds, airborne fibers, and respirable suspended particulate matter are also likely contributors to allergic reactions and other forms of chronic poor health. Tobacco smoking contributes to the indoor air burden of some of the suspect chemicals but ETS is seldom found to be the cause of sick building complaints. This is because other sources and factors usually unique to the environment or situation are required to produce acute effects. An obvious exception is that ETS can clearly be the cause of illness for those who are allergic to tobacco smoke.

Materials viewed as posing a cancer threat in indoor air are (Gold et al. 1990) asbestos, radon-222, ETS, a large variety of organic chemicals, and several heavy metals. Chemicals of particular concern include formaldehyde, benzene, methylene chloride, selected polycyclic aromatic hydrocarbons, selected pesticides, N-nitrosamines, cadmium, and nickel. Tobacco smoking is not a source of asbestos and is either not a source or is an insignificant source of radon-222, halocarbons, and pesticides. Tobacco smoking can be a significant (measurable contribution to the total quantity) source of most other chemical carcinogens of current common concern in indoor air.

The common principal sources (Table 2.1) of indoor air contaminants are outdoor air, the building itself, and occupant activities. Incursion of outdoor air includes that entering the indoor environment through the ventilation system, windows, doors, and the many imperfectly sealed points typical of common construction. Residential attached garages and commercial underground parking garages are important sources of indoor air contaminants and might be considered a special case of "outdoor" air inclusion. Another special case is those materials that enter the indoor environment on soil tracked indoors (Kliest et al. 1989).

Considering the building itself as a source includes contributions from the construction materials (Table 2.2), furnishings and decorations, heating and cooling systems, water heaters, refrigerator/freezers and humidification devices. Special combinations of circumstances can make the building an especially important source. Perhaps the best example is the combination of generally minimal ventilation and heavy use of particle board and urea-foam insulation in the past in mobile homes. This leads to uncommonly high concentrations of formaldehyde.

Occupant activity (Clobes and Anath 1992) is the third general source of indoor air contaminants. While occupants themselves contribute to indoor air constituents (e.g., carbon dioxide, other exhaled breath constituents, and dander), it is their activities that are the principal contributors because most activities involve the use of consumer products (Tables 2.2, 2.3). The most obvious example is tobacco smoking, which leads to visibly obvious ETS. Another example is painting or wood staining where emissions are obviously detected by their odor. Maintenance and repair activities (e.g., painting,

Table 2.1 Indoor Air Constituents of Common Concern and Their Major Sources

| Constituents | Outdoor Incursion | | Combustion Appliances[a] | Building and Furnishings | | | Occupant Activities | | |
	Outdoor Air	Attached Garage		AC/Humidity Appliances[b]	Construction Materials[c]	Furnishings[a]	Maintenance & Repair[d]	Personal Care[e]	Tobacco Smoking
Substances									
Microorganisms[f]	✓			✓					
Asbestos	✓				✓		✓		
Radioactivity/radon	✓				✓				
Environmental tobacco smoke	✓								✓
Respirable particulate matter	✓	✓	✓	✓			✓	✓	✓
Volatile organic compounds	✓		✓	✓	✓	✓	✓	✓	✓
Specific Chemical(s)									
Carbon monoxide	✓	✓	✓						✓
Nitrogen oxides	✓		✓						✓
Benzenes	✓				✓	✓	✓	✓	✓
Chlorinated hydrocarbons				✓	✓	✓	✓	✓	
Formaldehyde		✓	✓		✓	✓	✓		
Pesticides	✓	✓	✓						
Polycyclic aromatic hydrocarbons	✓								✓
Heavy metals	✓		✓						✓

[a] Particle board, urea-formaldehyde foam insulation, and textiles are major formaldehyde sources.
[b] Vacuuming, cleaning, polishing, painting, remodeling, etc.
[c] Air fresheners, laundry, shower, aerosols, deodorants, etc.
[d] Bacteria, fungi, molds.
[e] Air conditioning and humidification appliances.
[f] Central and space heating, cooking, water heaters (also see Table 2.6).

Table 2.2 Example Emissions of Volatile Organic Chemicals

Material	Major Constituents Reported
Silicone caulk[a]	Methylethylketone, butylpropionate, 2-butoxyethanol, butanol, benzene, toluene
Water based floor adhesive[a]	Nonane, decane, undecane, dimethyloctane, 2-methylnonane, dimethylbenzene
Particle board[a]	Formaldehyde, acetone, hexanal, propanol, butanone, benzaldehyde, benzene
Floor wax[a]	Nonane, decane, undecane, dimethyloctane, trimethylcyclohexane, ethylmethylbenzene
Wood stain[a]	Nonane, decane, undecane, methyloctane, dimethylnonane, trimethylbenzene
Latex paint[a]	2-Propanol, butanone, ethylbenzene, propylbenzene, 1,1'-oxybisbutane, butylpropionate, toluene
Furniture polish[a]	Trimethylpentane, dimethylhexane, trimethylhexane, trimethylheptane, ethylbenzene, limonene
Polyurethane floor finish[a]	Nonane, decane, undecane, butanone, ethylbenzene, dimethylbenzene
Room freshener[a]	Nonane, decane, undecane, ethylheptane, limonene, substituted aromatics
Moth crystals[a]	p-Dichlorobenzene
Carpet[b]	Benzene, 4-phenylcyclohexene, ethanol, acetone, ethylacetate, ethylbenzene, tetrachloroethylene
Drapery lining[b]	Acetone, benzene, chloroform, ethylacetate, 3-hexanone, methylene chloride

[a]Tichenor and Mason, 1988.
[b]Bayer and Papanicolopoulos, 1990.

Table 2.3 Representative Sources of Indoor
Volatile Organic Compounds (VOCs)

Building Materials	Consumer Products	Human Activity
Particle board	Furniture polish	Tobacco smoking
Adhesives	Floor polish	Painting
Caulks	Floor wax	Cleaning
Paints	Room fresheners	Arts and crafts
Stains	Moth crystals	Showering
Floor finishes	Paints	Cooking
Vinyl tiles	Stains	Photocopying
	Cleaning agents	
	Disinfectants	
	Deodorants	
	Hairspray	
	Furniture	
	Shoe polish	
	Pesticides	
	Gasoline	
Building Systems	**Textiles**	**Miscellaneous**
Space heaters	Drapery	Outdoor air
Central heating	Carpet	Attached garage
Room humidifiers	Furniture	Dry cleaning
Fireplace	Bedding	
Water heater		

cleaning, vacuuming, and remodeling) and daily activities (e.g., cooking) all contribute various constituents in varying quantities to indoor air.

In studies of various constituents and sources, numerous unexpected (or at least not intuitively obvious) associations have been identified. Examples are the release of chloroform by showering (Pellizzari et al. 1989), of respirable suspended particulate matter by ultrasonic room humidifiers operating on tap water (Highsmith et al. 1988), and of 2-ethylhexanol from the hydrolysis of phthalate plasticizer in carpet backing (McLaughlin and Aigner 1990).

Many factors affect the presence and persistence of indoor air contaminants. Factors important to all constituents include the presence of sources, their emission rates, ventilation conditions and air exchange rate, the presence of "sinks" (materials that absorb the constituent), and indoor temperature and humidity. Constituents such as formaldehyde and solvents, which are due principally to off-gassing from materials, are typically found at the highest concentrations in newly completed buildings or newly furnished or remodeled areas (Table 2.4; Fortmann and Roache 1998; Weislander and Norback 1997; Dietert and Hodge 1996). Concentrations due to such sources decrease with time but those due to some constituents require a year or more to reach true background levels (Schriever and Marutzky 1990, Baldwin and Farant 1990). Formaldehyde emissions, for example, occur continuously over

Table 2.4 Spatial and Temporal Variation of Indoor VOC Concentrations in an Office Building[a]

	First Floor Interior Office	First Floor Secretarial	Second Floor Office	Second Floor Secretarial	Third Floor Copier
Benzene					
July 1983	4.9	5.4	5.1	4.6	3.2
Sept 1983	3.1	7.7	4.1	4.5	6.0
Dec 1983	6.3	3.6	2.0	8.0	13.0
Styrene					
July 1983	0.4	7.7	7.1	10.5	6.7
Sept 1983	4.0	9.7	5.9	7.4	8.3
Dec 1983	3.6	3.6	5.8	3.7	2.4
Ethylbenzene					
July 1983	89	83	73	88	90
Sept 1983	4.4	6.3	4.8	7.2	8.8
Dec 1983	4.8	4.9	4.0	4.2	5.9
m+p-Xylene					
July 1983	160	130	110	100	140
Sept 1983	16	20	16	19	23
Dec 1983	9.7	11	7.4	8.6	8.6
n-Decane					
July 1983	NR	430	320	400	380
Sept 1983	39.7	50	39	25	37
Dec 1983	4.8	7.1	4.7	2.4	2.4
1,1,1-Trichloroethane					
July 1983	398	323	418	409	346
Sept 1983	75	139	82	99	106
Dec 1983	45	55	48	67	30

Average Concentration in $\mu g/m^3$

[a]Sheldon et al. 1988a,b. Building constructed/occupied July, 1983.
NR=not reported.

a long period while petroleum solvents (benzene and its derivatives) and chlorocarbon solvents are more quickly depleted. Constituents such as particulate matter and carbon monoxide, which are largely due to combustion processes, are especially influenced by their concentrations in outdoor air and by the presence of an active indoor source.

Temperature and humidity affect different contaminants in different ways. Elevated temperatures increase the emission rates of volatile organic compounds from building materials and furnishings. This is now used to advantage (Godish 1989) by performing a building "bake-out" prior to occupancy. Humidity especially affects the growth of microorganisms (Reponen et al. 1994, Ahearn and Crow 1996) and associated volatiles emissions and may have an effect on particulate matter aggregation and composition. The influence on particulates is in part because surface charge is affected (Samnaljarvi et al. 1990) by the moisture content of the air.

The rate of air exchange with outdoor air is also an important determinant of presence and persistence but its effect is principally important at low rates of exchange. VOC concentrations, for example, are closely related (Molhave 1984) to exchange rates below approximately 1 air change per hour. At higher rates, the presence of sources and their emission rates are more important. For constituents whose primary source is outdoor air, high air exchange rates can lead to higher-than-usual indoor air concentrations. Material (e.g., particulate matter) reduced by indoor sinks is replenished by infiltrating outdoor air.

An important characteristic of indoor air contaminants is that their concentrations vary both spatially and temporally to a greater extent than is common outdoors. This is due to the large variety of sources present, the intermittent operation of some of those sources, and the different types and quantities of "sinks" present at various locations. One might intuitively conclude that the compartmental nature of indoor environments would have a major influence on the spatial distribution of contaminants but its effect is largely limited to providing short-term, localized elevated concentrations in the immediate vicinity of a source. Mechanical ventilation and natural diffusion tend to distribute contaminants from a given source throughout the larger indoor environment. Most residences and other buildings contain sources of common contaminants throughout, thus further leveling spatial distributions.

Concentrations of contaminants such as carbon monoxide, nitrogen oxides, and particulate matter that arise principally from combustion sources are subject to very large temporal variation. As an example, the concentration of nitrogen oxides in an unoccupied research home has been found to rise from a baseline level of 20–30 ppb to a peak of approximately 140 ppb in the morning and a second peak of 180 ppb in the early evening (Leslie and Billick 1990). The peaks were achieved and depleted over a period of 2 to 3 h each and were due to the operation of various gas-fired appliances (e.g., range, oven, dryer). As a further example (Akland et al. 1995), personal exposures of a population of Denver residents to carbon monoxide varied from

approximately 0.7 ppm to almost 4 ppm over 1-h periods during a weekday. The hourly outdoor concentrations of carbon monoxide varied from approximately 1 ppm to approximately 5.8 ppm. As a final example, the average particulate matter concentration in the smoking sections of commercial aircraft is reported (Nagda et al. 1990) to increase from a background level of 34.8 μg/m³ to 75.8 μg/m³ during smoking. Nicotine concentrations rose from ≤0.05 μg/m³ to 13.43 μg/m³ in the same study. Peak concentrations can be much greater (Chap. 7). Related results are reported by Dechow and Sohn 1997.

The high temporal variability of combustion product concentrations reflects the intermittent nature of their source(s). Heating, cooking, smoking, etc., introduce contaminants into the environment only while they are occurring. Combustion product concentrations vary temporally, however, even in the absence of indoor sources. This is because the incursion of outdoor air dominates the concentrations of indoor air combustion products in the absence of indoor sources. Outdoor air concentrations of combustion products vary throughout the day (vehicle exhaust being an especially dominant variable).

Formaldehyde, benzene, and other volatile organic compounds are also subject to temporal variation in their indoor air concentrations. Because off-gassing from construction materials and furnishings is an important source of such constituents, newly constructed (Table 2.4), remodeled, or furnished environments would be expected to produce more highly contaminated indoor air. The elevated contamination is expected to peak more slowly and exist much longer than is typical of single-event combustion-derived contaminants. Episodic releases of the volatile organic compounds due to human activities such as painting and showering yield temporal variations similar to those from combustion sources. Tobacco smoking releases VOCs and can be a major source of temporal variations in their indoor air concentrations.

Spatial variations in contaminant concentrations within a given micro-environment can also be significant but are generally less dramatic than are temporal variations. Data for selected VOCs at various points in an office building (Table 2.4) and 2 rooms in a residence (Table 2.5) illustrate spatial variability for 12-h sampling periods. Average concentrations at a given time at various locations in the office building generally differ only by a factor of 2 to 3. The mean ratio of living room to bedroom concentrations in the residence studied was close to unity. The ranges (Table 2.5) of living room to bedroom concentrations varied by factors of 3 to greater than 10, again illustrating the importance of temporal variations. Similar observations have been reported for carbon monoxide and nitrogen oxides in a residence (Leslie and Billick 1990).

Spatial variations can be very large if steps are taken to control the environment. Perhaps the best example is the large spatial variation observed aboard passenger aircraft during smoking. In a study by Nagda et al. (1990) of 92 randomly selected flights, the average nicotine concentration in the smoking

Table 2.5 Ratio of Living Room to Bedroom Area
Volatile Organic Compound Concentrations[a]

Compound	No. of Samples	Ratio Arithmetic Mean	Ratio Standard Error	Ratio Range
1,1,1-Trichloroethane	74	1.05	0.03	0.57–1.77
Tetrachloroethylene	74	0.97	0.04	0.32–2.53
p-Dichlorobenzene	28	1.03	0.03	0.61–1.35
Benzene	75	1.24	0.26	0.46–20.5
Toluene	74	1,25	0.25	0.45–19.1
Ethylbenzene	74	1.00	0.03	0.49–2.22
m,p-Xylene	74	1.00	0.03	0.46–2.55
n-Octane	71	0.98	0.03	0.43–1.90
n-Nonane	74	1.24	0.27	0.47–21.1
n-Decane	62	0.97	0.03	0.38–2.18

[a]Pellizzari et al. 1989.

section was found to be 13.4 $\mu g/m^3$, dropping to 0.3 $\mu g/m^3$ at the smoking/nonsmoking boundary and reaching background concentrations of 0.05 $\mu g/m^3$ at the middle of the aircraft. Respirable particulate matter concentrations similarly dropped from 76 $\mu g/m^3$ to 54 $\mu g/m^3$ to a background concentration of 31 $\mu g/m^3$. Other studies (Chap. 7) show less dramatic but clearly measurable spatial isolation of ETS constituents aboard aircraft. This is because the passenger cabin ventilation system is designed to provide maximally individual micro-environments. Separately ventilated rooms or areas in buildings can provide similar isolation and resulting large spatial variations in contaminant concentrations.

The length of the sampling period is an important determinant of the concentration and variability found in any study. Short sampling periods (minutes to a few hours) tend to result in higher mean concentrations and greater variability. An example is the work of Akland et al. (1985) who find consistently higher and more variable exposure to carbon monoxide using 1-h sampling than 8-h sampling. Short sampling periods are more highly influenced by the particular circumstances prevailing at a given point at a given time and thus detect greater variability. Higher mean concentrations often reflect the contributions of one or more short-term high concentration results to the arithmetic mean.

'High arithmetic means and the maximum values of ranges reported in the literature are usually the result of spatial or temporal effects. Either an environment known to be particularly contaminated has been sampled or an episodic release has been detected. Median and geometric means are almost invariably lower than arithmetic means and better reflect concentrations encountered most frequently.

COMBUSTION PRODUCTS

Combustion processes are commonly recognized as the principal source of particulate matter, carbon monoxide, nitrogen oxides, and polycyclic aromatic hydrocarbons in indoor air. Combustion processes also, however, contribute formaldehyde and a wide range of other volatile and semi-volatile organic compounds to indoor air (Traynor et al. 1990, Rickert 1995). Including tobacco smoking as a combustion process further increases the variety of indoor air contaminants due to combustion.

Table 2.6 lists example sources of various combustion products along with a qualitative assessment of their contributions. All combustion appliances (ranges, ovens, water heaters, space heaters, etc.) contribute air contaminants while they operate but several factors are especially important. The most important factors are the nature of the fuel and whether the appliance is vented to the outdoors. The next most important factor (Godish 1989) is the state of maintenance of the appliance and the condition of its flue. Wood and coal combustion typically produces greater quantities of emissions than does kerosene or oil combustion. Kerosene and oil combustion generally produce greater quantities of emissions than gas combustion. Electric heating produces fewer indoor emissions than other combustion sources but the emissions introduced into the outdoor air by power plants are eventually transported indoors to some degree. Electric appliances actually contribute more to the indoor air contaminant burden than might be expected. Electric "convector" space heaters with unprotected coils (Samnaljarvi et al. 1990) and electric range burners (Krafthefer and MacPhaul 1990) produce respirable suspended particulate matter from contact with household dusts, pollen, and dander. High temperature (1800°F for electric ranges) sources produce greater quantities of pyrolytic decomposition products than do low temperature (300°F electric baseboard heaters) sources (Krafthefer and MacPhaul 1990).

Wood-burning fireplaces, stoves, and space heaters are especially important sources of indoor contaminants (Traynor et al. 1987, Perritt et al. 1990). The slow and incomplete combustion associated with wood-burning produces large quantities of particulate matter and a large variety of volatile and semi-volatile organic contaminants. Emissions are especially extreme (Table 2.5) for nonairtight stoves under extreme fire conditions and when fireplaces or stoves are being stoked or are otherwise directly open to the indoor environment. Geographic areas and cultures which rely heavily on wood and other biomass (e.g., oil shale and dung) for heating and cooking typically are exposed to both outdoor and indoor concentrations of particulate matter, polycyclic aromatic hydrocarbons, and carbon monoxide 1 to 3 orders of magnitude greater than those commonly tabulated and discussed here.

Poor maintenance, improper tuning (fuel/air mix), or inadequate ventilation can result in an otherwise minor source of indoor contaminants

Table 2.6 Example Sources of "Combustion Products" in Indoor Air

Sources	Constituents						
	RSP	CO	NO$_x$	SO$_2$	PAH	Form.	VOC
Tobacco smoking	ab	bc	Bc	c	ab	c	ab
Wood-burning stoves	ab	ab	ab	bc	ab	c	ab
Wood-burning fireplaces	ab	ab	ab	bc	ab	c	ab
Unvented kerosene space heaters	ab	ab	ab	ab	a	c	b
Radiant electric space heaters	b	c	c	c	c	c	c
Unvented gas space heaters	c	ab	c	bc	c	c	c
Gas water heaters	c	ab	ab	c	c	c	c
Gas cooking stoves/range	bc	ab	ab	c	c	c	c
Gas cooking stoves/oven	c	ab	ab	c	c	c	c
Coal-fired central heating	ab	ab	ab	ab	ab	c	ab
Oil-fired central heating	c	ab	ab	ab	c	c	ab
Gas-fired central heating	c	ab	ab	c	c	c	c
Outdoor air	a	a	a	a	a	c	a
Attached garage	c	ab	ab	ab	ab	c	ab
Room humidifier	ab	c	c	c	c	c	b
Carpet vacuuming	ab	c	c	c	c	c	b
Particle board	c	c	c	c	c	ab	ab

[a]Commonly important contributor (when operating/occurring for episodic sources).
[b]Can be responsible for short-term or localized high concentrations.
[c]Generally minor or no contribution relative to other common sources.

Respirable Suspended Particulate Matter (RSP), Carbon Monoxide (CO), Nitrogen Oxides (NO$_x$), Sulfur Dioxide (SO$_2$), Formaldehyde (Form.), Volatile Organic Chemicals (VOC's)

becoming an important source. Mal-tuned kerosene space heaters (Traynor et al. 1990), nonairtight wood-burning stoves (Traynor et al. 1987), and ventilated gas space heaters (Leslie and Billick 1990) are examples of readily measurable sources of indoor air contaminants. The importance of even partial ventilation is illustrated by the finding that concentrations of carbon monoxide in a living area are reduced from 10.8 ppm to 4.5 ppm and of nitrogen dioxide from 403 ppb to 220 ppb by using the ventilation hood of a gas-fired range/oven operated in the kitchen (Leslie and Billick 1990).

Emission rates of several combustion appliances are given in Table 2.7, and the impact of several combustion sources on indoor air concentrations may be determined from data given in Tables 2.8, 2.9, and 2.10. Emission rates are highly influenced by the individual model of the appliance tested, the degree to which it is tuned for optimal performance, and the manner in which the emission rate is experimentally determined. These factors, plus the absence of accounting for ventilation and for indoor sinks, make emission rates very poor predictors of resulting indoor air contaminant concentrations. Experimental emission rates are very important, however, for identifying significant sources of indoor contaminants and for identifying those characteristics of appliance design or operation that produce relatively increased or decreased emission. Results indicate, for example, that gas-fired space heaters and ovens/ranges can be a much more significant source of carbon monoxide than is a

Table 2.7 Example Combustion Emission Rates

Type	Milligrams per hour							
	Particulate Matter	Carbon Monoxide	Nitric Oxide	Nitrogen Dioxide	Sulfur Dioxide	Formaldehyde	Volatile Organics	Polycyclic Aromatics
Unvented Kerosene Space Heater								
Well-tuned radiant[a]	0.001	588	4.8	34.3			16.2[b]	0.06[c]
Mal-tuned convective[a]	0.28	138	148	45.5			24.7[b]	0.20[c]
Gas-Fired Space Heater[d]								
	0.38[e]	2659				41		
Gas Range[d]								
Oven	0.13[e]	1898	55.5	85.2	0.9	22.9		
Top burner (2)	9.20[e]	3680	178	272	2.9	31.3		
Wood-burning stoves[f]								
Airtight models	2.5–8.7[g]	11–160						≤ 0.0001[h]
Airtight as fireplace	5.2[g]	92						0.0008[h]
Non-airtight models	16–73	252–618						0.002–0.008[h]
Non-airtight extreme fire	230	2061						0.057[h]

[a]Traynor et al. 1990.
[b]Phthalates, aliphatic hydrocarbons, aliphatic alcohols, aliphatic ketones, and hydronaphthalenes predominate.
[c]Summation of naphthalene, $C_2 + C_3$ naphthalene, phenanthrene, fluoranthrene, anthracene, chrysene, and indenopyrene. Naphthalenes predominate.
[d]Girman et al. 1982.
[e]< 5 μm estimated by electric mobility analyzer.
[f]Traynor et al. 1987.
[g]Total suspended particulate matter.
[h]Benzo(a)pyrene.

Table 2.8 Example Non-ETS Particulate Matter Concentrations

Study	Location	Studied Source[a]	Total Suspended Particulate Mean	Total Suspended Particulate Range	Respirable Suspended Particulate Mean	Respirable Suspended Particulate Range
Outdoor						
Pacific Northwest commercial buildings (Turk et al. 1987)	Portland, OR[b]	—	58	26–113	32	28–44
	Salem, OR[b]	—	28	18–77	—	—
	Spokane, WA[b]	—	130	84–173	13	5–34
	Building sites[c]	—	—	—	19[b]	—
	Building sites[c]	—	—	—	14	—
Residential wood-burning stoves (Traynor et al. 1987)	Truckee, CA	—	16	7–31	—	—
Indoor						
Pacific Northwest commercial buildings (Turk et al. 1987)	Buildings	Background	—	—	19	—
	Buildings	Background	—	—	15[d]	—
Wood-burning stove residential (Traynor et al. 1987)	Residence	w/o stove	21	19–24	6	4–8
		Airtight stove	40	24–71	24	11–36
		Nonairtight stove	—	100–420	—	—
NY State indoor combustion sources (Perritt et al. 1990)	Residences, Onondaga City, NY	Background	—	—	25.7[d]	—
		w/kerosene heater	—	—		27.4[d]
	Residences, Suffolk City, NY	Background	—	—		34.2[d]
		w/kerosene heater	—	—		56.3[d]
Mobile home kerosene heaters (Mumford et al. 1990)	Mobile homes, Apex, NC	Background	—	—		56.1
		w/kerosene heater	—	—		73.7

[a]Emissions source studied. Includes unknown sources.
[b]Regional environmental monitoring data.
[c]Investigators outdoor air data in connection with indoor air study.
[d]Geometric mean. Arithmetic mean if not specified.

Table 2.9 Example Non-ETS Carbon Monoxide and Nitrogen Oxides Concentrations

Study	Location	Studied Source[b]	Carbon Monoxide (ppm) Mean	Carbon Monoxide (ppm) Range	Nitrogen Oxides[a] (ppb) Mean	Nitrogen Oxides[a] (ppb) Range
Outdoor						
Pacific Northwest commercial buildings (Turk et al. 1987)	Portland, OR[c]	—	3.3	2–5	3.8	3.6–4.0
	Salem, OR[c]	—	1.3	1–2	—	—
	Spokane, WA[c]	—	2.9	1–5	—	—
	Building sites	—	—	—	23	6–40
	Building sites	—	—	—	20[d]	6–40[d]
Residential wood-burning stoves (Traynor et al. 1987)	Truckee, CA	—	0.5	ND–1.1	—	—
Unoccupied home/gas appliances (Leslie and Billick, 1990)	Chicago, IL	—	—	1.8–5.8[e]	—	54–97[e]
Pacific Northwest commercial buildings (Turk et al. 1987)	Buildings OR, WA	Background	—	—	20	5–34
		Background	—	—	18	5–34
Residential wood-burning stoves (Traynor et al. 1987)	Residence, Truckee, CA	Wood stove	1.2	0.4–2.8	—	—
Indoor combustion sources (Perritt et al. 1990)	Residences, Onondaga City, NY	Background	1.3	—	11.1	—
		Kerosene heater	2.2	—	26.1	—
		Background	1.4	—	11.8	—
		Wood stove/fireplace	0.9	—	8.4	—
		Background	1.0	—	6.5	—
		Gas stove	1.7	—	21.9	—
		Background[f]	—	—	6.4	—
		Gas stove[f]	—	—	28.0	—

Table 2.9 continued

Study	Location	Studied Source[b]	Carbon Monoxide (ppm) Mean	Carbon Monoxide (ppm) Range	Nitrogen Oxides[a] (ppb) Mean	Nitrogen Oxides[a] (ppb) Range
Indoor combustion sources (Pernitt et al. 1990)	Residences, Suffolk County, NY	Background	1.6	–	11.8	–
		Kerosene heater	3.4	–	32.7	–
		Background	1.7	–	13.6	–
		Wood stove/fireplace	1.7	–	10.4	–
		Background	1.5	–	8.2	–
		Gas stove	2.0	–	25.1	–
		Gas central heating[g]				
		Living room	3.5[e]	–	63[e]	–
		Bedroom	3.5[e]	–	60[e]	–
		Basement	4.1[e]	–	45[e]	–
		Gas space heater[h]				
		Living room	4.2[e]	–	186[e]	–
		Bedroom	4.1[e]	–	187[e]	–
		Basement	3.8[e]	–	26[e]	–
		Gas range and oven[i]				
		Living room	4.5[e]	–	220[e]	–
		Bedroom	4.4[e]	–	191[e]	–
		Basement	1.3[e]	–	24[e]	–

[a]NO$_2$ unless otherwise specified.
[b]Known emission source.
[c]Regional environmental monitoring data. Other outdoor data acquired by reference investigators during IAQ study.
[d]Geometric mean. Arithmetic mean if not specified.
[e]Measured in kitchen rather than living areas.
[f]Maximum (peak) concentrations.
[g]Unoccupied one-level home with basement.
[h]Gas furnace in basement.
[i]Unvented gas-fired fan-forced space heater in living room.
[j]With ventilation hood operating.
ND=not detected.

Table 2.10 Example Non-ETS Indoor Air Concentrations of Polycyclic Aromatic Hydrocarbons

	Concentration in ng/m³								
	Pyrene	Bz Anth	Chrysene	Bz(b)Fl	Bz(k)Fl	Bz(a)Py	Dibzanth	Bz Peryl	Inden Pyr
Salem, Oregon (Turk et al. 1987)									
Buildings									
Urban multiuser (8)[a]	—	—	0.13	0.34	0.17	0.13	0.19	1.03	0.55
Urban office (9)[a]	—	—	0.09	0.24	0.06	0.09	0.11	0.73	0.35
Urban office (10)[a]	—	—	0.5	1.2	0.49	0.56	0.86	1.41	0.41
Urban office (15)[a]	—	—	0.12	0.26	0.11	0.13	0.21	0.62	0.29
Truckee, California (Traynor et al. 1987)									
Buildings									
Indoor background	—	—	—	0.06	0.04	0.10	—	0.25	0.27
Airtight stove A	—	—	—	0.56	0.31	0.88	—	1.1	1.7
Airtight stove C	—	—	—	0.17	0.07	0.34	—	0.37	1.1
Nonairtight stove D[b]	—	—	—	11	5.4	13	—	14	20
Stove A as fireplace	—	—	—	3.8	1.9	3.5	—	3.7	6.1
Stove D, extreme fire[b]	—	—	—	320	150	370	—	340	560
Stove D, extreme fire[c]	—	—	—	16	7.5	19	—	18	29
Apex, North Carolina (Mumford et al. 1990)									
Mobile Home/Kerosene Space Heater									
Indoor background	9.7	.72	1.5	0.45[d]	—	0.24	—	0.22	0.15
Kerosene heater	12.8	2.8	3.1	5.5[d]	—	2.0	—	3.7	1.3
Kawasaki, Japan (Matsushita et al. 1990)									
Kitchen	0.91	0.54	—	—	1.18	2.28	—	3.89	—
Living room	0.92	0.48	—	—	1.13	2.20	—	3.85	—

Table 2.10 continued

	Pyrene	Bz Anth	Chrysene	Bz(b)Fl	Bz(k)Fl	Bz(a)Py	Dibzanth	Bz Peryl	Inden Pyr
					Concentration in ng/m³				
Manila, Phillipines									
Kitchen	1.64	1.48	–	–	1.95	3.70	–	7.95	–
Living room	1.67	1.47	–	–	2.13	3.34	–	8.14	–

[a]Single no smoking observation (building designation).
[b]Average over-burn period.
[c]Single observation.
[d]Total benzofluoranthene.
Pyrene, Benz(a)anthracene (Bz Anth), Chysene, Benzo(b)fluoranthene (Bz(b)Fl), Benzo(k)fluoranthene (Bz(k)Fl), Benzo(a)pyrene (Bz(a)Py), Dibenz(a,h)anthracene (Dibzanth), Benzo(g,h,i)perylene (Bz Peryl), Ideno(cd)pyrene (Inden Pyr)

wood-burning stove even under conditions of an extreme fire in the stove (Table 2.7). Results also suggest that gas-fired appliances can produce particulate matter emissions quantitatively comparable to those produced by more visibly evident wood-burning stove emissions.

The principal findings of combustion appliance emission rate studies (Godish 1989) are that the appliances produce a wide variety of emissions and that the quantities of the emissions are highly dependent on the manner in which the appliance is used and maintained. As regards studies intended to address environmental tobacco smoke (ETS), combustion appliances are particularly confounding sources for measures of carbon monoxide and nitrogen oxides. They are also significant (often major) contributors to "background" concentrations of respirable suspended particulate matter, polycyclic aromatic hydrocarbons, and a wide variety of volatile (including formaldehyde) and semivolatile organic chemicals.

Respirable suspended particulate matter (RSP) is a particularly important indoor air contaminant because at least part of that which is inhaled can be irreversibly deposited in the respiratory tract. RSP can be hazardous because of its inherent properties (e.g., asbestos) or because it contains or carries hazardous constituents (e.g., polycyclic aromatic hydrocarbons, radon daughters). RSP is especially important in assessing the risk to ETS exposure because ETS can be a major contributor to ambient RSP concentrations and because ETS-specific RSP is considered by some to be related to mainstream cigarette smoke "tar." Cigarette smoke "tar" is known to contain a variety of chemicals classified as carcinogenic.

An obvious but seemingly often overlooked fact is that RSP, unlike individual chemicals, is not a single entity. Particulate matter mass collected or measured using any common technique includes materials as diverse as dust, mineral matter, dander, and smoke. Since health effects are highly dependent on the physicochemical nature of the particulate matter, it is very important to determine their nature in addition to their mass concentrations. Considerable effort is under way to identify constituents unique to ETS in the hope that their measurement will allow a measure of the fraction of total RSP that is due to ETS (Chaps. 4 and 6). As mentioned above, solanesol, scopoletin, long-chain alkanes, and cadmium are candidate markers.

As mentioned earlier, RSP is contributed to indoor air by noncombustion sources as well. Carpet vacuuming, for example, can be an especially important source because filters employed with vacuum cleaners are generally inefficient and frequently improperly installed by the user (Smith et al. 1990). A particularly dramatic example of noncombustion releases is the report of an RSP concentration of 6307 $\mu g/m^3$ in a bedroom due to operating an ultrasonic room humidifier using tap water rather than distilled water as was specified by the manufacturer (Highsmith and Rodes 1988). Polycyclic aromatic hydrocarbons (PAHs) are an important class of trace-level indoor contaminants and are derived principally from combustion (Chuang et al. 1999). They are generally present at ng/mL concentrations but are important because many of their class are known or suspected human carcinogens. Tables 2.10, 2.11, and 2.12 present results for benzo(a)pyrene (the most common PAH carcinogen of concern) and numerous other PAHs in outdoor and in nonsmoking indoor air.

Benzo(a)pyrene (BaP) is most commonly found at concentrations ranging from approximately 0.1–0.5 ng/m^3 but can be found at concentrations up to approximately 3 ng/m^3 without the presence of an obvious source (Tables 2.10 and 2.11). The presence of an obvious source (e.g., a wood-burning stove) can result in at least short-term concentrations of several hundred nanograms per cubic meter. It is probably safe to assume that the total PAH concentration is at least a factor of 10 greater than the BaP concentration at any given time. Summation of the few PAH concentrations given in Tables 2.10, 2.11, and 2.12 approach or exceed 10 times the BaP concentration yet they represent only a small number of PAHs actually present.

PAHs are found in both the particle phase and the gas phase of indoor air with the phase distribution depending on their molecular weight (see Chap. 12). PAHs consisting of 2 fused benzene rings (e.g., naphthalenes) are found principally in the gas phase, those consisting of 5 rings (e.g., benzo[a]pyrene) are found predominantly in the particle phase, and those of intermediate size are distributed between the phases. Particle-phase PAHs are dissolved in or adsorbed on particulate matter generated by the source and/or on ambient particulate matter from other sources.

VOLATILE ORGANIC COMPOUNDS

Indoor air contaminants classified as VOCs (volatile organic compounds or volatile organic chemicals) include a bewildering array (Shah and Singh 1988) of individual chemicals. Their only common characteristic is that they are sufficiently volatile to enter the air to some degree by evaporation at room temperature. A principal source of VOCs is evaporation ("out-gassing" and "off-gassing" are frequently applied terms) from materials but they are also contributed by combustion and by processes (showering, humidifiers, use of aerosol products, etc.), which enhance transport of the chemicals into the vapor phase (Howard-Reed and Corsi 1999).

Results (Table 2.13) of a survey of indoor environments for VOCs illustrate the variety of individual chemicals encountered. The study (Sheldon et al. 1988a,b) detected in excess of 200 individual chemicals or groups of chemicals representing 11 different chemical classes. Many additional VOCs have been reported (Chap. 11) in other studies.

The data in Table 2.13 illustrate two other important points concerning VOC contamination. First, while approximately 235 chemicals or chemical

Table 2.11 Background Polycyclic Aromatic Hydrocarbon (PAH) Concentrations — Pacific Northwest Buildings Study[a]

Building Type (Sample Site/Open or Restricted Policy)	Location	Benzo(a)pyrene Outdoor	Benzo(a)pyrene Indoor[c]	PAH[b] Outdoor	PAH[b] Indoor[c]	Ratio (I/O) BaP	Ratio (I/O) PAH
Urban multiuse (8/O)	Salem, OR	0.22	0.18	2.89	2.67	0.82	0.92
Urban office (9/O)	Salem, OR	0.27	0.09	3.44	1.67	0.30	0.49
Urban office (10/O)	Salem, OR	0.04	0.56	1.53	4.43	14	3.55
Urban office (11/R)	Salem, OR	0.09	0.17	1.92	2.16	1.89	1.13
Urban office (15/O)	Salem, OR	0.34	0.13	3.38	2.01	0.38	0.59
Suburban school (16/R)	Spokane, WA	0.29	0.23	NR	NR	0.79	—
Suburban library (23/R)	Spokane, WA	0.11	0.10	1.75	1.55	0.91	0.89
Suburban office (25/O)	Spokane, WA	0.12	0.10	1.94	NM	0.83	—
Urban office (26/O)	Spokane, WA	0.29	0.23	NR	NR	0.79	—
Urban office (27/O)	Spokane, WA	0.38	0.26	2.03	NM	0.68	—
Urban multiuse (28/O)	Spokane, WA	0.23	0.30	1.63	1.30	1.30	0.80
Suburban school (30/R)	Spokane, WA	2.84	1.35	19.66	NM	0.47	—
Suburban multiuse (31/R)	Spokane, WA	0.58	0.25	5.64	NM	0.43	—
Urban office (34/O)	Spokane, WA	0.73	0.24	10.32	NM	0.33	—
Urban office (35/O)	Spokane, WA	1.25	1.10	14.86	NM	0.88	—
Urban office (36/R)[d]	Salem, OR	0.40	0.19	4.85	NM	0.48	—
Urban office (37/O)	Portland, OR	0.42	0.98	6.26	NM	2.33	—
Urban office (39/O)	Portland, OR	0.47	NM	NR	NR	—	—

Table 2.11 continued

Building Type (Sample Site/Open or Restricted Policy)	Location	Concentration ng/m³						Ratio (I/O)	
		Benzo(a)pyrene		PAH[b]				BaP	PAH
		Outdoor	Indoor[c]	Outdoor	Indoor[c]				
Arithmetic mean		0.52	0.39	5.47	2.40			–	–
Arithmetic standard development		0.66	0.40	5.41	1.41			–	–
Geometric mean		0.31	0.27	3.84	2.15			–	–
Geometric standard development		2.74	2.36	2.29	1.60			–	–

[a]Table text derived from Turk, B. H., Brown, J. T., Geisling-Sobotka, K., Froehlich, D. A., Grimsrud, D. T., Harrison, J., Koonce, J. F., Prill, R. J., & Revzan, K. L. (1987) *Indoor Air Quality and Ventilation Measurements in 38 Pacific Northwest Commercial Buildings, Volume I—Measurement Results and Interpretation* (LBL 22315 1/2), *Volume II—Appendices* (LBL 22315 2/2), Lawrence Berkeley Laboratory, Berkeley, CA 94720.

[b]Chrysene, benzo(b)fluoranthene, benzo(k)fluoranthene, benzo(a)pyrene, dibenz(a,h)anthracene, benzo(ghi)perylene, plus indeno(1,2,3-cd)pyrene.

[c]Nonsmoking observations.

[d]Repeat of Building 11 sampling.

NM=not measured.

NR=not reported.

Table 2.12 Outdoor Air Concentrations of Polycyclic Aromatic Hydrocarbons

Location	Pyrene	Bz Anth	Chrysene	Bz(b)Fl	Bz(k)Fl	Bz(a)Py	Dibzanth	Bz Peryl	Inden Pyr
Salem, OR[a] (Turk et al. 1987)									
Buildings									
8 Urban multiuse	–	–	0.18	0.56	0.25	0.22	0.43	0.91	0.33
9 Urban office	–	–	0.27	0.71	0.32	0.27	0.47	1.00	0.40
10 Urban office	–	–	0.04	0.26	0.08	0.04	0.14	0.67	0.29
15 Urban office	–	–	0.27	0.81	0.34	0.34	0.60	0.90	0.1
Truckee, CA[a] (Traynor et al. 1987)									
Residence/Wood Stove									
Airtight stove A	–	–	–	0.56	0.21	0.47	–	0.96	1.3
Airtight stove C	–	–	–	0.8	0.30	0.55	–	0.92	1.2
Stove A as fireplace	–	–	–	0.52	0.23	0.51	–	0.59	1.3
Stove D extreme fire	–	–	–	2.0	1.2	1.7	–	1.7	3.9
Kawasaki, Japan (Matsushita et al. 1990)	0.75	0.48	–	–	1.19	1.84	–	4.02	–
Manila, Philippines	1.58	1.36	–	–	1.88	2.54	–	6.69	–

[a]Keyed to Indoor Air Concentrations in Table 2.9.

Bz anth=Benzo(a)anthracene,
Bz(b)Fl=Benzo(b)fluoranthene,
Bz(k)Fl=Benzo(k)fluoranthene,
Bz(a)Py=Benzo(a)pyrene,
Dibzanth=Dibenz(a,h)anthracene,
Bz Peryl=Benzo(g,h,i)perylene,
Inden Pyr=Indeno(cd)pyrene.

Table 2.13 Volatile Organic Compounds Detected in Public Buildings Study[a]

	Frequency Observed					
	Total[b] 16[c]	Out[b] 4[c]	HmN[b] 5[c]	HmS[b] 2[c]	Off[b] 3[c]	Sch[b] 2[c]
Aliphatic Hydrocarbons						
Decane[d]	16	4	5	2	3	2
Undecane[d]	16	4	5	2	3	2
Dodecane[d]	16	4	5	2	3	2
Aromatic Hydrocarbons						
Benzene	16	4	5	2	3	2
Toluene	16	4	5	2	3	2
m,p-Xylene	16	4	5	2	3	2
Styrene	16	4	5	2	3	2
Ethylbenzene	16	4	5	2	3	2
Phenylacetylene	3	1	0	0	2	0
Isopropylbenzene	8	1	3	2	0	2
n-Propylbenzene	13	2	4	2	3	2
Ethylmethylbenzenes	16	4	5	2	3	2
Trimethylbenzenes	16	4	5	2	3	2
Indene	1	0	0	1	0	0
2-Methylphenylacetylene	2	0	1	0	1	0
sec-Butylbenzene	2	0	1	0	1	0
t-Butylbenzene	3	0	1	0	2	0
Propylmethylbenzenes	14	4	4	1	3	2
Dimethylethylbenzenes	15	4	5	1	3	2
Diethylbenzenes	12	4	3	1	3	1
Tetramethylbenzenes	8	2	2	1	1	2
Ethenylethylbenzene	7	2	2	0	2	1
Methylpropenylbenzene	4	1	2	0	1	0
Diethenylbenzene	2	1	1	0	0	0
Naphthalene	15	3	5	2	3	2
Methylindane	2	1	0	0	0	1
Pentylbenzenes	8	3	2	0	2	1
Dimethylpropylbenzene	4	0	2	1	0	1
Pentamethylbenzenes	1	0	0	1	0	0
Ethyltrimethylbenzenes	4	2	1	0	0	1
Ethylpropylbenzenes	2	1	0	0	0	1
Methylbutenylbenzenes	2	0	2	0	0	0
Methylnaphthalenes	15	3	5	2	3	2
Methyldihydronaphthalene	2	1	0	1	0	0
Methyltetralins	5	1	2	1	0	1
Dimethylindans	3	2	0	0	0	1
(3,3-Dimethylbutyl)benzene	1	0	0	0	1	0
Hexamethylbenzene	1	0	0	1	0	0
Triethylbenzenes	4	1	1	1	0	1
Diisopropylbenzene	1	1	0	0	0	0
Diethyldimethylbenzene	1	1	0	0	0	0
Dimethyl(methylpropyl)benzene	1	1	0	0	0	0
Cyclohexenylbenzene	3	1	0	1	1	0

Table 2.13 continued

	Frequency Observed					
	Total[b] 16[c]	Out[b] 4[c]	HmN[b] 5[c]	HmS[b] 2[c]	Off[b] 3[c]	Sch[b] 2[c]
Butenyldimethylbenzene	2	0	0	1	0	1
Dimethyl(methylpropyl)benzene	1	1	0	0	0	0
Cyclohexenylbenzene	3	1	0	1	1	0
Butenyldimethylbenzene	2	0	0	1	0	1
Triemthyl(methylethenyl)benzene	1	0	0	0	0	1
Di(methylethenyl)benzene	1	0	0	0	1	0
Dimethylnaphthalenes	7	1	2	1	1	2
Ethylnaphthalene	2	1	0	0	0	1
Diemthyltetralins	1	1	0	0	0	0
Ethlytetralins	2	1	0	1	0	0
Biphenyl	5	1	1	1	2	0
Trimethylnaphthalene	1	0	1	0	0	0
Ethylphenylbenzene	1	1	0	0	0	0
Butylhexylbenzene	1	0	0	0	1	0
Halogenated Hydrocarbons						
Chloromethane	6	2	1	2	0	1
Dichloromethane	11	3	5	2	0	1
Chloroform	10	3	3	1	2	1
Carbontetrachloride	8	3	2	0	2	1
Trichlorofluoromethane	12	3	5	2	0	2
Dichlorodifluoromethane	3	1	2	0	0	0
Dichlorofluoromethane	1	0	1	0	0	0
1,1,1-Trichloroethane	15	4	5	1	3	2
Trichloroethylene	14	3	5	2	3	1
Tetrachloroethylene	16	4	5	2	3	2
1,2-Dichloroethane	3	2	1	0	0	0
Vinylidene chloride	2	0	1	1	0	0
1,1,2-TriCl-1,2,2-TriFluroethane	4	2	1	1	0	0
1,1-DiBr-2-Cl-2-Fl-cyclopropane	1	1	0	0	0	0
Chlorobenzene	6	3	1	0	1	1
m-Dichlorobenzene	13	3	4	2	2	2
p-Dichlorobenzene	12	4	4	2	1	1
o-Dichlorobenzene	11	3	4	1	2	1
Trichlorobenzene	2	0	2	0	0	0
3-Chloro-1-phenyl-2-butene	1	0	0	0	1	0
Esters						
Ethylformate	1	0	1	0	0	0
1-Methylethylformate	1	0	0	0	1	0
n-Butylformate	1	0	0	0	1	0
Ethylacetate	8	1	3	2	2	0
2-Propylacetate	1	0	0	0	1	0
Butylacetate	1	0	1	0	0	0
Diethylacetate	3	1	1	0	0	1
3-Methyl-1-butylacetate	1	0	1	0	0	0
1-Phenylacetate	1	0	1	0	0	0
n-Hexylacetate	1	0	1	0	0	0
n-Decylacetate	1	0	1	0	0	0
α-Ethylmethylbenzeneacetate	1	1	0	0	0	0
n-Pentylbutanoate	1	0	1	0	0	0
1-Methylethylbutanoate	2	0	1	0	0	1
n-Hexylbutanoate	4	1	0	0	2	1

Table 2.13 continued

	Frequency Observed					
	Total[b] 16[c]	Out[b] 4[c]	HmN[b] 5[c]	HmS[b] 2[c]	Off[b] 3[c]	Sch[b] 2[c]
Methylbutylbenzoate	1	0	1	0	0	0
Dimethylphthalate	1	1	0	0	0	0
Diethylphthalate	1	1	0	0	0	0
Dibutylphthalate	1	1	0	0	0	0
Alcohols						
Ethanol	1	0	1	0	0	0
2-Nitroethanol	1	0	1	0	0	0
2-Propanol	3	1	1	0	1	0
2-Methyoxyethanol	1	1	0	0	0	0
1,2-Propanediol	1	0	1	0	0	0
n-Butanol	3	0	2	1	0	0
Cylcobutanol	1	1	0	0	0	0
1,2-Butanediol	1	1	0	0	0	0
n-Pentanol	2	1	0	0	1	0
4-Penten-2-ol	1	0	0	0	1	0
n-Hexanol	8	2	3	2	1	0
2-Cyclohexene-1-ol	2	1	1	0	0	0
2-Methylpentanol	1	0	1	0	0	0
2-Ethylbutanol	1	0	1	0	0	0
3-Methylpentanol	1	0	0	0	0	1
2-(1,1-Dimethylethoxy)ethanol	2	0	1	1	0	0
2-Butoxyethanol	2	0	1	0	1	0
n-Heptanol	3	0	2	1	0	0
5-Methylhexanol	1	0	0	1	0	0
3-Methylhexanol	1	0	1	0	0	0
2,2-Dimethyl-1-pentanol	1	0	1	0	0	0
1-Methyl-2-cyclohexene-1-ol	1	0	1	0	0	0
3-Methylcyclohexanol	2	1	0	0	1	0
2-Ethyl-4-methyl-1-pentanol	4	0	0	2	1	1
2-Ethyl-1-hexanol	9	3	2	0	2	2
3,5-Dimethylcyclohexanol	2	0	0	1	0	1
2,2,4-Trimethyl-3-penten-ol	3	0	0	1	1	1
4-Methylbenzenemethanol	1	0	1	0	0	0
2-Propyl-1-hexanol	1	0	0	0	1	0
4,5-Dimethyl-2-hepten-3-ol	1	0	1	0	0	0
α-Ethylbenzenemethanol	1	0	1	0	0	0
2,2-Dimethyl-1-octanol	1	0	1	0	0	0
2-(1,1-Dimethylethyl)cyclohexanol	1	0	1	0	0	0
4-(1,1-Dimethylethyl)cyclohexanol	1	0	1	0	0	0
n-Decanol	3	1	1	0	1	0
Dimethyloctanol	1	0	0	0	0	1
2-Propyl-1-heptanol	7	2	1	0	2	2
n-Undecanol	1	0	0	0	0	1
n-Dodecanol	6	1	2	1	0	2
2-Butyloctanol	7	2	2	1	2	0
2,2-Dimethyl-1-decanol	1	0	0	1	0	0
6-Ethyl-4,5-decanediol	1	0	0	0	1	0
2-Methyl-1-dodecanol	1	0	1	0	0	0
n-Hexadecanol	1	0	0	0	1	0
n-Heptadecanol	1	0	0	0	1	0

Table 2.13 continued

	Frequency Observed				
	Total[b] 16[c]	Out[b] 4[c]	HmN[b] 5[c]	HmS[b] 2[c]	Off[b] 3[c]	Sch[b] 2[c]
Phenols						
Phenol	1	0	1	0	0	0
2-Methylphenol	1	0	0	1	0	0
2,4-Dimethylphenol	2	0	1	1	0	0
2,3-Dimethylphenol	1	0	1	0	0	0
3,4-Dimethylphenol	2	1	1	0	0	0
2,5-Dimethylphenol	3	2	1	0	0	0
3-Ethylphenol	2	1	0	0	1	0
2-t-Butylphenol	1	0	1	0	0	0
2-Methyl-4-t-butyl-phenol	1	0	0	1	0	0
Ketones						
Acetone	16	4	5	2	3	2
Ketones continued						
Methylethylketone	2	1	0	1	0	0
4-Pentene-2-one	1	0	0	1	0	0
2-Methyl-1-propene-1-one	1	0	1	0	0	0
Dihydro-2(3H)-furanone	1	0	0	1	0	0
3-Methyl-2-butanone	1	0	1	0	0	0
3-Hexanone	2	1	1	0	0	0
4-Hexene-3-one	1	1	0	0	0	0
5-Hexene-2-one	1	1	0	0	0	0
3-Methyl-2(5)-furanone	1	1	0	0	0	0
2-Methyl-3-pentanone	1	0	1	0	0	0
4-Methyl-2-pentanone	2	0	0	0	2	0
2-Hexanone	1	0	0	0	0	1
2-Heptanone	3	1	2	0	0	0
Acetophenone	4	1	2	1	0	0
3,4-Dihydro-1(2H)naphtalenone	2	1	0	1	0	0
Indan-1-one	2	1	0	1	0	0
7-Methyl-3-octane-2-one	1	1	0	0	0	0
5-Methyl-1(3H)-isobenzofuranone	1	1	0	0	0	0
2,6-Dimethyl-4-hepten-3-one	1	0	0	0	1	0
5-Nonen-4-one	1	0	0	0	1	0
Methylacetophenone	2	1	1	0	0	0
1-Phenyl-1-propanone	1	0	0	1	0	0
1-Ethylacetophenone	1	1	0	0	0	0
4-Phenyl-3-butene-2-one	1	1	0	0	0	0
1-(2-Methylphenol)ethanone	1	1	0	0	0	0
5-Methyl-2-(1-methylethenyl)-cyclohexane	1	0	0	0	1	0
Cyclohexanone	1	0	0	1	0	0
2-Decen-2-one	1	0	0	1	0	0
2,4-Dimethylacetophenone	4	3	0	1	0	0
1,7,7-Trimethylbicyclo[2,2,1]-Heptanone	4	3	0	1	0	0
1-Phenyl-2-pentanone	1	0	0	1	0	0
1-Phenyl-1-pentanone	1	1	0	0	0	0
2,3-Dihydro-4,7-dimethyl-1H-indene-1-one	2	1	1	0	0	0
3,3-Dimethyl-1-indene-1-one	1	1	0	0	0	0

Table 2.13 continued

	Frequency Observed					
	Total[b] 16[c]	Out[b] 4[c]	HmN[b] 5[c]	HmS[b] 2[c]	Off[b] 3[c]	Sch[b] 2[c]
4(1-Methylethyl)acetophenone	1	1	0	0	0	0
2,6-bis(1,1-Dimethylethyl)-2,5-cyclohexadien-1,4-dione	1	0	0	0	1	0
2,5-cyclohexadien-1,4-dione	1	0	0	0	1	0
Benzophenone	2	1	1	0	0	0
5-Methyl-1-phenyl-hexene-3-one	1	1	0	0	0	0
6,10-Dimethyl-5,9-undecadiene-2-one	1	0	0	0	1	0
Aldehydes						
2-Oxopropanol	1	0	1	0	0	0
n-Butanal	2	1	1	0	0	0
2-Methylpropanal	1	0	1	0	0	0
2-Methyl-2-butenal	1	0	1	0	0	0
2-Furaldehyde	4	1	1	2	0	0
n-Hexanal	1	0	1	0	0	0
n-Heptanal	4	1	2	1	0	0
3,3,-Dimethylhexanal	3	1	0	2	0	0
Benzaldehyde	4	1	2	1	0	0
n-Octanal	1	0	1	0	0	0
n-Nonanal	13	3	3	2	3	2
α-Methylbenzaldehyde	3	2	0	1	0	0
4-Methylbenzaldehyde	1	1	0	0	0	0
3-Phenyl-2-propanal	1	1	0	0	0	0
Ethylbenzaldehyde	4	2	1	0	1	0
n-Decanal	10	1	3	2	3	1
n-Undecanal	1	1	0	0	0	0
4-(Methylethyl)benzaldehyde	1	0	0	0	0	1
n-Tetradecanal	1	0	1	0	0	0
Ethers						
Methoxyethane	1	1	0	0	0	0
Ethoxyethane	1	0	1	0	0	0
Ethoxy-1-propene	1	0	1	0	0	0
Methylbutylhydroperoxide	1	1	0	0	0	0
Hexylpentylether	1	0	1	0	0	0
Tetrahydrofuran	3	1	1	1	0	0
2-Butyltetrahydrofuran	1	0	0	0	1	0
Benzofuran	1	0	0	1	0	0
4,7-Dimethylbenzofuran	2	0	2	0	0	0
Dihydrobenzopyran	1	0	1	0	0	0
Epoxides						
Ethylene oxide	4	2	1	0	1	0
Trimethyloxirane	1	1	0	0	0	0
Ethyloxirane	2	2	0	0	0	0
Methyloxirane	1	0	1	0	0	0
2,2-Dimethyloxirane	1	1	0	0	0	0
3,3-Dimethyloxirane	1	1	0	0	0	0
Carboxylic Acids						
Acetic acid	10	3	3	0	2	2
Propanoic acid	1	0	0	1	0	0
Propanedioic acid	2	1	1	0	0	0

Table 2.13 continued

	Frequency Observed					
	Total[b] 16[c]	Out[b] 4[c]	HmN[b] 5[c]	HmS[b] 2[c]	Off[b] 3[c]	Sch[b] 2[c]
Pentanoic acid	1	0	1	0	0	0
Sulfur-Compounds						
Carbon disulfide	2	2	0	0	0	0
6-Methylbenzo(b)thiophene	1	1	0	0	0	0
Benzothiazole	2	0	1	0	1	0
Thiopropanoicacid-3-sec-butylester	1	0	0	0	1	0
2-Ethylthiophenol	1	0	0	0	1	0
Nitrogen-Compounds						
Propionamide	1	1	0	0	0	0
Benzonitrile	2	1	0	0	0	1
3-Methylpyrrolide	1	1	0	0	0	0
3-Methylcinnoline	2	2	0	0	0	0
Pyridine	1	1	0	0	0	0
Decylhydroxylamine	3	0	0	0	2	1

[a]Sheldon et al. 1988a,b.
[b]Total sites; Outdoor near air intake; Home for Elderly, nonsmoking (NS) and smoking (S) environment; Office; School, respectively.
[c]Number of sampling sites.
[d]Not tabulated. Text states decane, undecane, dodecane were ubiquitous and major.

groups are reported, a relatively small number are detected at high frequency. In each environment studied, 12 of the chemicals were detected and an additional 16 were detected in 10 or more of the 16 samplings performed. Second, the presence of a unique source (home for the elderly with a smoking resident) is not necessarily reflected in indoor air contaminant distribution at any given time. Table 2.13 indicates the presence of some constituents in the smoking residence not detected in other environments but that tend to be detected in low frequency and are not major constituents of cigarette smoke.

At least three classes of chemicals appear to be ubiquitous in indoor air. They are found whenever they are specifically sought and they are often reported as having been detected by investigators performing studies of other VOCs or of VOC sources. These are n-alkanes (particularly n-nonane, n-decane, n-undecane, and n-dodecane), benzene and its derivatives (particularly toluene, ethylbenzene, styrene, and the xylenes), and halogenated hydrocarbons (e.g., 1,1,1-trichloroethane, trichloroethylene, and tetrachloroethylene. These chemicals are common petroleum and/or industrial solvents and are likely to be major constituents of many household and maintenance products as well as to be residual constituents of many finished products (e.g., furniture, textiles). Table 2.14 summarizes the predominant VOCs identified in various studies and Table 2.2 summarizes major chemicals in emissions from several specific materials. The same or similar chemicals appear on many of the lists. The most common are petroleum-derived n-alkanes and benzene derivatives.

Common sources of VOCs in indoor air are given in Table 2.3. They include the materials of construction of the building itself, the building heating and air conditioning system, consumer and commercial products used for building maintenance and for personal care, building furnishings and decorations, and the incursion of outdoor air. Specific activities such as tobacco smoking, showering, and photocopying contribute an additional VOC burden. Each of these sources can be important contributors to many of the commonly detected VOCs in indoor air. This very large variety of sources coupled with each source releasing very many constituents (Table 2.2) is the reason that both the quantities and the nature (the number of constituents) of indoor VOCs are highly variable and are difficult to apportion by source. Surrogates for source apportionment are unavailable and need to be developed.

A few VOCs are highly related to one or a few major sources (although others may be found as studies progress). A most important example is formaldehyde. The major source (Godish 1989) of formaldehyde is wood products bonded with or finished with urea-formaldehyde resins. These include particle board, hardwood plywood, medium density fiberboard, and finishes on furnishings, cabinetry, and wood floors. Urea-formaldehyde foam insulation (Chap. 10) is also an especially important source of formaldehyde. While other sources (e.g., combustion) contribute to the indoor air burden of formaldehyde, the extensive use of urea-formaldehyde bonded wood products in construction and furnishings generally make these the primary source of indoor formaldehyde concentrations.

Other VOCs have been associated with one or a few related sources as well. Chloroform is largely introduced into the indoor air as a result of showering or otherwise dispersing or heating tap water. p-Dichlorobenzene is largely the result of using moth crystals, and tetrachloroethylene is especially prevalent after returning dry-cleaned textiles to the home. Liquid process photocopiers release (Kerr and Sauer 1990) isodecanes into the air. Glycols 2[2-ethoxyethoxy]-ethanol) are released (Plehn 1990, Clausen et al. 1990) by"low pollutant" water-based paints. Polyurethane wood sealants (Schriever and Marutzky 1990) release butylacetates, ethylacetate, and 2-methoxyethanol.

As for combustion sources discussed above, emission rates for VOCs are poor predictors of their ambient concentrations. Emission rate studies are very important, however, for identifying those factors that determine the extent of release and for identifying sources of release. This may be more important (e.g., 1,2-ethanediol, 1,2-propanediol) and glycol ethers (e.g., 2-butoxyethanol for VOCs than for major combustion products because of the greater variety of potential sources and the greater influence of ambient conditions on off-gassing. Indoor temperature and humidity influence the release of VOCs resulting from evaporative sources (Godish 1989).

Table 2.14 Predominant Volatile Organic Chemicals Detected in Various Studies

Office Buildings[a] Baldwin et al. '90	Building Materials[a] Molhave, '82	Building Materials[b] Molhave, '82	Public Buildings[a,b] Sheldon et al. '88	Residential[b,d] Pellizzari et al. '87
n-Octane	Toluene	Toluene	n-Decane	m+p-Xylene
n-Nonane	m-Xylene	n-Decane	n-Undecane	Tetrachloroethylene
n-Decane	$C_{10}H_{16}$ Terpene	1,2,4-TriMebenzene	n-Dodecane	Benzene
n-Undecane	n-Butylacetate	n-Undecane	Benzene	Ethylbenzene
n-Dodecane	n-Butanol	m-Xylene	Toluene	o-Xylene
Toluene	n-Hexane	o-Xylene	m+p-Xylenes	Styrene
Ethylbenzene	p-Xylene	n-Propylbenzene	Styrene	1,1,1-Trichloroethane
m+p-Xylenes	Ethoxyethylacetate	Ethylbenzene	Ethylbenzene	m,p-Dichlorobenzene
o-Xylene	n-Heptane	n-Nonane	Ethylmethylbenzenes	Trichloroethylene
Cumene	o-Xylene	1,3,5-TriMebenzene	Trimethylbenzenes	Chloroform
Mesitylene			Tetramethylbenzenes	Carbon tetrachloride
p-Cymene			Tetrachloroethylene	
Butylbenzene			Acetone	
Styrene			Dimethylethylbenzenes	
			Naphthalene	
			Methylnaphthalenes	
			1,1,-Trichloroethane	
			Propylmethylbenzenes	
			Trichloroethylene	
			n-Propylbenzene	
			Dichlorobenzenes	
			n-Nonanal	

[a]Predominant by quantity.
[b]Predominant by frequency of observation.
[c]Homes for elderly, school, office, outdoors.
[d]Overnight indoor air, Bayonne, NJ.

The work of Tichenor and Mason (1988) on emissions from building materials is a good example of the general experimental design and of valuable information generated in emission rate studies. The investigators employed the now common approach of placing a sample of the material of interest in a test chamber and monitoring the resulting emissions. The air exchange rate and product loading (typically in square meters of surface area per cubic meters of chamber volume or sweep volume) are employed along with the monitoring result to yield emission rates in mass per square meter of surface per hour. Using this approach, the investigators find emission rates of formaldehyde from a sample of particle board aged 8 months prior to the test to range from 95-230 $\mu g/m^2h$ depending on the air exchange rate and product loading used. Emission rates for acetone and hexanal were less variable (37-41 $\mu g/m^2h$ for acetone, 15-26 $\mu g/m^2h$ for hexanal) under the conditions used. Emission rates for formaldehyde, acetone, and hexanal were found to be 140, 37, and 24 $\mu g/m^2h$, respectively, under a product loading (0.39 m^2/m^3) and air exchange rate (0.54 ach) typical of residences. In another part of the same study, the investigators examined the rate of total VOC (TVOC) release from a freshly applied adhesive and a silicone caulk. The TVOC emission rate was 1700 mg/m^{2}h 30 min after application of the adhesive to the test substrate and dropped to 100 mg/m^{2}h at 5 h. The emission rate from the silicone caulk dropped from 50 mg/m^{2}h to 2 mg/m^{2}h in the same time frame. Such information on source strengths and characteristics are especially important for locating sources of specific emissions and ranking the relative importance of various sources and conditions to the indoor air burden of VOCs.

"Whole house" source strengths have been determined for a variety of VOCs in residences (Pellizzari et al. 1989). Indoor and outdoor concentrations are simultaneously measured and combined with data on the air exchange rate to compute the quantity of the VOC due to indoor sources. As illustrated by the results in Table 2.15, 12-h, whole-house contributions generally range from negative (outdoor sources predominate) to a few milligrams per hour for most constituents measured. Higher source strengths, e.g., 20–40 mg/h for tetrachloroethylene, toluene, and limonene in Table 2.15, are likely associated with specific sources being present during the period sampled. Dry cleaning, an attached garage, and recent renovations are possible sources of the elevated tetrachloroethylene, toluene, and limonene concentrations. The results clearly demonstrate considerable home-to-home and temporal (12-h period) variation in "background" VOC emissions by the overall indoor environment itself.

Concentrations of common VOCs in representative nonsmoking environments are given in Table 2.16. Common aliphatic, aromatic, and halogenated hydrocarbons are most frequently present at individual concentrations of 1-20 $\mu g/m^3$. The concentration of any given chemical can range, however, from less than 0.1 $\mu g/m^3$ to several hundred micrograms per cubic meter at any given time and place. High concentrations, e.g., 72.7 $\mu g/m^3$ of 1,1,1-trichloroethane and 213 $\mu g/m^3$ of toluene generally indicate that an unusual source is present (Table 2.16). These high concentrations were found (Pellizzari et al. 1989) in a hardware store where one might expect a prevalence of petroleum distillates and industrial solvents.

Exposure to VOCs is not limited to the time spent in buildings. Considerable exposure occurs during commuting (Lawryk and Lioy 1995 and Table 2.17) and outdoor activities in certain environments. Concentrations of VOCs in automobiles are often greater than those in buildings. Concentrations are further increased in the winter months when the heater is used and are generally greater for urban driving than rural driving. Benzene and its alkyl derivatives and aliphatic hydrocarbons make up the bulk of VOC exposure when the vehicle is powered by gasoline or other petroleum distillates.

While individual chemicals are commonly present at concentrations of 1-20 $\mu g/m^3$, the TVOC concentration commonly ranges from 100-3000 $\mu g/m^3$ and has been observed to exceed 10 $\mu g/m^3$ (Table 2.18). A study of personal exposure levels in residences in Baltimore, California, and New Jersey found mean TVOC to be 1.6, 1.4, and 3.8 $\mu g/m^3$, respectively (Wallace et al. 1990). Outdoor means were 0.4 $\mu g/m^3$ for the California study and 0.5 $\mu g/m^3$ for the New Jersey study. TVOC concentrations have also been determined in a survey (Turner and Binnie 1990) of 26 buildings in 20 Swiss cities. TVOC concentrations ranged from 0.06 to 1.4 mg/ m 3 for the 6 buildings where ETS was not present (as determined by undetectable nicotine concentrations). A recent study (Baldwin and Farant 1990) of office buildings in Montreal reports TVOC concentrations for a given building as 0.6 mg/m^3 when unoccupied, 1.8 mg/m^3 when unoccupied but in the presence of construction activities, 4.4 mg/m3 when later occupied, and 13.5 mg/m^3 when occupied and undergoing renovations. A second building with stable occupancy exhibited concentrations ranging from 0.4 to 1.4 mg/m^3 TVOC.

Results of a typical survey (Tsuchiya and Stewart 1990) for TVOC concentrations are given in Table 2.18. Approximately 28 buildings in Ottawa and other Canadian cities were surveyed with a particular interest in the contribution of photocopiers to the TVOC concentration. The photocopier contribution was determined by comparing the gas chromatographic profile of the ambient sample with the relatively distinctive profiles of photocopier emissions. The study reports generally observed TVOC concentrations of 0.1-4 mg/m^3 indoors and 0.1 mg/m^3 outdoors. Ten of the 28 buildings exhibited TVOC concentrations greater than 10 mg/m^3 at one or more samplings and the investigators report short excursions of as much as 80-100 mg/m^3. High percentage contributions by the photocopier were frequently associated with high mean concentrations of TVOC but other sources predominated in most cases.

Table 2.15 Nonsmoking Whole-House Source Strengths for Selected Volatile Organic Compounds

Chemical	Range of 12-h Integrated Emissions (mg/hr)[a]		
	Residence G[b]	Residence J[c]	Residence K[d]
Chloroform	0.7–2.9	0–0.1	-0.1–0.1
1,1,1-Trichloroethane	14.4–24.1	-3.0–1.0	-0.2–0.8
Carbon Tetrachloride	-0.04–0.05	-0.5–0	-0.01–-0.1
Trichloroethylene	1.0–2.2	-0.2–0.1	-0.01–0.06
Tetrachloroethylene	0.3–23.8	-3.0–1.0	-0.4–0.3
Chlorobenzene	-0.1–0.01	0–<0.1	-0.01–0.06
m-Dichlorobenzene	0–0.2	0–<0.1	-0.01
p-Dichlorobenzene	0.6–2.7	0.4–2.8	0.3–0.5
Benzene	<0.1–0.6	-2.4–2.6	-0.4–0.2
Toluene	11.9–37.2	-4.1–8.2	-0.4–0.8
o-Xylene	0.5–2.2	-3.2–1.0	-1.0–0.2
m+p-Xylene	-0.2–6.7	-4.0–3.6	-2.3–0.5
Ethylbenzene	<0.1–1.3	-0.3–0.8	-0.5–0.2
Styrene	0–0.4	-0.04–0.2	-0.05–0.1
n-Octane	0.2–1.4	0.5–2.0	0.02–0.4
n-Nonane	0.1–2.0	-1.0–1.3	-0.19–0.4
n-Decane	0.2–0.4	-1.4–1.2	-0.10–0.8
n-Dodecane	0.4–0.9	0.7	-0.10–0.3
Limonene	12.4–101	3.2–20.2	0.8–2.2
1,4-Dioxane	0.2–1.1	-0.02–<0.01	-0.01–0.3

[a]Pellizzari et al. 1989.
[b]Attached garage, mothball/deodorizer use, recent dry cleaning, recent home improvements, 10 12-h samplings.
[c]Recent home improvements, chosen as comparative home due to low VOC content found in earlier studies, 6 12-h samplings.
[d]Mothball/deodorizer use, chosen as comparative home due to low VOC content found in earlier studies, 6 12-h samplings.

Table 2.16 Example Non-ETS Indoor Air Concentrations of Common VOCs

			Concentration in $\mu g/m^3$					
	School[a]	Unoccupied Apartment[b]	Residences[c]		Business[d]		In Vehicle[e]	
Chloroform	0.08–0.29	1.2	0.6	1.1	1.1	0.43	—	—
1,2-Dichloroethane	0.02–0.05	0.17	—	—	—	—	—	—
1,1,1-Trichloroethane	17.0–29.0	5.0	22.1	21.0	72.7	25.3	—	—
Carbon Tetrachloride	0.71–0.97	1.6	—	—	0.69	0.69	—	—
Trichloroethylene	0.61–0.93	0.8	0.8	1.9	0.91	0.64	—	—
Tetrachloroethylene	5.0–7.8	5.0	6.6	6.6	12.3	7.73	—	—
Chlorobenzene	0.11–0.17	0.06	—	—	0.52	0.14	—	—
m+p-Dichlorobenzene	1.6–2.2	0.56	11.1	5.0	108.5	26.5	—	—
o-Dichlorobenzene	0.10–0.13	0.07	—	—	—	—	—	—
1,1,2,2-Tetrachloroethane	0.02–0.02	0.02	—	—	—	—	—	—
Benzene	6.3–12.0	8.2	18.7	11.5	13.1	6.5	—	—
Toluene	—	—	—	—	213	56.7	43.1	10.4
o-Xylene	4.2–5.9	4.9	19.6	15.2	37.1	8.6	10.6	3.6
Ethylbenzene	1.7–4.6	4.7	11.1	8.3	18.7	4.5	8.1	2.5
Styrene	0.98–3.2	0.73	2.7	2.4	5.4	1.2	—	—
n-Decane	0.61–1.1	2.0	—	—	152	31.7	6.7	3.3
n-Undecane	0.69–0.92	2.1	—	—	45.8	4.3	5.8	4.4
n-Dodecane	0.72–0.83	1.6	—	—	—	—	—	—

[a] Sheldon et al. 1988, range of average concentrations over three locations.
[b] Sheldon et al. 1988, unoccupied apartment in a home for the elderly, single observation.
[c] Pellizzari et al. 1989, overnight living room, single observation fall 1987, nonsmoking low emissions homes.
[d] Pellizzari et al. 1989, hardware stores, average of two 12-h samplings.
[e] Cooper et al. 1990, urban and rural driving in Raleigh, NC.

The finding of a single or a clearly predominant source of TVOC (as is found for many samplings in the photocopier study) is unusual and often reflects an experimental design optimized to detect a particular source being studied. TVOC concentrations determined in studies without a particular source in mind generally reflect the contributions of many individual chemicals from many sources. Individual chemicals of particular concern such as benzene and formaldehyde are frequently very minor contributors to the TVOC concentration. TVOC measurements remain very important, however, because they provide a measure of the material present, which is unaccounted for by determinations of preselected chemicals. They are also important for assessing the possible contribution of VOCs to illness and discomfort commonly associated with the Sick Building Syndrome. Studies of human sensitivity to indoor environments have shown (Molhave 1984) that TVOC concentrations of less than 0.2 mg/m^3 are generally tolerated without discomfort, that concentrations equal to or greater than 2 mg/m^3 are frequently associated with subjective complaints of discomfort, and that TVOC concentrations of 5 mg/m^3 routinely result in eye and nose irritation.

A review of the data presented here yields several important conclusions and at least one important caution. Important conclusions are that the TVOC content of indoor air is made up of contributions from many individual chemicals, that a very large number of chemicals are likely to be present at any given time and place, that the chemicals arise from many different sources, that their concentrations vary with time and space, and that a relatively small number of chemicals or chemical classes predominate either in frequency of detection or in quantity present. The caution is that much of the data have been generated using a single measurement technique, i.e., gas chromatography (GC) with one form of detection or another. This may contribute to the high frequency of detection of aliphatic, aromatic, and halogenated hydrocarbons and the less common reports of chemicals not so amenable to detection by GC. Chemicals more difficult to detect or more difficult to identify using commonly employed GC methods include carbonyl compounds (e.g., formaldehyde, acrolein); acidic compounds (e.g., carboxylic acids and phenols); and polyfunctional compounds (e.g., diols, diol ethers). It is possible that such chemicals will be found to be common components of the VOC content of indoor air if they are specifically sought using methods optimized for their detection.

Table 2.17 Exposure to Aromatic Hydrocarbons by Commuting Mode[a]

	Concentration (μg/m³)				
	Benzene	Toluene	Ethyl Benzene	m+p Xylene	o-Xylene
Mean Exposure					
In vehicle (40)[b]	17.0	33.1	5.8	20.9	7.3
Subway (37)[b]	6.9	30.8	2.5	9.8	3.6
Walking (31)[b]	10.6	19.8	3.0	12.6	4.1
Biking (11)[b]	9.2	16.3	2.4	10.0	3.0
Sidewalk[c] (10)[b]	11.8	17.7	3.9	10.4	5.3
Maximum Observed					
In vehicle	64.0	105.1	21.6	74.6	26.1
Subway	13.2	151.7	5.8	21.6	7.8
Walking	24.2	44.3	6.8	32.9	8.9
Biking	28.0	45.1	7.1	28.3	8.9
Sidewalk[c]	17.6	29.1	4.9	13.6	11.6

[a]Chan et al. 1990. Boston, MA, winter 1989.
[b]Number of samples.
[c]Outdoor stationary sample.

Table 2.18 Photocopier Contribution to Total Volatile Organic Compound Concentrations in Buildings

Building	No. Samples	Concentration, mg/m^3		Percent Copier Contribution
		Average	Maximum	
Mall	3	0.3	0.4	0
Hospital 1	5	0.7	1.3	0
Hospital 2	3	0.3	0.4	0
Hospital 3	3	20	98	16
Hospital 4	17	1.3	4.6	0
Hospital 5	16	2.2	23	0
Hospital 6	17	6.5	83	0
Library 1	24	11	79	> 90
Library 2	3	8	13	> 90
School 1	8	2	5.9	80
School 2	12	2.4	5.3	> 90
Office 1	7	0.4	13	> 90
Office 2	14	4	19	> 90
Office 3	2	0.5	3	70
Office 4	4	0.1	0.4	0
Office 5	4	0.8	2.4	32
Office 6	5	0.7	2.8	80
Office 7	4	17	23	67
Office 7a	4	8	88	> 90
Office 8	1	2.7	2.7	28
Office/lab 1	4	1.6	10	> 90
Office/lab 3	1	0.1	0.1	50
Office/lab 4	1	0.2	0.2	0
Office/lab 5	3	1.2	6.2	30
Office/print shop 1	3 •	3	6.4	> 90
Office/print shop 1a	6	3.4	3.9	90
Office/print shop 1b	4	26	29	> 90
Office/factory	6	0.4	0.9	0
Warehouse	11	1.8	3.9	27
Conference center	3	2.5	4.2	0

Tsuchiya and Stewart 1990.

REFERENCES

Ahearn, D. G. & Crow, S. A. (1996) Fungal ionization of fiberglass insulation in the air distribution system of a multi-story office building: VOC production and possible relationship to a sick building syndrome. *J. Ind. Microbiol.*, 16, 280–285.

Akland, G. G., Hartwell, T. D., et al. (1985) Measuring human exposure to carbon monoxide in Washington, D.C. and Denver, CO during the winter of 1982-1983. *Environ. Sci. Technol.,* 19, 911-918.

Baldwin, M. E. & Farant, J.-P. (1990) Study of selected volatile organic compounds in office buildings at different stages of occupancy. *Proceedings of the 5th International Conference on Indoor Air Quality and Climate, Volume 2,* Toronto, Canada, pp. 665-670.

Bayer, C. W. & Papanicolopoulos (1990) Exposure assessments to volatile organic compound emissions from textile products. *Proceedings of the 5th International Conference on Indoor Air Quality and Climate, Volume 3,* Toronto, Canada, pp. 725-730.

Bloeman, H. J. T., Kliest, J. J. G., & Bos, H. P. (1990) Indoor air pollution after the application of moisture repellents. *Proceedings of the 5th International Conference on Indoor Air Quality and Climate, Volume 3,* Toronto, Canada, pp. 569-574.

Brown, V. M., Cockram, A. H., Crump, D. R., & Gardiner, D. (1990) Investigations of the volatile organic compound content of indoor air in homes with an odorous damp proof membrane. *Proceedings of the 5th International Conference on Indoor Air Quality and Climate, Volume 3,* Toronto, Canada, pp. 575-580.

Chan, C. D., Spengler, J. D., Ozkaynak, H., & Lefkopoulou, M. (1990) Commuter exposure to volatile organic compounds. *Proceedings of the 5th International Conference on Indoor Air Quality and Climate, Volume 2,* Toronto, Canada, pp. 627-632.

Chuang, J. C., Callahan, P. J., Lyu, C. W., & Wilson, N. K. (1999) Polycyclic aromatic hydrocarbon exposures in low-income families. *J. Expo. Anal. Environ. Epidemiol.*, 9, 85–98.

Clausen, P. A., Wolkoff, P., & Nielsen, P. A. (1990) Long term emission of volatile organic compounds from waterborne paints in environmental chambers. *Proceedings of the 5th International Conference on Indoor Air Quality and Climate, Volume 3,* Toronto, Canada, pp. 557-562.

Clobes, A. L. & Anath, G. P. (1992) Human activities as sources of volatile organic compounds in residential environments. *Ann. NY Acad. Sci.*, 641, 79–86.

Cooper, S.D., Thomas, K. W., Raymer, J. H., & Pellizzari, E. D. (1991) The microenvironment as a source of volatile organic compounds in air. *Environ. Int.,* submitted.

Daisey, J. M. (1999) Tracers for assessing exposure to environmental tobacco smoke: what are they tracing? *Environ. Health Perspect.*, 107(Suppl. 2), 319–327.

Dechow, M. & Sohn, H. (1997) Concentrations of selected contaminants in cabin air of airborne aircrafts. *Chemosphere*, 35, 21–31.

Dietert, R. R. & Hodge, A. (1996) Toxicological considerations in evaluating indoor air quality and human health: impact of new carpet emissions. *Crit. Rev. Toxicol.*, 26, 633–707.

Flannigan, B. (1990) Symposium on bioaerosols. *Proceedings of the 5th International Conference on Indoor Air Quality and Climate, Volume 2,* Toronto, Canada, pp. 1-121.

Fortmann, R. & Roache, N. (1998) Characterization of emissions of volatile organic compounds from interior alkyd paint. *J. Air & Waste Manage. Assoc.*, 48, 931–940.

Godish, T. (1989) *Indoor Air Pollution Control,* Lewis Publishers, 121 South Main Street, Chelsea, MI 48118, 401 pp.

Gold, K. W., Naugle, D. F., Berry, M. A. (1990) *Indoor Concentrations of Environmental Carcinogens,* Research Triangle Institute, Report Number 4479/07-F, P. O.Box 12194, Research Triangle Park, NC 27709-2194, April 1990, 40 pp.

Greenwood, M. R. (1990) The toxicity of isoparaffinic hydrocarbons and current exposure practices in the non-industrial (office) indoor air environment. *Proceedings of the 5th International Conference on Indoor Air Quality and Climate, Volume 5,* Toronto, Canada, pp. 169-175.

Heavner, D. L., Morgan, W. T., & Ogden, M. W. (1996) Determination of volatile organic compounds and respirable suspended particulate matter in New Jersey and Pennsylvania homes and workplaces. *Environ. Int.*, 22, 159–183.

Highsmith, V. R., & Rodes, C. E. (1988) Indoor particle concentrations associated with use of tap water in portable humidifiers. *Environ. Sci. Technol.,* 22, 1109-1112.

Howard–Reed, C. & Corsi, R. L. (1999) Mass transfer of volatile organic compounds from drinking water to indoor air: the role of residential dishwashers. *Environ. Sci. Technol.*, 33, 2266–2272.

Jenkins, R. A. & Counts, R. W. (1999) Occupational exposure to environmental tobacco smoke: Results of two personal exposure studies. *Environ. Health Perspect.*, 107, 341–348.

Kavouras, I. G., Statigakis, N., & Stephanou, E. G. (1998) Iso- and ante-iso-alkanes:specific tracers of environmental tobacco smoke in indoor and outdoor particle-size distributed urban aerosols. *Environ. Sci. Technol.,* 32, 1369–1377.

Kerr, G. & Sauer, P. (1990) Control strategies for liquid process photocopier emissions. *Proceedings of the 5th International Conference on Indoor Air Quality and Climate, Volume 3,* Toronto, Canada, pp. 759-764.

Kliest, J., Fast, T., Boley, J. S. M., van de Weil, H., & Bloeman, H. (1989) The relationship between soil contaminated with volatile organic compounds and indoor air pollution. *Environ. Int.,* 15, 419-425.

Klus, H., Begutter, H., Ball, M., & Intorp, M. (1987) Environmental tobacco smoke in real life situations. *Proceedings of the 4th International Conference on Indoor Air Quality and Climate, Volume 2,* Berlin (West), pp. 137-141.

Krafthefer, B. C. & MacPhaul, D. (1990) Ultrafine particle emission from baseboard and other resistance-type heaters. *Proceedings of the 5th International Conference on Indoor Air Quality and Climate, Volume 3,* Toronto, Canada, pp. 659-664.

Lawryk, N. J. & Lioy, P. J. (1995) Exposure to volatile organic compounds in the passenger compartment of automobiles during periods of normal and malfunctioning operation. *J. Expo. Anal. Environ. Epidemiol.,* 5, 511–531.

Lebowitz, M. D., Holberg, C. J., O'Rourk, M. K., Corman, G., & Dodge, R. (1983) Gas stove usage CO and TSP and respiratory effects. *76th Annual Meeting of Air Pollution Control Association (APCA),* Atlanta, GA, June 19-24, 1983.

Leslie, N. P. & Billick, 1. H. (1990) Examination of combustion products in an unoccupied research house. *Proceedings of the 5th International Conference on Indoor Air Quality and Climate, Volume 2,* Toronto, Canada, pp. 349-354.

Matsushita, H., Tanabe, K., Koyano, M., Laquindanum, J., & Lim-Sylianco, C. Y. (1990) Automatic analysis for polynuclear aromatic hydrocarbons indoors and its application to human exposure assessment. *Proceedings of the 5th International Conference on Indoor Air Quality and Climate, Volume 2.* Toronto, Canada, pp. 287-292.

McLaughlin, P. & Aigner, R. (1990) Higher alcohols as indoor air pollutants: source, cause, mitigation. *Proceedings of the 5th International Conference on Indoor Air Quality and Climate, Volume 3,* Toronto, Canada, pp. 587-591.

Molhave, L. (1982) Indoor air pollution due to organic gases and vapours of solvents in building materials. *Environ. Int.* 8, 117-127.

Molhave, L. (1984) Volatile organic compounds as indoor air pollutants. In: Gainmage, R. B. & Kaye, S. V., eds., *Indoor Air and Human Health,* Lewis Publishers, Inc., 121 S. Main Street, Chelsea, MI 48118, pp. 403-417.

Molhave, L. (1986) Indoor air quality in relation to sensory irritation due to volatile organic compounds *ASHRE Trans.,* 92(IA), 306-316.

Moschandreas, D. J., Winchester, J. W., Nelson, J. W., & Burton, R. M. (1979) Fine particle residential indoor air pollution. *Atmos. Environ.*, 13, 1413-1418.

Mumford, J. L., Lewtas, J., Burton, R. M., Svendsgaard, D. B., Houk, V. S., Williams, R. W., Walsh, D. B. & Chuang, J. C. (1990) Unvented kerosene heater emissions in mobile homes: studies on indoor air particles, sernivolatile organics, carbon monoxide, and mutagenicity. *Proceedings of the 5th International Conference on Indoor Air Quality and Climate, Volume 2*, Toronto, Canada, pp. 257-262.

Nagda, N., Fortmann, R., Koontz, M., & Konheirn, A. (1990) Investigation of cabin air quality aboard commercial airliners. *Proceedings of the 5th International Conference on Indoor Air Quality and Climate, Volume 2*, Toronto, Canada, pp. 245-250.

Pellizzari, E. D., Thomas, K. W., Smith, D. J., Perritt, R. L., & Morgan, M. A. (1989) *Total Exposure Assessment Methodology (TEAM): 1987 Study in New Jersey, Volume I, RTI/108/00*, Research Triangle Institute, P. O. Box 12194, Research Triangle Park, NC 27709-2194.

Pellizzari, E. D., Perritt, K., Hartwell, T. D., Michael, L. C., Sparacino, C. M., Sheldon, L. S., Whitmore, R., Leninger, C., Zelon, H., Hardy, R. W., & Smith, D. (1987) *Total Exposure Assessment Methodology (TEAM) Study. Elizabeth and Bayonne, New Jersey, Devils Lake, North Dakota and Greensboro, North Carolina, Volume II, Part II, Final Report*, EPA/600/6-87/002b, p. 328.

Perritt, R. L., Hartwell, T. D., Sheldon, L. S., Cox, B. G., Smith, M. L., & Rizzuto, J. E. (1990) Distribution of N02, CO, and respirable suspended particulates in New York state homes. *Proceedings of the 5th International Conference on Indoor Air Quality and Climate, Volume 2*, Toronto, Canada, pp. 251-256.

Plehn, W. (1990) Solvent emission from paints. *Proceedings of the 5th International Conference on Indoor Air Quality and Climate, Volume 3*, Toronto, Canada, pp. 563-568.

Reponen, T. R., Hyvärinen, A., RuushKannen, J., Raunemaa, T., & Nevalainen, A. (1994) Comparison of concentrations and size distributions of fungal spores in buildings with and without mould problems. *J. Aerosol. Sci.*, 25, 1595–1603.

Rickert, W. S. (1995) An assessment of the relative contributions of burning wood, candles, lamps, incense, and cigarettes to levels of particulates, nicotine, benzo(a)pyrene, carbonyls, solanesol, hydrogen cyanide, benzene, nitric oxide, and carbon monoxide in ambient air. Tobacco Chemists' Research Conference, Lexington, KY, September 24–27 (Labstat Incorporated, 2652 Manitou Dr., Kitchener, Ontario, Canada, N2C 1L3).

Rogge, W. F., Hildemann, L. M., Mazurek, M. A., Cass, G. R., & Simoneit, B. R. (1994) Sources of fine organic aerosol. 6. Cigarette smoke in the urban atmosphere. *Environ. Sci. Technol.*, 28, 1375–1388.

Samnaljarvi, E., Laaksonen, A., & Raunemaa, T. (1990) Aerosol and reactive gas effects by electric heating units. *Proceedings of the 5th International Conference on Indoor Air Quality and Climate, Volume 3*, Toronto, Canada, pp. 653-658.

Schriever, E. & Marutzky, R. (1990) VOC emissions of coated parqueted floors. *Proceedings of the 5th International Conference on Indoor Air Quality and Climate, Volume 3*, Toronto, Canada, pp. 551-556.

Shah, J. J. & Singh, H. B. (1988) Distribution of volatile organic chemicals in outdoor and indoor air: a national VOCs database. *Environ. Sci. Technol.*, 22, 1381–1388.

Sheldon, L. S., Handy, R. W., Hartwell, T. D., Whitmore, R. W., Zelon, H. S., & Pellizzari, E. D. (1988a) *Indoor Air Quality in Public Buildings, Volume I*, EPA/ 660/6-88/009a.

Sheldon, L. S., Zelon, H., Sickles, J., Eaton, C., & Hartwell, T. D. (1988b) *Indoor Air Quality in Public Buildings, Volume II*, EPA/660/6-88/009b.

Smith, D. D., Donovan, R. P., Ensor, D. S., & Sparks, L. E. (1990) Quantification of particulate emission rates from vacuum cleaners. *Proceedings of the 5th International Conference on Indoor Air Quality and Climate, Volume 3*, Toronto, Canada, pp. 647-652.

Tichenor, B. A. & Mason, M. A. (1988) Organic emissions from consumer products and building materials to the indoor environment. *JAPCA*, 38, 264-268.

Traynor, G. W. , Apte, M. G., Sokol, H. A., Chuang, J. C., Tucker, W. G., & Mumford, J. L. (1990) Selected organic pollutant emissions from unvented kerosene space heaters. *Environ. Sci. Technol.*, 24, 1265-1270.

Traynor, G. W., Apte, M. G., Carruthers, A. R., Dillworth, J. F., Grimsrud, D. T., & Gundel, L. A. (1987) Indoor air pollution due to emissions from wood-burning stoves. *Environ. Sci. Technol.*, 21, 691-697.

Tsuchiya, Y. & Stewart, J. B. (1990) Volatile organic compounds in the air of Canadian buildings with special reference to wet process photocopying machines. *Proceedings of the 5th International Conference on Indoor Air Quality and Climate, Volume 2*, Toronto, Canada, pp. 633-638.

Tu, K. W. & Hinchliffe, L. E. (1983) A study of particle emissions from portable space heaters. *J. Am. Ind. Hygiene Assn.*, 44, 857-862.

Turk, B. H., Brown, J. T., Geisling-Sobotka, K., Froehlich, D. A., Grimsrud, D. T., Harrison, J., Koonce, J. F., Prill, R. J., & Revzan, K. L. (1987) *Indoor Air Quality and Ventilation Measurements in 38 Pacific Northwest Commercial Buildings, Volume I—Measurement Results and Interpretation (LBL 22315 ½), Volume II—Appendices (LBL 22315 2/2)*, Lawrence Berkeley Laboratory, Berkeley, CA 94720.

Turner, S. & Binnie, P. W. H. (1990) An indoor air quality survey of twenty-six Swiss office buildings. *Proceedings of the 5th International Conference on Indoor Air Quality and Climate, Volume 4,* Toronto, Canada, pp. 27-32.

Virelizier, H., Gaudin, D., Anguenot, F., & Aigueperse, J. (1990) An assessment of the organic compounds present in domestic aerosols. *Proceedings of the 5th International Conference on Indoor Air Quality and Climate, Volume 3,* Toronto, Canada, pp. 737-741.

Wallace, L., Pellizzari, E. D4, & Wendel, C. (1990) Total organic concentrations in 2500 personal, indoor, and outdoor air samples collected in the USEPA TEAM studies. *Proceedings of the 5th International Conference on Indoor Air Quality and Climate, Volume 2,* Toronto, Canada, pp. 639-644.

Wieslander, G. & Norback, D. (1997) Asthma and the indoor environment: the significance of formaldehyde and volatile organic compounds from newly painted indoor surfaces. *Int. Arch. Occup. Environ. Health,* 69, 115–124.

3

Mainstream and Sidestream Cigarette Smoke

INTRODUCTION

Mainstream cigarette smoke is the material drawn from the mouth-end of a cigarette during a puff. Sidestream cigarette smoke is the material released directly into the air from the burning tip of the cigarette plus that which diffuses through the cigarette paper. The material issuing from the mouth-end of the cigarette between puffs is sometimes considered a component of sidestream smoke. Most sidestream smoke is emitted between puffs, but some is also released during a puff.

Cigarette smoke is among the most extensively studied materials. This is especially the case for mainstream smoke, and increased attention has been given to sidestream smoke. Early work focused on identifying the constituents of mainstream smoke that were important to product quality and to determining the relationship of those constituents to tobacco content and to cigarette design. An early review (Kosak 1954) listed fewer than 100 components identified in mainstream smoke. Later work included extensive studies of mainstream smoke in search of constituents that might be responsible for adverse health impacts attributed to smoking. With the introduction of new and more sensitive analytical methods, an ever-increasing number of smoke constituents were reported. By 1982 approximately 4000 individual constituents had been identified (Dube and Green 1982). An updated review by Roberts (1988) and recent summaries (Anonymous 1997, Hoffmann and Hoffmann 1998) continue to report the presence of approximately 4000 constituents. It has been suggested that the total number of mainstream smoke components may be ten to twenty times the number of identified components; i.e., mainstream cigarette smoke might include as many as 100,000 components (Rodgman 1991). It is to be noted, however, that the approximately 4000 mainstream smoke components identified to date account for more than 95% of mainstream smoke weight. Thus, most of the thousands of mainstream components that may not yet have been identified must be delivered at extremely low levels (nanograms, picograms, etc., per cigarette).

Major attention to sidestream smoke is more recent, but results to date suggest that it is equally complex. Sidestream smoke composition is qualitatively similar to mainstream smoke composition in that it contains the same constituents. It is quantitatively different, however, in that the relative quantities of many individual constituents present are different from those found in mainstream smoke. Also, the absolute quantities of most constituents released in sidestream smoke differ from those delivered in mainstream smoke.

The cigarette itself and factors that influence its deliveries have also been extensively studied. Fundamental characteristics such as temperature distributions, gas distributions, and air flux in and around the firecone (burning tip) have been measured. The influences of tobacco type, tobacco processing, paper porosity, filtration, filter ventilation, and major additives, such as humectants, have been studied. The influence of smoking parameters, such as puff volume, puff duration, puff frequency, butt length, and ambient environmental conditions, have been considered for at least major constituents of interest, such as tar, nicotine, and carbon monoxide. Most studies have centered on mainstream smoke, but many also consider sidestream smoke.

Much of the literature on cigarette smoke chemistry appears in commonly available peer-reviewed journals. Much, however, is found in specialty journals, proceedings of specialty symposia, and in institutional or government reports. Particularly important sources are (a) *Beitrage zur Tabakforschung International, Verband der Cigarettenindustrie,* Koenigswinterer Strasse 550, D-5300 Bonn, Germany; (b) *Tobacco Science,* annual compilation of peer-reviewed scientific articles published in the weekly journal *Tobacco International,* 130 W. 42nd St., Suite 2200, New York, NY 10036 through calendar year 1990 and published in *Tobacco Reporter,* 3000 Highwoods Blvd., Suite 3000, Raleigh, NC 27625 beginning in calendar year 1991; (c) *Proceedings of Symposia on Recent Advances in Tobacco Science,* published as part of the annual Tobacco Scientists' (formerly Tobacco Chemists') Research Conference; and (d) The annual *Surgeon General's Report on Smoking and Health,* Department of Health and Human Services, Office on Smoking and Health, Rockville, MD 20857. Reviews and treatises providing additional information on cigarette smoke chemistry referenced in this chapter include Wynder and Hoffmann 1967; Baker 1980a; Dube and Green 1982; Guerin 1980, 1991; Guerin et al. 1987; Johnson et al. 1973; Klus and Kuhn 1982; Klus 1990; National Research Council (NRC) 1986; O'Neill et al. 1987; Surgeon General 1986; Anonymous 1997; Hoffmann and Hoffmann 1998; and Hoffmann et al. 1997. Reports referenced in this work that are not available from the publisher may be available through the authors.

A definitive treatment of cigarette smoke chemistry is beyond the scope of a single chapter. The intent of this chapter is to summarize findings to date related to sidestream smoke as a source term for environmental tobacco smoke (ETS) and to draw attention to analytical conventions and to the extensive literature related to cigarette smoke chemistry.

TERMS AND CONVENTIONS

The manner in which a cigarette is smoked greatly influences its mainstream delivery and also influences sidestream emissions. It has thus been necessary to standardize smoking conditions for experimental and testing purposes in order to ensure that results can be compared. Standard smoking conditions in the United States require the use of a 35 ± 2-cc puff volume of 2 ± 0.2-s duration taken once per minute to a 23-mm butt for nonfilter cigarettes and to 3 mm from the filter overwrap for filter cigarettes. These conditions are employed in U.S. Federal Trade Commission (FTC) testing of the tar, nicotine, and carbon monoxide mainstream deliveries of commercial cigarettes (Federal Trade Commission 1997). The international convention is similar but employs a 30-mm butt length rather than a 23-mm length.

Standard smoking conditions were developed from the work of Bradford et al. (1936) and were widely adopted in the early 1960s. It was recognized early on that individuals smoked cigarettes in different and varying ways (Wakeham 1972 reported human puff volumes to range from 17–73 cc, durations from 0.9–3.2 s, and intervals from 22–72 s) but the standard conditions were felt to reflect common smoking conditions of the time. The most popular cigarettes in today's market, however, are filtered rather than nonfiltered cigarettes and the market share of low-tar (10–15 mg tar or less) and very-low-tar (less than 5 mg tar) is increasing. Smokers tend to compensate for the lower tar, lower nicotine deliveries by taking larger puffs, puffing more frequently, or smoking more cigarettes (e.g., Russell 1977; Herning et al. 1981; Djordjevic, et al. 1995, 1997). The puff duration and butt length of standard smoking conditions remain reasonable, but puff volumes of 45–55 cc and puff frequencies of 2–3 per minute may better reflect human smoking conditions for today's lower tar and nicotine products. This suggests that using the standard smoking parameters on today's more popular products will provide a significant underestimate of mainstream deliveries and at least a measurable overestimate of sidestream emissions as compared to human smoking.

Much attention is being given to developing a standard machine-smoking protocol that better reflects current cigarette smoking practices (e.g., National Cancer Institute 1996; Eberhardt and Scherer 1995; Borgerding 1997). The FTC standard (e.g., Rodgman 1977) was adopted for comparing brands and not for measuring human exposure; however, FTC results have become used as an indication of relative exposure by consumers. Major differences between perceived (from FTC results) and actual (human-smoking) exposures can result from the differences in smoking practices. Differences are especially great for cigarettes that deliver very small quantities of tar and nicotine by the FTC test (e.g., Rickert et al. 1983, Djordjevic et al. 1995, National Cancer Institute 1996). Laboratory studies will likely continue to employ FTC smoking conditions in order to allow comparison with the existing literature.

Several less-obvious parameters must also be controlled for experimental smoking. The two most important are the moisture content of the cigarette and the air flow around the cigarette. Moisture content is controlled by the standard practice of conditioning the test cigarettes for at least 48 hours in an environment maintained at $60 \pm 2\%$ relative humidity and $24 \pm 1\,°C$ prior to smoking. Ideally, the cigarettes are also smoked in a room maintained at the same conditions, but adequate results for most purposes can be obtained for cigarettes smoked under ambient conditions if they are transferred to the machine shortly after removal from the humidifier.

Air flow across the cigarette is important because it affects the speed with which a cigarette is consumed between puffs and thus affects the number of standard puffs that can be taken from the cigarette. High air flows result in rapid consumption (a high "static burn rate"). The optimal flow is one that vents the sidestream smoke while minimally affecting the static burn rate. This is generally experimentally determined by comparing the number of standard puffs generated by a reference cigarette in the apparatus used with the number of puffs reported by others or determined by the investigators using standard smoking machines employed for FTC testing. Comparisons with FTC tar, nicotine, and/or carbon monoxide results are used when the number of puffs is not available.

Several other parameters have been found to influence deliveries of constituents determined by machine smoking. These are the "puff profile" (the pressure/air-flow change over the course of the 2-s puff), the configuration of the cigarette holder, and whether the cigarette butt is open to the atmosphere ("free smoking") or closed from the atmosphere ("restricted smoking") between puffs. The preferred conditions are a Gaussian puff profile, a cigarette holder that secures the cigarette without leakage but does not physically indent the butt, a cigarette held in the horizontal position or slightly elevated position as opposed to a vertical position, and "free smoking". Most commercial and investigator-constructed smoking machines meet all these requirements except for free smoking. Many machines isolate the butt end of the cigarette between puffs.

The question of cigarette-to-cigarette variability must also be addressed in experimental smoking. Although cigarettes of any given brand produced with a given formulation are remarkably uniform and are essentially equivalent to the consumer, experimental measurement methods are sufficiently sensitive to be affected by subtle differences between individual cigarettes. Such effects are important because the results of testing only a very few cigarettes are extrapolated to the product as a whole and to associated human exposures. This problem has been addressed in various ways. Products for FTC testing (now carried out by the Tobacco Institute Testing Laboratory) of commercial cigarettes are acquired by purchasing two packages of each variety at each of 50 locations throughout the U.S. Two cigarettes per pack are selected at random to provide a sample size of 100 cigarettes for testing. An alternative approach is to select cigarettes for testing from a smaller sample based on weight and pressure drop ("resistance-to-draw"). Early studies for the National Cancer Institute that considered specially manufactured experimental cigarettes selected only those cigarettes for testing that fell within ± 20 mg of the average weight of 200 cigarettes of each type and ± 10% of their average pressure drop. Pressure drop is defined as the difference in pressure between the butt end and the cigarette tip while air is drawn from the butt end at a rate of 17.5 cc/s. Principal concerns for research purposes are that the cigarettes studied not be physically damaged, not be overly dry or wet, and not be at the extremes of weight and pressure drop of the sample set selected for study.

The introduction of reference cigarettes (Diana and Vaught 1990) has been a major contribution to the standardization of tobacco smoke chemistry studies. The Tobacco and Health Research Institute of the University of Kentucky at Lexington provides reference nonfilter, filter, and special-interest products to the scientific community. The cigarettes are formulated and manufactured to represent commercial products omitting only those ingredients that give each brand its unique characteristics. Kentucky Reference Cigarettes have become the benchmark for development and standardization of analytical methods. Table 3.1 summarizes the results of a study by the R. J. Reynolds Tobacco Company (RJR 1988) that used the Kentucky Reference IR4F cigarette as a reference product for comparison with a sidestream-free experimental cigarette. The Kentucky Reference 1R1 (second generation, 2R1) nonfilter cigarette has not been as systematically studied for sidestream delivery.

The air dilution filter, a now common component of cigarette design, has confounded some investigations of cigarette smoke chemistry. Air dilution filters, introduced in the late 1960s (Grise 1984), use barely visible perforations toward the tip-end of the cigarette filters and are designed to dilute smoke from the firecone with ambient air during a puff. Some products have employed highly air-permeable filter wraps to allow a similar dilution of smoke. The mainstream deliveries of such products are highly dependent on the air-dilution mechanism being open to ambient air both for its diluting effect and because it influences pyrosynthesis of certain constituents (Norman 1974). Obstruction of the air-dilution region by the cigarette holder results in a greatly increased mainstream cigarette smoke delivery. Such obstruction by the mouth has been found to be a means of smoker compensation to ultra-low-tar cigarettes (e.g., Koslowski et al. 1980, National Cancer Institute 1996), and some degree of obstruction is incorporated in smoking protocols proposed (National Cancer Institute 1996) or used (BCM 1998) to simulate human smoking conditions.

Mainstream cigarette smoke is commonly described as comprising a particulate phase and a gas phase. The particle phase is experimentally defined as those materials that are collected on standard Cambridge glass fiber filters (Cambridge Filter Corporation, East Syracuse, NY); the gas phase is experimentally defined as those materials that pass through the filters. Cambridge filters collect particulate matter of 0.2-μm diameter or larger and have an efficiency of greater than 99%.

Table 3.1 Mainstream and Sidestream Deliveries of the Kentucky Reference 1R4F Cigarette[a]

	Per Cigarette		Ratio SS/MS
	Mainstream	Sidestream	
FTC Tar (mg)	-	16.9	-
Total Particulate Matter (mg)	0.79	5.6	7.1
Nicotine (mg)	11.3	54.1	4.8
Carbon Monoxide (mg)	41.9	474	11.3
Carbon Dioxide (mg)	0.23	0.9	3.9
Nitrogen Oxides (mg)	0.02	9.1	455
Ammonia (mg)	0.02	0.73	36.5
Formaldehyde (mg)	0.63	4.2	6.7
Acetaldehyde (mg)	0.07	1.3	18.6
Acrolein (mg)	-	0.9	-
Propionaldehyde (mg)	89.0	53.3	0.6
Hydrogen Cyanide (μg)	45.2	299	6.6
Benzene (μg)	10.5	197.5	18.8
Benz[a]anthracene (ng)	9.2	147.9	16.0
Benzo[a]pyrene (ng)	101.0	171	1.7
NNN[b] (ng)	84.0	419	5.0
NNK[b] (ng)	114.0	110	1.0
NAT[b] (ng)	18.0	13	0.7
NAB[b] (ng)	ND	298	-
DMNA (ng)	ND		-
EMNA[b] (ng)	ND	-	-
DENA[b] (ng)	ND	-	
NPYR[b] (ng)	14.0	182	13

[a]R. J. Reynolds, 1988, unless otherwise noted.
[b]N-nitrosonornicotine (NNN), 4-Methylnitrosoamino-1-(3-pyridinyl)-1-butanone (NNK), N-nitrosoantabine (NAT), N-nitrosoanabasine (NAB), Dimethylnitrosamine (DMNA), Ethylmethylnitrosamine (EMNA), Diethylnitrosamine (DENA), Nitrosopyrrolidine (NPYR).
ND=not detected.

The distinction between particle-phase and gas-phase constituents is appropriate for those constituents that are clearly nonvolatile (e.g., high molecular weight multifunctional organics and most metals) and those that are clearly gases (e.g., carbon monoxide). Constituents with appreciable vapor pressures (i.e., most of the constituents of tobacco smoke) can be found in both the particle phase and the gas phase of cigarette smoke. The term "semivolatiles" has been used to describe such constituents. These compounds distribute between the particle and gas phases to a degree determined by their volatility and the circumstances of their environment. Highly concentrated smokes, such as those inhaled by a smoker, find these constituents preferentially distributed in the particle phase. Highly diluted smokes, such as those encountered by the passive smoker, find these constituents preferentially distributed in the gas phase. Experimental studies find these constituents variously distributed between the particle phase and vapor phase, depending on the conditions employed for sampling and analysis.

Most individual constituents of tobacco smoke are reported as quantities of the individual constituents present. The terms "tar" and "total particulate matter" (TPM) have been used to describe materials in tobacco smoke whose compositions are not fully defined. The term "TPM" is used to mean the total material (weight-gain) collected on standard glass fiber Cambridge filters. The term "tar" is used to mean mainstream TPM minus the quantities of water and nicotine in the TPM. Material collected in cold traps is commonly referred to as cigarette smoke condensate (CSC) and is variously reported as CSC (including water) and dry CSC (corrected for water content).

Conventions for sidestream smoke analysis are not as well defined as those for mainstream smoke. Devices used to generate and collect sidestream smoke for analysis are generally optimized to provide comparable mainstream deliveries to those provided by standard smoking machines while allowing the simultaneous collection of sidestream emissions (Dube and Green 1982; McRae 1990; Proctor et al. 1988; Perfetti et al. 1998; see Fig. 3.1). Sidestream collection systems account for the material emitted at the firecone and through the cigarette paper but not the material emitted from the butt of the cigarette between puffs. A standard method for the generation and collection of sidestream smoke has not yet been defined.

It is important to recognize that mainstream and sidestream "tar" and particulate matter bear only an operational relationship. Mainstream particulate matter exists as an entity during active smoking and enters the respiratory tract largely intact. Sidestream particulate matter is available for inhalation only after dilution in ambient air and whatever physical and chemical composition changes that attend that dilution. The principal objective of sidestream smoke analysis is to determine the quantities of individual constituents released into the air. The distribution of the constituents between the particle phase and vapor phase in indoor air and their actual concentrations depends on the

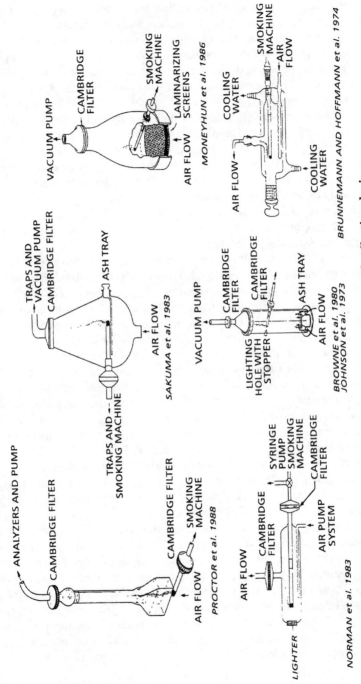

Fig. 3.1. Sidestream smoke generation and collection devices.

degree of vapor phase in indoor air of dilution, on ventilation, and on the nature of surfaces available for adsorption and desorption, and on the materials in indoor air from sources other than tobacco smoke. It is because of this that an increasing number of studies employs dilution chambers (e.g., Mahanama and Daisey 1996) or use actual indoor environments.

FORMATION AND PROPERTIES

Physical and chemical processes that result in the formation of cigarette smoke have been studied in considerable detail. The classic review by Baker (Baker 1980b) details firecone temperature and gas flux distributions and their contributions to mainstream and sidestream smoke formation and delivery. A more recent review (Baker and Robinson 1990) updates the findings and includes the results of more recent studies on the influence of perforated cigarette papers and other variables (e.g., Baker 1989). Studies involving the addition of radiologically labeled tobacco constituents to cigarettes followed by detailed analyses of the fate and distribution following smoking have contributed much to the current understanding of chemical factors that contribute to smoke composition (e.g., Jenkins et al. 1970, 1971, 1990; Bass et al. 1989). Techniques for measuring the yield of nicotine over the course of a puff (Crooks and Lynn 1992) and for time-resolved measurement of sidestream particulate emissions (Dittmann et al. 1992) promise to provide further insights.

During a puff, gas-phase temperatures reach 850°C at the core of the firecone and solid-phase temperatures reach 800°C at the core and 900°C or greater at the char line (the line of paper burn). Core temperatures are high enough to carbonize the tobacco and thus produce a region of the firecone that is highly impermeable to air. This region is thus oxygen-deficient and contributes to the formation of smoke constituents through reductive processes. The relative impermeability of the firecone coupled with the high porosity of the cigarette paper results in most of the puffing air in the region of the firecone to enter near the char line. Both the gas-phase and solid-phase temperatures drop from the range of 800–900°C to the range of 200–400°C within 2 mm of the char line and to 30°C within 1–2 cm of the char line. The extremely high temperature of the firecone coupled with the steep temperature gradient produces smoke. Gases formed by combustion and pyrolysis condense as they leave the firecone region and are cooled. The constituents are further cooled and are diluted by air entering through the cigarette paper.

The region of the cigarette immediately behind the char line is of importance beyond its role as a condensation region. Temperatures of 200°C to 400°C are insufficient for combustion but are sufficient for distillation (and, to a lesser degree, pyrolysis). Distillation behind the firecone is enhanced by the localized high concentrations of water (Johnson et al. 1973). The process may be likened to steam distillation. At least 1200 thermally stable constituents of tobacco (e.g., nicotine, n-paraffins, some terpenes) are transferred intact into the smoke stream by distillation from the tobacco rod behind the firecone (Rodgman 1991). The degree of transfer depends on the volatility, stability, and chemistry of the constituent.

Conditions producing sidestream smoke differ from those producing mainstream smoke in at least two important ways. First, firecone temperatures are lower (reaching approximately 600°C) during the period between puffs than during active puffing. Second, the flow of air is convectively driven in the reverse direction of air flow accompanying a puff. Convection driven by the high temperature of the firecone is sufficiently strong that sidestream gases are emitted from the firecone even during a puff.

Mainstream and sidestream smokes are produced by the same fundamental processes operating on the same fuel and thus consist of the same constituents. However, differences in their conditions of formation result in quantitative differences in their compositions (e.g., Klus and Kuhn 1982, Guerin et al. 1987). The relative quantities of individual constituents in mainstream and sidestream smoke can differ greatly depending on their mechanisms of formation (Table 3.1). Sidestream smoke is slightly more alkaline than is mainstream smoke. It contains much greater quantities of ammonia and generally greater quantities of organic bases, it is relatively depleted in acidic constituents, and it contains lesser quantities of hydrogen cyanide than does mainstream smoke.

Both sidestream and mainstream smoke are highly complex and dynamic aerosols. Both consist of submicron (0.1-0.3 μm diam) electrically charged liquid particles in a complex mixture of air, combustion gases, and a wide variety of gas- and vapor-phase constituents. Mainstream smoke contains approximately 5×10^9 particles per cubic centimeter. The degrees to which the constituents of both mainstream and sidestream smoke distribute between the particle phase and vapor phase depends on their individual physical and chemical properties and their environments. Constituents of sidestream smoke are especially subject to changes in phase distribution because they are rapidly and extensively diluted in ambient air.

MAINSTREAM AND SIDESTREAM CONSTITUENTS

The deliveries of mainstream and sidestream smoke constituents are most commonly determined by machine smoking under standard FTC conditions. Mainstream deliveries are determined with investigator-constructed smoking machines capable of achieving the standard experimental smoking parameters or with commercial smoking machines (e.g., Fidus 1998). Sidestream deliveries are generally determined with investigator-constructed sidestream collection devices in conjunction with mainstream smoking machines. Figure 3.1 illustrates a number of the sidestream collection devices used to date. Most devices are designed to collect the sidestream smoke of single cigarettes, but multi-port smoking machines have also been modified to allow the study or collection of larger quantities of sidestream smoke (e.g., Ueno and Peters 1986, Chortyk and Schlotzhauer 1986).

Sidestream collection devices must be capable of collecting sidestream smoke without influencing smoking characteristics. That is, the number of standard puffs required to consume a cigarette and the mainstream deliveries of tar, nicotine, and carbon monoxide should be the same in the presence of the sidestream apparatus as they are in its absence. It is also important that the cigarette be smoked in an environment of fresh air comparable with that associated with human smoking and standard FTC machine smoking. Devices that highly confine the burning cigarette (e.g., Norman et al. 1983, Brunnemann et al. 1977) have been found to result in smoking environments highly elevated in temperature and moisture content and are thus falling from favor relative to other devices (see Fig. 3. 1). The "fishtail" sampler (Proctor et al. 1988; Perfetti et al. 1998; see Fig. 3.1) was designed to maximally collect sidestream smoke while minimally affecting smoking conditions and is currently being considered as a possible standard collection apparatus.

Table 3.2 reproduces mainstream cigarette smoke deliveries and sidestream/mainstream constituent ratios given by the National Academy of Sciences in its review of environmental tobacco smoke (NRC 1986). The

**Table 3.2 National Research Council Summary
of Deliveries by Nonfilter Cigarettes**

Constituent	MS per Cigarette	SS/MS Ratio
Mainstream vapor phase		
Carbon monoxide	10–23 mg	2.5–4.7
Carbon dioxide	20–40 mg	8–11
Carbonyl sulfide	18–42 μg	0.03–0.13
Benzene	12–48 μg	5–10
Toluene	100–200 μg	5.6–8.3
Formaldehyde	70–100 μg	0.1–~50
Acrolein	60–100 μg	8–15
Acetone	100–250 μg	2–5
Pyridine	16–40 μg	6.5–20
3-Methylpyridine	12–36 μg	3–13
3-Vinylpyridine	11–30 μg	20–40
Hydrogen cyanide	400–500 μg	0.1–0.25
Hydrazine	32 ng	3
Ammonia	50–130 μg	40–170
Methylamine	11.5–28.7 μg	4.2–6.4
Dimethylamine	7.8–10 μg	3.7–5.1
Nitrogen oxides	100–600 μg	4–10
N-Nitrosodimethylamine	10–40 ng	20–100
N-Nitrosodiethylamine	ND–25 ng	<40
N-Nitrosopyrrolidine	6–30 ng	6–30
Formic acid	210–490 μg	1.4–1.6
Acetic acid	330–810 μg	1.9–3.6
Methyl chloride	150–600 μg	1.7–3.3

Table 3.2 continued

Constituent	MS per Cigarette	SS/MS Ratio
Mainstream particle phase		
Particulate matter	15–40 mg	1.3–1.9
Nicotine	1–2.5 mg	2.6–3.3
Anatabine	2–20 μg	<0.1–0.5
Phenol	60–140 μg	1.6–3.0
Catechol	100–360 μg	0.6–0.9
Hydroquinone	110–300 μg	0.7–0.9
Aniline	360 ng	30
2-Toluidine	160 ng	19
2-Naphthylamine	1.7 ng	30
4-Aminobiphenyl	4.6 ng	31
Benz[a]anthracene	20–70 ng	2–4
Benzo[a]pyrene	20–40 ng	2.5–3.5
Cholesterol	22 μg	0.9
γ-Butyrolactone	10–22 μg	3.6–5.0
Quionoline	0.5–2 μg	8–11
Harman[a]	1.7–3.1 μg	0.7–1.7
N'-Nitrosonornicotine	200–3,000 ng	0.5–3
NNK[b]	100–1,000 ng	1–4
N-Nitrosodiethanolamine	20–70 ng	1.2
Cadmium	100 ng	7.2
Nickel	20–80 ng	13–30
Zinc	60 ng	6.7
Polonium–210	0.04–0.1 pCi	1.0–4.0
Benzoic acid	14–28 μg	0.67–0.95
Lactic acid	63–174 μg	0.5–0.7
Glycolic acid	37–126 μg	0.6–0.95
Succinic acid	110–140 μg	0.43–0.62

[a]I-Methyl-9H-pyrido(3,4-b)-indole.
[b]NNK = 4-(N-methyl-N-nitrosamino)-1-(3-pyridyl)-I-butanone.
NRC 1986.

tabulation is limited to the range of deliveries provided by commercial nonfilter cigarettes and to those constituents most expected to contribute to health impacts. Results are highly representative of the nature of mainstream smoke, however, in that they illustrate the variety of constituents present, their variability even within a narrow range of products, and their widely differing quantities. A relatively few constituents (e.g., particulate matter, nicotine, and carbon monoxide) are delivered at milligram-per-cigarette quantities. Many more constituents are delivered at microgram-per-cigarette quantities and even more at nanograms-per-cigarette quantities. In terms of numbers of constituents, most constituents of biological concern are delivered at microgram- and nanogram-per-cigarette quantities in mainstream smoke. Reviews of the overall chemical composition of mainstream smoke and treatments detailing a wide variety of individual constituents of mainstream smoke may be found in the references given in the introduction to this chapter.

The National Academy presentation includes the common practice of relating mainstream and sidestream deliveries by means of sidestream/mainstream (SS/MS) ratios (NRC 1986) (see Table 3.2). The finding is that except for hydrogen cyanide and a few phenolic or carboxylic acid constituents, a much greater quantity of each smoke constituent is released in sidestream smoke than is transported in mainstream smoke.

The use of SS/MS ratios is reasonable for the NRC presentation because it is restricted to a small subset (commercial nonfilter) of cigarettes. The use of SS/MS ratios to indicate SS deliveries is very misleading, however, for filter cigarettes (currently by far the most popular products) and very-low-tar cigarettes (increasingly popular products). This is because product innovations to date such as filtration, ventilation, and tobacco processing have focused on reducing mainstream deliveries. Mainstream deliveries have been dramatically reduced while sidestream deliveries have stayed the same or have been only slightly changed. The result is that SS/MS ratios increase only because mainstream deliveries decrease.

An early study (Table 3.3) of the influence of puff volume and filter ventilation on sidestream and mainstream deliveries illustrates the misleading nature of SS/MS ratios. The investigators found the expected increase in mainstream delivery of particulate matter and carbon monoxide with an increase in puff volume and the expected decrease with increasing filter ventilation. SS/MS ratios vary by factors of 2 to 10, depending on the conditions and constituents considered, but the actual sidestream deliveries are almost constant (Browne et al. 1980).

The difficulty of simplifying the relationship between sidestream and mainstream deliveries is perhaps best illustrated by a study of commercial low-tar cigarettes (Table 3.4). The investigators determined the sidestream and mainstream deliveries of 19 brands of commercial cigarettes rated as delivering (FTC mainstream) from 10 to 1 mg/cigarette of tar (Chortyk and Schlotzhauer 1989). The results confirm the general trend of greatly increasing SS/MS ratios with relatively little change in sidestream deliveries as products of lower mainstream delivery are considered. Sidestream deliveries of low tar (low FTC mainstream tar) products were found, however, to rival or exceed deliveries of the nonfilter control cigarette. Individual products were found to vary greatly in SS/MS ratios. Absolute sidestream deliveries were found to vary to a much smaller degree except for a few product-constituent combinations.

A summary of the reported sidestream measurements of constituents considered in this book is given in Table 3.5. Although the data are rather limited, they clearly suggest that sidestream deliveries are relatively constant across products. Differences in sidestream deliveries measured under standard smoking conditions generally fall within factors of 2 to 3 rather than the factors of 10 or more suggested by SS/MS ratios. The results of Sakuma et al. (1983 1984a, 1984b) for other alkaline, acidic, and semivolatile constituents suggest a similar relatively constant sidestream delivery. The lack of a major

Table 3.3 Influence of Puff Volume and Filter Ventilation on Mainstream (MS) and Sidestream (SS) Particulate Matter and Carbon Monoxide Deliveries[a]

Variable[b]	No. Puffs	Milligrams per cigarette and SS/MS ratio					
		Particulate Matter			Carbon Monoxide		
		MS	SS	SS/MS	MS	SS	SS/MS
Puff volume							
None, free burn	0	-	23	-	-	58	-
17.5 cc	9.6	29	23	0.8	9	63	7
35 cc	8.7	46	20	0.4	19	50	2.6
50 cc	7.4	55	21	0.4	20	56	2.8
Filter Ventilation[c]							
0%	8.7	46	20	0.4	19	50	2.6
33%	8.8	32	21	0.6	13	49	3.8
48%	9.8	21	21	1.0	7	58	8.3
83%	10.6	12	21	1.8	2	56	28

[a]Browne et al. 1980.
[b]USA blend cigarette, FTC smoking conditions unless otherwise noted.
[c]Percent of mainstream puff air entering through periphery of filter.

variation in sidestream deliveries across products is consistent with the fact that sidestream deliveries are primarily related to the weight of tobacco and paper consumed during the smolder period. Deliveries are little influenced by cigarette design factors such as filtration that do profoundly influence mainstream smoke delivery.

With very few exceptions (e.g., hydrogen cyanide and organic acids) greater quantities of smoke constituents are released in sidestream smoke on a per-cigarette basis than in mainstream smoke. The concentrations of the constituents are many times greater in mainstream smoke than in environmental tobacco smoke, however. The increased sidestream delivery is massively attenuated by dilution in ambient air.

A number of chemicals known or suspected to contribute to adverse health effects associated with other materials has been specifically sought in cigarette smoke. Mainstream vinyl chloride is reported to be delivered at 12.4 ng/cigarette by the Kentucky Reference nonfiltered 1R1 cigarette and at quantities ranging from 1.3-15.8 ng/cigarette for commercial cigarettes (Hoffmann et al. 1976). Polychlorodibenzodioxins (PCDDs) and polychlorodibenzofurans (PCDFs) have recently been sought in the ten best-selling brands of German cigarettes (Ball et al. 1990). None of the cigarettes were found to contain 2,3,7,8-tetrachlorobenzodioxin. The total mainstream delivery of tetra-octachlorodibenzodioxins expressed as TCDD equivalents ranged from 0.05-0.17 pg/cigarette. The total PCDD deliveries ranged from 4.4-10.3 pg/cigarette: and PCDF deliveries from 1.4 to 5.2 pg/cigarette. Nitropolycyclic aromatics have been sought in the mainstream smoke of the Kentucky Reference 1R1 cigarette and commercial cigarettes and found to be undetectable (< 10 ng/cigarette for 1-nitronaphthalene and 1-nitropyrene and < 1 ng/cigarette for 6-nitrochrysene) (El-Bayoumy et al. 1985. Hydrogen sulfide and sulfur dioxide have been determined in mainstream smoke and found to range from 25 to 89 μg H_2S per cigarette and 3 to 4 μg SO_2 per cigarette (Horton and Guerin 1974). Various trace elements have been determined in tobacco and tobacco smoke (see Table 12.7). The free-radical content of mainstream tar (Pryor et al. 1983) and of mainstream and sidestream gas phase and tar (Pryor, Prier, and Church 1983) has been studied. Radioactive constituents of tobacco and tobacco smoke have been considered (e.g., Martell 1975, 1982).

More recent work has centered on developing methods for better defining the quantities of suspect tobacco-related bioactive constituents in smoke and in assessing the effect of human-smoking conditions on the quantities delivered. Examples include the determination of polycyclic aromatic hydrocarbons (Gmeiner et al. 1997), aromatic amines (Eberhardt and Scherer 1995), benzene (Darrall et al. 1998) and enantiomeric forms of nicotine and secondary alkaloids (Perfetti and Coleman 1998a, 1998b; Perfetti et al. 1998). N-nitrosamines are receiving particular attention (e.g., Song and Ashley 1999,

Table 3.4 Mainstream and Sidestream Deliveries of Low-Tar Cigarettes[a]

Constituent		NF Cntrl	Low-Tar Cigarette (Microgram Per Cigarette)																		
FTC MS Tar (mg/cig.):		23	10	10	9	9	9	10	7	6	6	5	4	3	3	3	3	2	1	1	1
Nicotine	MS	1680	902	1052	695	387	641	655	596	584	216	554	368	343	355	363	86	79	180	138	97
	SS	9075	909	1046	8218	9141	8154	1120	6242	8946	7041	8057	8137	7763	7488	7644	8474	6701	6928	5719	6348
	SS/MS	5.4	10.1	9.9	11.8	14.2	12.7	17.1	10.5	15.3	32.6	14.5	22.1	22.6	21.1	21.1	98.5	84.8	38.5	41.4	65.4
Limonene	MS	56	22	34	30	15	22	25	28	40	20	16	12	23	19	24	2	4	16	10	10
	SS	226	72	65	85	107	69	86	62	467	411	45	45	74	275	405	45	54	404	375	39
	SS/MS	4.0	3.3	1.9	2.8	7.1	3.1	3.4	2.2	11.7	20.1	2.8	3.8	3.2	14.5	16.9	22.5	13.5	25.3	37.5	3.9
Phenol	MS	73	13	13	17	20	15	15	21	70	31	14	8	20	24	44	1	3	7	1	5
	SS	121	94	85	71	111	59	83	56	124	121	78	92	359	261	307	49	44	371	110	70
	SS/MS	1.7	7.2	6.5	4.2	5.6	3.9	5.5	2.7	1.8	3.9	5.6	11.5	18.0	10.9	7.0	49	14.7	53	110	14
Catechol	MS	121	36	44	40	19	34	25	33	27	9	25	16	15	19	21	—	—	—	—	—
	SS	125	162	168	160	101	73	118	82	86	61	93	106	61	75	66	189	189	57	54	64
	SS/MS	1.0	4.5	3.8	4.0	5.3	2.1	4.7	2.5	3.2	6.8	3.7	6.6	4.1	3.9	3.1	—	—	—	—	—
Naphthalene	MS	14	1	4	4	8	6	7	7	9	6	9	6	5	4	7	<1	1	4	2	5
	SS	53	79	73	69	177	117	87	120	170	80	121	69	133	118	90	70	61	113	70	148
	SS/MS	3.8	79	18.3	17.3	22.1	19.5	12.4	17.1	18.9	13.3	13.4	11.5	26.6	29.5	12.9	-	61	28.3	35	29.6
Neophytadiene	MS	83	19	18	17	18	19	14	20	22	10	19	9	18	11	12	3	4	12	8	7
	SS	166	150	143	111	153	117	125	127	150	122	144	114	147	121	115	98	136	144	117	160
	SS/MS	2	7.9	7.9	6.5	8.5	6.2	8.9	6.4	6.8	12.2	7.6	12.7	8.2	11.0	9.6	32.7	34.0	12.0	14.6	22.8
Palmitic acid	MS	156	57	61	52	48	53	46	58	44	24	51	42	33	41	30	16	10	25	16	24
	SS	143	166	188	153	135	136	148	170	92	92	151	128	90	109	86	139	69	84	70	89
	SS/MS	0.9	2.9	3.1	2.9	2.8	2.6	3.2	2.9	2.1	3.8	3.0	3.0	2.7	2.7	2.9	8.7	6.9	3.4	4.46	3.7

[a] Chortyk and Schlotzhauer, 1989.

Table 3.5 Example Sidestream Cigarette Smoke Deliveries[a]

Constituent	Kentucky Reference[b]	Experimental[c]	Commercial
Milligrams/cigarette			
Condensate			36-67
TPM	16.9		16-36, 20-23
Nicotine	5.6	3.2-5.8, 5.9-7.0	5.7-11.2, 2.7-6.1
Carbon Monoxide	54	49-63	41-67
Carbon Dioxide	474	422-598	
Nitrogen Oxides	0.9	1.8-2.7	
Ammonia	9.1	5.7-9.4	
Formaldehyde	0.7		
Acetaldehyde	4.2		
Acrolein	1.3, 1.4		0.7-1.0
Propionaldehyde	0.9		
Benzene	0.3, 0.4, 0.7		0.3-0.5
Toluene	0.8, 1.3		0.8-1.1
Styrene			
Pyrrole	0.4	0.2-0.3	
Pyridine	0.3	0.1-0.2	
3-Vinylpyridine		0.3-0.4	
3-Hydroxypyridine		0.1-0.2	
Limonene	0.3	0.2-0.4	<0.1-0.4
Neophytadiene		0.2-0.4	0.1-0.2
Isoprene	2.5, 6.1		4.4-6.5
nC_{27} - n-C_{33}	0.2-0.8		
Acetonitrile	1.0, 0.8[d]	0.7-1.0	
Acrylonitrile	0.2		
Micrograms/cigarette			
Hydrogen Cyanide	53, 17[d]	53-108, 14-25	
Phenol		69-241	44-371
o-Cresol		17-24	24-98
m + p-Cresol		49-70	59-299
Catechol		138-292	46-189
Hydroquinone		95-295	26-256
Naphthalene			53-177
Phenanthrene			2.4
Anthracene			0.7
Fluoranthene			0.7
Pyrene			0.5
Benz[a]anthracene	0.2		0.2
Benzo[a]pyrene	0.1		0.1
NNN[e]	0.2	1.7-6.1	1.7
NNK[e]	0.4	0.5-0.7	0.4
NAT[e]	0.1		
NAB[e]	<0.1		
DMNA[e]	0.3		0.7-1.0
EMNA[e]			<0.1
DENA[e]			<0.1-0.1
NPYR[e]	0.2		0.2-0.4
2-Naphthylamine			<0.1-1[f]
4-Aminobiphenyl			<0.1-0.2[f]
Nickel		<0.1–0.5	

Table 3.5 continued

Constituent	Kentucky Reference[b]	Experimental[c]	Commercial
Cadmium		<0.1–0.5	
Lead		<0.1–1.0	
Chromium		<0.1	

[a]From Browne et al. 1980; Brunnemann et al. 1977, 1978, and 1990; Chortyk and Schlotzhauer
1989; Grimmer et al. 1987; Guerin 1991; Higgins et al. 1987; Johnson et al. 1973; O'Neill et al.
1987; R. J. Reynolds 1988; Rickert et al. 1984; Sakuma et al. 1983, 1984a, 1984b; Norman et
al. 1983.
[b]Filter 1R4F unless otherwise specified.
[c]Cigarettes containing a single common tobacco type (e.g., Burley, Bright, Turkish, etc.) with or
without additives or containing blends of common tobaccos.
[d]Nonfilter 1R1.
[e]N-Nitrosonornicotine (NNN), 4-Methylnitrosoamino-I-(3-pyridinyl)-I-butanone(NNK),
–Nitrosoanatabine (NAT), N-Nitrosoanabasine (NAB), Dimethyinitrosamine (DMNA), Ethyl
methyl1nitrosamine (EMNA), Diethy1nitrosamine (DENA), N-Nitrosopyrrolidine (NPYR).
[f]Calculated from NRC 1986, SS/MS ratio.

Mitacek et al. 1999) based on the pioneering work of D. Hoffmann and colleagues of the American Health Foundation, Valhalla, NY (e.g., Brunnemann et al. 1978). There is now an Internet site (BCM 1998) that provides both mainstream and sidestream deliveries of 44 constituents by commercial Canadian cigarettes under both standard and simulated human-smoking conditions.

Table 3.6 lists the biologically active constituents of mainstream smoke as given by Hoffmann and Hoffmann 1998. These are the constituents most likely to be sought in sidestream and environmental tobacco smoke.

SUMMARY

Mainstream and sidestream cigarette smokes are chemically and physically complex mixtures consisting of electrically charged submicron liquid particles at very high concentration in a vapor phase consisting of permanent gases, reactive gases, and a large variety of organic chemicals. The composition of the smoke and especially the total quantities of individual constituents delivered are dependent on the conditions of smoke generation. Standard conditions for mainstream smoke generation have been established to allow interlaboratory comparison. Modified smoking conditions are being proposed to better reflect current human-smoking conditions. Sidestream smoke collection has not yet been standardized.

The mechanisms of smoke formation and the chemical composition of mainstream smoke have been studied in great detail. Sidestream smoke has received increased attention as a source term for environmental tobacco smoke. Sidestream smoke is similar to mainstream smoke in that they both contain the same constituents. Sidestream smoke is different from mainstream smoke in that the relative quantities of the constituents present differ.

Table 3.6 Biologically Active Agents in the Mainstream Smoke of Nonfilter Cigarettes[1]

Smoke Compounds	Amount/Cigarette	Smoke Compounds	Amount/Cigarette
PAHs		**N-Nitrosamines**	
Benz[a]anthracene (BaA)	20–70 ng	N-Nitrosodimethylamine (DMNA)	2–180 ng
Benzo[b]fluoranthene (BbFL)	4–22 ng	N-Nitrosoethylmethylamine (EMNA)	3–13 ng
Benzo[j]fluoranthene (BjFL)	6–21 ng	N-Nitrosodiethylamine (DENA)	ND–2.8 ng
Benzo[k]fluoranthene (BkFL)	6–12 ng	N-Nitroso-di-n-propylamine (DPNA)	ND–1.0 ng
Benzo[a]pyrene (BaP)	20–40 ng	N-Nitroso-di-n-butylamine (DBNA)	ND–30 ng
Dibenz[a,h]anthracene (DBA)	4 ng	N-Nitrosopyrrolidine (NPYR)	3–110 ng
Dibenzo[a,l]pyrene (DB,al,Py)	1.7–3.2 ng	N-Nitrosopiperidine (NpiP)	ND–9 n
Dibenzo[a,e]pyrene (DB,ae,Py)	Present	N-Nitrosodiethanolamine (NDELA)	ND–68 ng
Indeno[1,2,3-cd]pyrene (IndPy)	4–20 ng	N'-Nitrosonornicotine (NNN)	120–3700 ng
5-Methylchrysene (5-MeChr)	0.6 ng	N'-Nitrosoanabasine (NAB)	ND–150 ng
		4-(Methylnitrosamino)-1-(3-pyridyl)-1-butanone (NNK)	80–770 ng
Volatile Hydrocarbons		**Aldehydes**	
1,3-Butadiene	20–75 μg	Formaldehyde	70–100 μg
Isoprene	450–1000 μg	Acetaldehyde	500–1400 μg
Benzene	20–70 μg	Acrolein	60–140 μg
Styrene	10 μg	Crotonaldehyde	10–20 μg
Heterocyclic Compounds		**N-Heterocyclic Amines**	
Pyridine	16–40 μg	2-Amino-9H-pyrido[2,3-b]indole (AaC)	25–260 ng
Nicotine	1.0–3.0 mg	2-Amino-3-methyl-9H-pyrido[2,3-b]indole (MeAaC)	2–37 ng
Quinoline	2–180 ng	2-Amino-3-methylimidazo[4,5-b]quinoline (IQ)	0.3 ng
Dibenz[a,h]acridine	0.1 ng	3-Amino-1,4-dimethyl-5H-pyrido[4,3-b]indole (Trp-P-1)	0.3–0.5 ng
Dibenz[a,j]acridine	3–10 ng	3-Amino-1-methyl-5H-pyrido[4,3-b]indole (Trp-P-2)	0.8–1.1. ng
7H-Dibenzo[c,g]carbazole	0.9 ng	2-Amino-6-methyl[1,2-a:3',2'-d]imidazole (Glu-P-1)	6.37–0.89 ng
Furan	18–30 ng	2-Aminodipyrido[1,2-a:3',2'-d]imidazole (Glu-P-2)	0.25–0.88 ng
Benzo[b]furan	Present	2-Amino-1-methyl-6-phenylimidazo[4,5-1]pyridine (PhIP)	11–23 ng

Table 3.6 continued

Smoke Compounds	Amount/Cigarette	Smoke Compounds	Amount/Cigarette
Miscellaneous Organic Compounds		**Inorganic Compounds**	
Methanol	100–250 μg	Carbon monoxide (CO)	10–23 mg
Acetamide	38–56 μg	Carbon disulfide	0.6–2.6 μg
Acrylamide	Present	Ammonia	10–130 μg
Acrylonitrile	3–15 μg	Nitrogen oxides[2] (NO$_x$)	100–600 μg
Vinyl chloride (VC)	11–15 ng	Hydrogen cyanide	400–500 μg
Ethylene oxide (EO)	7 μg	Cyanogen	7–11 μg
Ethyl carbamate	20–38 ng	Hydrogen sulfide	10–90 μg
1,1-Dimethylhydrazine	Present	Hydrazine	24–34 μg
Maleic hydrazide (MH)	1.16 μg	Arsenic	40–120 ng
Methyl isocyanate	1.5–5 μg	Beryllium	0.3 μg
2-Nitropropane	0.2–2.2 μg	Nickel	ND–600 ng
Nitrobenzene	25 μg	Cobalt	0.13–0.2 ng
Phenol	80–160 μg	Chromium	4–70 ng
Catechol	200–400 μg	Lead	34–85 ng
Di-(2-ethylhexyl)phthalate	20 μg	Mercury	4 ng
1,1,1-Trichloro-2,2-bis (4-chlorophenyl)ethane (DDT)	800–1200 ng	Polonium-210	0.03–1.0 pCi
1,1-Dichloro-2,2-bis (4-chlorophenyl)ethylene (DDE)	200–370 ng		
		Aromatic Amines	
		Aniline	360–655 ng
		2-Toluidine	30–337 ng
		2-Naphthylamine	1–334 ng
		4-Aminobiphenyl	2–5.6 ng

ND = Not detected
[1]Hoffmann and Hoffmann (1998)
[2]Unaged smoke contains primarily NO and only traces of NO$_2$ (< 10 μg).

REFERENCES

Anonymous (1997) Tobacco smoke components. *Beitr. Tabakforsch. Int.*, 17, 61–66.

Baker, R. R. (1980a) Mechanisms of smoke formation and delivery. In: *Recent Advances in Tobacco Science, Vol. 6, Symposium on Chemical, Physical, and Production Aspects of Tobacco and Smoke.* 34th Tobacco Chemists' Research Conference, Richmond, VA, pp. 184-224.

Baker, R. R. (1980b) Variation of sidestream gas formation during the smoking cycle. *Beitr. Tabakforsch. Int.*, 11, 181-193.

Baker, R. R. (1989) The viscous and inertial flow of air through perforated papers. *Beitr. Tabakforsch. Int.*, 14(5), 253-260.

Baker, R. R. & Robinson, D. P. (1990) Tobacco combustion-the last ten years. In: *Recent Advances in Tobacco Science, Vol. 16. Symposium on the Formation and Evolution of Cigarette Smoke.* 44th Tobacco Chemists' Research Conference, Winston-Salem, NC, pp. 3-72.

Ball, M. Papke, O., & Lis, A. (1990) Polychlorinated dibenzodioxins and dibenzofurans in cigarette smoke. *Beitr. Tabakforsch. Int.*, 14(6), 393-402.

Bass, R. T., Brown, L. E., Hassam, S. B., Newell, Jr., G. C., & Newman, R. H. (1989) Cigarette smoke formation studies, the transfer of added (18-C^{14}) octatriacontane. *Beitr. Tabakforsch. Int.*, 14(5), 289-296.

BCM (1998) British Columbia Ministry of Health and Ministry Responsible for Seniors. *1998 Reports on Cigarette Additives and Ingredients and Smoke Constituents*, 12/6/98, http://www.cctc.ca/bcreports/.

Borgerding, M. F. (1997) The FTC method in 197—what alternative smoking condition(s) does the future hold? In: *Recent Advances in Tobacco Science, Vol. 23, Symposium Proceedings on Smoke, Smoking, and Smokers*, 51st Tobacco Chemists' Research Conference, Winston-Salem, NC, September 14–17, pp. 75–151.

Bradford, J. A., Harlan, W. R., & Hanmer, H. R. (1936) Nature of cigaret smoke. Technic of experimental smoking. *Ind. Eng. Chem.*, 28, 836-839.

Browne, C. L., Keith, C. H., & Allen, R. E. (1980) The effect of filter ventilation on the yield and composition of mainstream and sidestream smoke. *Beitr. Tabakforsch. Int.*, 10, 81-90.

Brunnemann, K. D., Adams, J. D. Ho, D. P, S., & Hoffmann, D. (1978) The influence of tobacco smoke on indoor atmospheres. II. Volatile and tobacco-specific nitrosamines in mainstream and sidestream smoke and their contribution to indoor pollution. *Proceedings, 4th Joint Conference on Sensing of Environmental Pollutants*, American Chemical Society, Washington, D.C., pp. 876-880.

Brunnemann, K. D. & Hoffmann, D. (1974) pH of tobacco smoke. *Food Cosmet. Toxicol.*, 12, 115-124.

Brunnemann, K. D., Kagan, M. R., Cox, J. E., & Hoffmann, D. (1990) Analysis of 1,3-butadiene and other selected gas-phase components in cigarette mainstream and sidestream smoke by gas chromatography-mass selective detection. *Carcinogenesis,* 11, 1863–1868.

Brunnemann, K. D., Yu, L., & Hoffmann, D. (1977) Assessment of carcinogenic volatile n-nitrosamines in tobacco and in mainstream and sidestream smoke from cigarettes, *Cancer Res.,* 37, 3218-3222.

Chortyk, O. T. & Schlotzhauer, W. S. (1986) Modifications of an automatic cigarette smoking machine for sidestream smoke collection. *Tob. Sci.,* 30, 122-126.

Chortyk, O.T. & Schlotzhauer, W. S. (1989) The contribution of low tar cigarettes to environmental tobacco smoke. *J. Anal. Toxicol.,* 13, 129–134.

Crooks, E. L. & Lynn, D. (1992). The measurement of intrapuff nicotine yield. *Beitr. Tabakforsch Int.,* 15, 75–86.

Darrall, K. G., Figgins, J. A., Brown, R. D., and Phillips, G. F. (1998) Determination of benzene and associated volatile compounds in mainstream smoke. *Analyst,* 123, 1095–1101.

Diana, J. N. & Vaught, A. (1990) The research cigarette. Tobacco and Health Research Institute, University of Kentucky, Lexington, KY (606) 257-2816).

Dittmann, R., Field, H.-J., Müller, B.-H., and Schneider, W. (1992) Time resolved emission of sidestream smoke particles. *Beitr. Tabakforsch. Int.,* 15, 53–58.

Djordjevic, M., Fan, J., Ferguson, S., & Hoffmann, D. (1995) Self-regulation of smoking intensity. Smoke yields of the low-nicotine, low-"tar" cigarettes. *Carcinogenesis,* 16, 2015–2021.

Djordjevic, M., Hoffmann, D., & Hoffmann, I. (1997) Nicotine regulates smoking patterns. *Prev. Med.,* 26, 435–440.

Dube, M. F. & Green, C. R. (1982) Methods of collection of smoke for analytical purposes. In: *Recent Advances in Tobacco Science, Vol. 8. Symposium on the Formation, Analysis, and Composition of Tobacco Smoke.* 3th Tobacco Chemists' Research Conference, Raleigh, NC, pp. 42-102.

Eberhardt, H.-J. & Scherer, G. (1995) Human smoking behavior in comparison with machine smoking methods: a summary of the five papers presented at the 1995 meeting of the CoRESTA smoke and technology groups in Vienna. *Betir. Tabakforsch. Int.,* 16, 131–140.

El-Bayoumy, K., O'Donnell, M., Hecht, S. S., & Hoffmann, D. (1985) On the analysis of 1-nitronaphthalene, 1-nitropyrene, and 6-nitrochrysene in cigarette smoke. *Carcinogenesis,* 6, 505-507.

Federal Trade Commission (1997) *Tar, Nicotine, and Carbon Monoxide of the Smoke of 1206 Varieties of Domestic Cigarettes.* Federal Trade Commission, Washington, D.C.

Fidus Instrument Corporation (Offices of Filtrona Instruments and Automation Ltd.) 7400 Whitepine Rd., Richmond, VA 23237-2219 (804) 275-7850); In: *Tobacco International 1998 Buyers Guide and Directory*, 7730 Whitepine Road, P.O. Box 34277, Richmond, VA 23237-2255, (804) 275-7186; also 130 W. 42nd St., New York, NY.

Gmeiner, G., Stehlik, G., and Tousch, H. (1997) Determination of seventeen polycyclic aromatic hydrocarbons in tobacco smoke condensate. *J. Chromatogr. A*, 767, 163–169.

Grimmer, G., Naujack, K.-W., & Dettbarn, G. (1987) Gas chromatographic determination of polycyclic aromatic hydrocarbons, aza-arenes, and aromatic amines in the particle and vapor phase of mainstream and sidestream smoke of cigarettes. *Toxicol. Lett.*, 35, 117-124.

Grimmer, G., Schneider, D., Naujack, K.-W., Dettborn, G., & Jacob, J. (1995) Intercept-reactant method for the determination of aromatic amines in mainstream tobacco smoke. *Beitr. Tabakforsch. Int.*, 16, 141–156.

Grise, V. N. (1984) Market growth of reduced tar cigarettes. *Recent Adv. Tob. Sci.*, 10, 4-14.

Groenen, P. J. & Van Gemert, L. J. (1971) Flame photometric detection of various sulfur compounds in smoke from various types of cigarettes. *J. Chromatogr.*, 57, 239-246.

Guerin, M. R. (1980) Chemical composition of cigarette smoke. *Banbury Report 3: A Safe Cigarette?*, Cold Spring Harbor Laboratory, pp. 191-204.

Guerin, M. R. (1991) Environmental tobacco smoke. In: Hanson, L. D. & Eatough, D. J., eds., *Organic Chemistry of the Atmosphere*, CRC Press, Boca Raton, FL pp. 70–119.

Guerin, M. R., Higgins, C. E., & Jenkins, R. A. (1987) Measuring environmental emissions from tobacco combustion: sidestream cigarette smoke literature review. *Atmos. Environ.*, 21, 291-297.

Herning, R. I., Jones, R. T., Bachmann, J., & Mines, A. H. (1981) Puff volume increases when low-nicotine cigarettes are smoked. *Brit. J. Med.*, 283, 187-189.

Higgins, C. E., Jenkins, R. A., & Guerin, M. R. (1987) Organic vapor phase composition of sidestream and environmental tobacco smoke from cigarettes. *Proceedings of the 1987 EPA IAPCA Symposium on Measurement of Toxic and Related Air Pollutants*, Research Triangle Park, NC, pp. 140-150.

Hoffmann, D., Djordjevic, M. V., & Hoffmann, I. (1997) The changing cigarette. *Prev. Med.*, 26, 427–434.

Hoffmann, D. & Hoffmann, I. (1998). Letters to the Editor, Tobacco Smoke Components. *Beitr. Tabakforsch. Int.*, 18, 49–52.

Hoffmann, D., Patrianakos, C., Brunnemann, K. D., & Gori, G. B. (1976) Chromatographic determination of vinyl chloride in tobacco smoke. *Anal. Chem.*, 48(1), 47-50.

Horton, A. D. & Guerin, M. R. (1974) Quantitative determination of sulfur compounds in the gas phase of cigarette smoke. *J. Chromatogr.,* 90, 63-70.

Houseman, T. H. (1973) Studies of cigarette smoke transfer using radioisotopically labelled tobacco constituents, part II. The transference of radiosotopically labelled nicotine to cigarette smoke. *Beitr. Tabakforsch. Int.,* 7, 142-147.

Jenkins, R. W., Chavis, M. K., Newman, R. H., & Morrell, F. A. (1971) The quantitative recovery of smoke from radioactivity labeled cigarettes. *Int. J. App. Rad. Isotopes,* 22, 691-697.

Jenkins, R. W., McRae, D. D., Johnson, R. O., Brenizer, J. S., & Williamson, T. G. (1990) Smoke entrainment as a major mechanism of sodium transfer from the 2R1 cigarette. *Tob. Sci.,* 34, 93-98.

Jenkins, R. W., Newman, R. H., Carpenter, R. D., & Osdene, T. S. (1970) Cigarette smoke formation studies. 1. Distribution and mainstream products from added ^{14}C-dotriacontane-16,17. *Beitr. Tabakforsch. Int.,* 5, 205-208.

Johnson, W. R., Hale, R. W., Nedlock, J. W., Grubbs, H. J., Powell, D. H. (1973) The distribution of products between mainstream and sidestream smoke. *Tob. Sci.,* 17, 141-144.

Keith, C. H. & Tesh, P. G. (1965) Measurement of the total smoke issuing from a burning cigarette. *Tob. Sci.,* 9, 61-64.

Klus, H. (1990) Distribution of mainstream and sidestream cigarette smoke components. In: *Recent Advances in Tobacco Science, Vol. 16. Symposium on the Formation and Evolution of Cigarette Smoke.* 4th Tobacco Chemists' Research Conference, Winston-Salem, NC, pp. 189-232.

Klus, H. & Kuhn, H. (1982) Distribution of various tobacco smoke components among mainstream and sidestream smoke (a survey) (In German). *Beitr. Tabakforsch. Int.,* 11, 229-265.

Kosak, A. I. (1954) The composition of tobacco smoke. *Experientia,* 10, 69-71.

Koslowski, L. T., Trecker, R. C., Khouw, V., & Pope, M. A. (1980) The misuse of "less hazardous" cigarettes and its detection: hole-blocking of ventilated filters. *Am. J. Public Health,* 70, 1202-1203.

Mahanama, K. R. R. & Daisey, J. M. (1996) Volatile N-nitrosamines in environmental tobacco smoke: sampling, analysis, emission factors, and indoor air exposures. *Environ. Sci. Technol.,* 30, 1477–1484.

Martell, E. A. (1975) Tobacco radioactivity and cancer in smokers. *Am. Sci.,* 63, 404-412.

Martell, E. A. (1982) Radioactivity in cigarette smoke. *New Eng. J. Med.,* 307, 309-310.

McRae, D. D. (1990) The physical and chemical nature of tobacco smoke. In: *Recent Advances in Tobacco Science, Vol. 16. Symposium on the Formation and Evolution of Cigarette Smoke.* 4th Tobacco Chemists' Research Conference, Winston-Salem, NC, pp. 233-323.

Mitacek, E. J., Brunnemann, K. D., Hoffmann, D., Limsila, T., Suttajit, M., Martin, N., & Caplan, L. S. (1999) Volatile and tobacco-specific nitrosamines in the smoke of Thai cigarettes: a risk factor for lung cancer and a suspected risk factor for liver cancer in Thailand. *Carcinogenesis,* 20, 133–137.

Moneyhun, J. H., Jenkins, R. A., Guerin, M. R. (1986) *Development of an improved sidestream smoke generation and collection apparatus.* 40th Tobacco Chemists Research Conference, Knoxville, TN.

Morrell, F. A. & Varsel, C. (1966). A total combustion product cigarette smoking machine -analysis of radioactive cigarette paper. *Tob. Sci.,* 10, 45-50.

National Cancer Institute (1996) The FTC cigarette test method for determining tar, nicotine, and carbon monoxide yields of U.S. cigarettes. *Smoking and Tobacco Control, Monograph 7,* Report of the NCI Expert Committee, DHHS, NIH, 275 pp.

(NRC) National Research Council (1986) *Environmental Tobacco Smoke. Measuring Exposures and Assessing Health Effects.* National Academy Press, 2101 Constitution Avenue, NW, Washington, D.C. 20418, 337 pp.

Norman, V. (1974) The effect of perforated tipping paper on the yield of various smoke components. *Beitr. Tabakforsch. Int.,* 7, 282-287.

Norman, V., Ihrig, A. M., Larson, T. M., & Moss, B. L. (1983) The effect of some nitrogenous blend components on NO/NO_x, and HCN levels in mainstream and sidestream smoke. *Beitr. Tabakforsch. Int.,* 12, 55-62.

O'Neill, I. K., Brunnemann, K. D., Dodet, B., & Hoffmann, D. (1987) *Environmental Carcinogens. Methods of Analysis and Exposure Assessment. Vol. 9- Passive Smoking (JARC Scientific Publication No. 81),* International Agency for Research on Cancer, Lyon, France, 372 pp.

Perfetti, T. A. & Coleman, III, W. M. (1998a). Chiral-gas chromatography-selected ion monitoring-mass selective detection analysis of tobacco materials and tobacco smoke. *Beitr. Tabakforsch. Int.,* 18, 15–34.

Perfetti, T. A. & Coleman, III, W. M. (1998b). Chiral-gas chromatography-selected ion monitoring-mass selective detection analysis of secondary alkaloids in tobacco and tobacco smoke. *Beitr. Tabakforsch. Int.,* 18, 35–42.

Perfetti, T. A., Coleman, III, W. M., & Smith, W. S. (1998) Determination of mainstream and sidestream cigarette smoke components for cigarettes of different tobacco types and a set of reference cigarettes. *Beitr. Tabakforsch. Int.,* 18, 95–114.

Proctor, C. J., Martin, C., Beven, J. L., & Dymond, H. F. (1988) Evaluation of an apparatus designed for the collection of sidestream tobacco smoke. *Analyst,* 113, 1509-1513.

Pryor, W. A., Hales, B. J., Premovic, P. I., & Church, D. F. (1983) The radicals in cigarette tar: their nature and suggested physiological implications. *Science,* 220, 425-427.

Pryor, W. A., Prier, D. G., & Church, D. F. (1983) Electron-spin resonance study of mainstream and sidestream cigarette smoke: nature of the free radicals in gasphase smoke and in cigarette tar. *Environ. Health Perspectives,* 47, 345-355.

(RJR) R. J. Reynolds Tobacco Company (1988) *New Cigarette Prototypes that Heat Instead of Burn Tobacco.* R J. Reynolds Tobacco Co., Winston-Salem, NC 27102, 743 pp.

Rickert, W. S., Robinson, J. C., & Collishaw, N. (1984) Yields of tar, nicotine, and carbon monoxide in the sidestream smoke from 15 brands of Canadian cigarettes. *Am. J. Public Health,* 74, 228-231.

Rickert, W. S., Robinson, J. C., Young, J. C., Collishaw, N. E., and Bray, D. F. (1983) A comparison of the yields of tar, nicotine, and carbon monoxide of 36 brands of Canadian cigarettes tested under three conditions. *Preven. Med.,* 12, 682–694.

Risner, C. H. (1988) The determination of benzo(a)pyrene in the total particulate matter of cigarette smoke. *J. Chromatogr. Sci.,* 26, 113-120.

Roberts, D. L. (1988) Natural tobacco flavor. *Rec. Advan. Tob. Sci.,* 14, 49-81.

Rodgman, A. (1991) Personal communication. R. J. Reynolds Tobacco Co., Winston-Salem, NC 27102.

Rodgman, A. (1997) FTC tar and nicotine in mainstream cigarette smoke: a retrospective. In: *Recent Advances in Tobacco Science, Vol. 23, Symposium Proceedings on Smoke, Smoking, and Smokers*, 51st Tobacco Chemists' Research Conference, Winston-Salem, NC, September 14–17, pp. 5–74.

Russell, M.A. H. (1977) Smoking problems: an overview. In: Jarvik, J. M. E., Cullen, J. W., Gritz, E. R., Vogt, T. M., & West, L. H., eds., *Research on Smoking Behavior,* National Institute Drug Abuse, *Res. Monogr.,* 17, 13-34.

Sakuma, H., Kusama, M., Munakata, S., Ohsumi, T., & Sugawara, S. (1983) The distribution of cigarette smoke components between mainstream and sidestream smoke. I. Acidic components. *Beitr. Tabakforsch. Int.,* 12, 63-71.

Sakuma, H., Kusama, M., Yamaguchi, K., Matsuki, T., & Sugawara, S. (1984a) The distribution of cigarette smoke components between mainstream and sidestream smoke. II. Bases. *Beitr. Tabakforsch. Int.,* 12, 199-209.

Sakuma, H., Kusama, M., Yamaguchi, K., Matsuki, T., & Sugawara, S. (1984b) The distribution of cigarette smoke components between mainstream and sidestream smoke. III. Middle and higher boiling components. *Beitr. Tabakforsch. Int.*, 12 , 251–258.

Song, S. & Ashley, D. L. (1999) Supercritical fluid extraction and gas chromatography/mass spectrometry for the analysis of tobacco-specific nitrosamines in cigarettes. *Anal. Chem.*, 71, 1303–1308.

Surgeon General (1986) *The Health Consequences of Involuntary Smoking. A Report of the Surgeon General.* DHHS, PHS, CDC, Office on Smoking and Health, Rockville, MD 20857, 359 pp.

Ueno, Y. & Peters, L. K. (1986) Size and generation rate of sidestream cigarette smoke. *Aerosol Sci. Technol.*, 5, 469-476.

Wakeham, H. (1972) Recent trends in tobacco and tobacco smoke research. In: Schmeltz, I., ed. *The Chemistry of Tobacco and Tobacco Smoke,* Plenum Press, New York, NY, pp. 1-20.

Wynder, E. L. & Hoffmann, D. (1967) *Tobacco and Tobacco Smoke. Studies in Experimental Carcinogenesis.* Academic Press, 111 Fifth Avenue, New York, NY 10003, 730 pp.

Young, J. C., Robinson, J. C., & Rickert, W. S. (1981) A study of chemical deliveries as a function of cigarette butt length. *Beitr. Tabakforsch. Int.*, 11, 87-95.

4

Properties and Measures
of Environmental Tobacco Smoke

INTRODUCTION

Environmental tobacco smoke (ETS) is considered by many authorities to be an important component of indoor air pollution in part because it is often viewed as being equivalent to mainstream cigarette smoke (MS). It has been clearly demonstrated that ETS is not the same as MS. Sidestream cigarette smoke (SS) is the major contributor to ETS. SS is generated under different conditions than MS, and as a result, has a different relative chemical composition (see Chap. 3). Exhaled MS, the second primary contributor to ETS, is a different material from that which leaves the cigarette butt and enters the lungs. Exhaled MS has been substantially depleted in vapor-phase constituents, and the particulate matter is likely to have increased its water content in the high-humidity environment of the respiratory tract. As the cigarette smoke, both SS and exhaled MS, enters the atmosphere, it is diluted by many orders of magnitude and subsequently undergoes both physical transformation and alterations in it chemical composition. Upon standing, or during air exchange from other sources, ETS continues to change. The purpose of this chapter is to discuss the physical characteristics of ETS, its chemical composition, the changes that it undergoes with time, and the utility of selected chemical markers for ETS. For additional perspectives and information, the reader is referred to reviews on the chemical composition of ETS by Guerin (1991), Baker and Proctor (1990), and Eatough et al. (1990) and to reviews on ETS markers by Goodfellow et al. (1990) and Ogden and Jenkins (1999).

PHYSICAL CHARACTERISTICS OF ETS

The largest physical differences between MS and ETS are due to the conditions under which the precursor SS is generated and the degree of dilution that it undergoes. In general, the phase distribution of the various components of ETS is shifted in favor of the vapor phase, and the individual particles that compose the particulate phase are smaller in diameter relative to those found in MS. First, because of the slower air flow into the firecone during smoldering, the production of more alkaline constituents is favored.

Thus, the pH at which the sidestream components are generated is higher than that of mainstream components. This causes nicotine to be present in its free base form. That nicotine is present predominantly (90–95%) in the vapor phase (Eudy et al. 1986) has some important ramifications for the ultimate relative chemical composition of ETS under field conditions. That is, because of its chemical properties relative to those of many materials, nicotine has a high affinity for surfaces. Nicotine's presence in the vapor phase means that it can diffuse more rapidly through the air than if it were present in the particulate phase and that it can strike surfaces and be adsorbed onto them. As such, it is depleted from the field environment more rapidly than are less absorptive constituents (e.g., Eatough et al. 1989a, Piadé et al. 1999), and significant changes in relative chemical composition of ETS can occur with aging. In addition, unprotonated nicotine in the vapor state is also more readily absorbed through the oral mucosa of ETS-exposed individuals than is protonated nicotine (Armitage and Turner 1970, Schievelbein and Eberhardt 1972). The shift in phase equilibrium has been illustrated for even higher-molecular-weight species. Ramsey et al. (1990) demonstrated that a significant fraction of straight-chain saturated hydrocarbons above C_{25} are distributed into the vapor phase. Also, studies employing ^{123}I-labeled iodohexadecane (boiling point: 380°C) as a tracer added to cigarettes indicated that while 95% of the tracer was present in the particle phase of SS when it was freshly generated, about 70% of the tracer was found in the vapor phase when the SS was diluted to particulate levels comparable with those of ETS (Black et al. 1987, Pritchard et al. 1988, Proctor et al. 1988). The influence of dilution and the loss of nicotine on ETS composition can be seen in the results of chamber studies presented in Table 4.1 (R. J. Reynolds 1988). The ratio of particulate matter to carbon dioxide drops by about 50% when SS is diluted to SS ETS, and the nicotine:CO_2 ratio decreases by about 85%, presumably due to losses of nicotine to the walls of the chamber and other surfaces.

Table 4.1 Weight Ratios of Selected Components[a]
Mainstream(MS)/Sidestream (SS)/ETS

Constituents	Kentucky Reference 1R4F Cigarettes			
	MS	SS	Pure SS ETS	True ETS (SS + Exhaled MS)
Carbon Monoxide/ Carbon Dioxide	0.270	0.114	0.1.35	0.150
Ammonia/ Carbon Dioxide	0.00043	0.0192	0.0243	0.013
Nicotine/Carbon Dioxide	0.0189	0.0110	0.00 17	0.00135
TPM/Carbon Dioxide	0.26	0.0357	0.0165	0.0249
Formaldehyde/Acrolein	0.233	0.576	2.94	2.03
Vinyl Pyridine/Nicotine	-	-	0.612	0.561

[a]Data from R. J. Reynolds 1988.

The influence of exhaled MS on the overall physical composition of real ETS is a factor that is often overlooked in many studies. Using human smokers to generate true ETS, Baker and Proctor (1990) demonstrated that exhaled MS can contribute between 15 and 43% of the particulate matter to ETS while contributing only small amounts of gas-phase constituents. The data in Table 4.1 support this finding. The ratio of total particulate matter (TPM) to carbon dioxide is considerably greater for "true ETS" (that generated by human smokers) than that of ETS generated only from SS. The greater contribution of particulates from exhaled MS may be due to greater losses to the respiratory tract of the more rapidly diffusing vapor-phase compounds. Given the degree of variation of human smoke-generation and inhalation patterns (e.g., Creighton and Lewis 1978), it seems reasonable to assume that the extent of the human contribution to ETS particulates will vary considerably.

A number of studies (Ueno and Peters 1986, Duc and Huynh 1987, Black et at. 1988, and Benner et al. 1989) have sought to address particle-size distribution and particle dynamics using diluted SS as a surrogate for ETS. Table 4.2 below summarizes the results of a study by Ingebrethsen and Sears (1985), which are particularly informative. These data were generated with a condensation nuclei counter and an optical particle counter for particle number

Table 4.2 Particle Characteristics as a Function of ETS Concentration[a,b]

Particulate Concentration (μg/m³)	Mass Median Diameter (μm)	Particles Smaller than 0.10 μm Diameter	
		Number Fraction of Total (%)	Mass Fraction of Total (%)
1.4	0.185	73	17
28	0.198	54	8.5
226	0.210	39	5.5

[a]From Ingebrethsen and Sears 1985.
[b]Measurements made on dilute sidestream(SS) smoke as an ETS surrogate, 1 h after generation, in an 18 m³ minimally [0.05 air change/hour (ACH)] ventilated chamber.

measurements and an electrical mobility analyzer, modified with an impactor to remove large particles, for determining size distribution. The data indicate that as the overall mass concentration in the chamber increases, the mass median diameter (MMD) increases (see Chap. 6 for a description of the terminology). The number fraction of particles smaller than 0.1 μm diam drops from 73% to 39% of the total. Clearly, the small particles compose a substantial fraction of the number of particles; however, they also compose one-sixth or less of the mass of particulate matter. Interestingly, the MMDs reported by Ingebrethsen and Sears (1985) are in good agreement with those reported by other investigators, despite substantially different conditions of measurement. Ueno and Peters (1986) used an electrical aerosol analyzer to determine particle size of freshly generated SS diluted about 1000-fold with the vapor phase of SS (to minimize evaporation of the particulates), and found an MMD of 0.16 μm. Black et al. (1988) used cascade impaction to determine the particle-size distribution of SS diluted into a 16-m^3 chamber; they reported an MMD of 0.25 μm. Benner et al. (1989) reported number mean diameter and an MMD of 0.107 and 0.255 μm, respectively, for freshly generated diluted SS in a 30-m^3 Teflon bag, determined using an optical particle counter. A recent study (Morawska et al. 1999b) reported a number mean diameter of 0.206 μm for simulated ETS particles contained in a 3-m^3 chamber. Densities of ETS particles have shown to be in the range of 1.12 to 1.18 g/cm^3 (Lipowicz 1988, Morawska et al. 1999a). In general, these data indicate that ETS particulate matter will be smaller than that of MS, which averages between 0.35 and 0.4 μm MMD (Ingebrethsen 1989). Given the range of particulate concentrations likely to be encountered in realistic environments, there will be some variation in ETS particle size and density. However, it also is clear that the greatest fraction of ETS particulates will be of a size range that can be inhaled by humans (Raabe 1980).

CHEMICAL COMPOSITION OF ETS

ETS is composed of aged exhaled MS and diluted SS. Thus, ETS most likely contains all the constituents present in its two precursor matrices, plus any constituents that may form during the aging process. Determining the composition of true ETS in field, or "real-world," conditions is difficult, partly because of the presence of confounding sources of ETS constituents. For this reason, many investigators have chosen to perform compositional and physical studies in chambers. The primary advantage to using these systems is the lack of other sources of atmospheric components. The incoming and/or dilution air can be cleaned by means of filters and sorbent media, so that the only source of ETS constituents is derived from tobacco. Because the term "chamber" denotes a specific type of enclosure, the term "controlled experimental atmosphere " (CEA) is probably more useful in describing the kinds of systems that have been used to study ETS under regulated conditions. Eatough et al. (1989b) used 10- and 30-m^3 Teflon bags to study the composition of vapor-phase and particle-phase ETS. Several investigators have used animal-inhalation exposure chambers for ETS studies, sized from 0.4 to 1.4 m^3 (Ingebrethsen and Sears 1985, Thompson et al. 1989). Larger, climate-controlled chambers have become increasingly popular for ETS studies (Duc and Huynh 1987, R. J. Reynolds 1988, Hammond et al. 1987, Löfroth et al. 1989, Tang et al. 1989, Rando et al. 1992, Van Loy et al. 1997, Nelson et al. 1998). Monitored but only semicontrolled rooms have also been used in several studies (Klus et al. 1985, Thompson et al. 1989). Although such CEAs have been used to study ETS, their use still presents some disadvantages. The primary disadvantages of CEAs are related to their relevance, or potential lack thereof, to the field. In some cases, ventilation rates of chambers are either very low [e.g., < 0.05 air change/hour (ACH) (Ingebrethsen and Sears 1985)] or very high [e.g., as much as 150 ACH (Thompson et al. 1989)] relative to those rates encountered in private residences (about 1 ACH) or public buildings (2–5 ACH). As such, the smoke can either be more or less "aged" than it might be under more realistic circumstances. Although many CEAs have as their primary surface of contact with the ETS Teflon, stainless steel, epoxy-based paint, or plastic, few field environments can make such a claim. Also, under field conditions, smoke is added to the environment at different times from different locations, giving rise to both temporal and spatial inhomogeneities. Several investigators have employed humans to smoke cigarettes for ETS generation; however, great care is typically taken to ensure that the humans smoke the cigarettes under reasonably defined conditions, a situation that does not exist in the field. Many studies have skirted both the human-studies issues and uncertainties of using human smokers and have used dilute SS generated by smoking machines as a surrogate for true ETS.

The number of comprehensive studies of the composition of ETS, either in CEAs or in the field, has been limited. There may be a number of reasons for this. First, the logistics of conducting a major field study can be expensive and often requires the cooperation of the building occupants. Conversely, the maintenance of specialized chamber facilities, smoking apparatuses, and monitoring equipment can be costly. The complexity of experimental design can be significant for any study seeking to examine more than a few constituents. (To what extent should the smoke be aged? To what extent should MS be added to the atmosphere? Under what conditions should the cigarettes be smoked? What is the appropriate ventilation rate? Should lights be present and if so, what kind? What kinds of surfaces will best mimic the real world?) In short, in contrast to the generation of mainstream cigarette smoke, where a battery of standard criteria exist, there are still no standardized criteria for the development of experimental atmospheres of ETS.

Despite the complexities and difficulties, a few studies of ETS composition [beyond the "common" constituents nicotine, carbon monoxide (CO), and respirable suspended particulates (RSPs)] have been performed. Selected results from some of the more informative studies are presented in Table 4.3. Several comments are in order. First, many investigators have focused on the vapor-phase constituents. That is probably because the vapor phase represents the bulk of the mass of ETS and because of the very low levels of RSP-related constituents. For example, if the RSP levels are in the range of 20 to 1000 $\mu g/m^3$, constituents of the particle phase present at concentrations of 1 to 100 ppm in the particulates themselves will be present at airborne concentrations of 20 pg/m^3 up to 100 ng/m^3. These are challenging concentrations for any sampling and analysis method. Also, because the number of constituents present in either phase runs into the thousands, it is difficult to find many studies that have determined the concentrations of the same constituents, obviating direct comparisons. For manipulated environments, in which no other confounding sources were present, the Higgins and Guerin (1988) study and the R. J. Reynolds (1988) study probably represent the extremes of ventilation and concomitant nicotine:RSP ratios likely to be encountered in realistic environments. The Guerin study was conducted in an office with an unusually high ventilation rate (ca. 15 ACH). As such, any ETS present was very "fresh"; the nicotine lacked the opportunity to be adsorbed on the surfaces present. As a consequence, the nicotine:RSP ratio was unusually high (ca. 0.36). In contrast, the R. J. Reynolds study was conducted in an 18-m³ chamber; the only air added was that required to replace the volume removed for sampling. The result was a ventilation rate of 0.05 ACH. The measured nicotine:RSP ratio was ca. 0.05. The remainder of the studies, conducted in field environments, show a considerable range of constituent concentrations. Interestingly, the nicotine:RSP ratio in a disco (Table 4.3) was fairly high (ca. 0.15). For those studies in which benzene levels were reported, the differences between the manipulated smoky office (Higgins and Guerin 1988) and the field studies were small, about a factor of two. This suggests that in some cases in the field studies, there may have been other contributors to the benzene levels. Indeed, in a field study of 49 homes, Heavner et al. (1995) estimated the fraction of benzene contributed from ETS to be 13%. Hodgson et al. (1996) estimated that ETS contributed between 37 and 58% of the benzene in smoking lounges. Scherer et al. (1995) estimated the ETS contribution to benzene levels to be no greater than 15% in a small study of ETS and benzene in both smoking and nonsmoking homes in Germany; however, the location of the residence (urban vs. suburban) was much more important to the benzene level than smoking status.

The two research groups that have performed the most extensive studies of the vapor-phase composition of ETS in recent years are those at the Lawrence Berkeley National Laboratory and the R. J. Reynolds Tobacco Company (e.g., Daisey et al. 1998, Martin et al. 1997. Although the focuses of the research teams are somewhat different, both have used static chambers to perform relatively detailed compositional studies. Comparative results for selected emission-factor studies are presented in Table 4.4. One common thread has been the use of large chambers in a static mode (no air ventilation or recirculation). One group used aged and diluted SS to determine commercial cigarette-emission factors, while the other used humans to generate true ETS. Agreement is remarkably good among these studies of commercial brands with the exception of emissions of RSP, nicotine, and 3-ethenyl pyridine (3-EP). Apparent emission factors are higher in the studies using humans to generate ETS, probably due to the contribution of exhaled particulate matter. Potential explanations for differences between emission factors for nicotine and 3-EP are less clear. Overall, however, these studies have indicated that it is possible to achieve, under static conditions in controlled atmospheres, reasonable consistency with regard to emission rates for volatile organic compounds (VOCs) from smoldering cigarettes. The translation of these

Table 4.3 Comparative ETS Composition Reported in Selected Studies

Constituent Concentrations (In units of $\mu g/m^3$, unless reported otherwise)

Constituent	Office with High Ventilation Rate[a]	18 m³ Chamber, 1R4F Cigarette[b]	18 m³ Chamber, 50 Top Selling US Commercial Brand Styles[c]	Tavern[d]	Mechanically Ventilated Building[e]	Disco[f]	Smoker's Office[g]	Smoker's Office[h]
RSP	330	449	1440	420	13	801	148	NR
Nicotine	120	24.4	90.8	71	4.0	120	18.1	14.5
CO, ppm	NR	2.37	5.09	3.85	3.6	22.1	1.4	NR
CO₂, ppm	NR	10.04[i]	NR	NR	454	NR	570	NR
Benzene	16	NR	30	27	NR	NR	8	19
Toluene	70	NR	54.5	NR	NR	NR	24	258
3-Ethenyl Pyridine	NR	13.7	37.1	NR	NR	18.2	NR	NR
Ethylbenzene	8.9	NR	8.50	NR	NR	NR	5	ND
Pyridine	11.7	30.5	23.8	NR	NR	17.6	NR	NR
Isoprene	47	NR	657	150	NR	NR	NR	NR
Formaldehyde	NR	3.5	143	104	12	NR	NR	NR
Limonene	16	NR	29.1	NR	NR	NR	7	NR

[a]Higgins and Guerin 1988
[b]R. J. Reynolds 1988
[c]Martin et al. 1997
[d]Löfroth et al. 1989
[e]Turner and Binnie 1990
[f]Eatough et al. 1989
[g]Proctor et al. 1989
[h]Bayer and Black 1987
[i]ppm above ambient level
NR= Not Reported
ND= Not Detected

Table 4.4 Chamber Studies of Cigarette Emissions: Dilute Sidestream or Environmental Tobacco Smoke Emissions per Cigarette (μg/cigarette)

Constituent	20 m³ Chamber Dilute SS from 6 Commercial Cigarette Brands Mean ± Std. Deviation	18 m³ Chamber Human Generated ETS from 50 Top US Commercial Brand Styles Weighted by Market Share Mean ± Std. Deviation	45 m³ Chamber Human Generated ETS from a Full Flavor Cigarette Brand Style Mean ± Std. Deviation
Acetaldehyde	2150 ± 477	2496 ± 55	2,347 ± 218
Benzene	406 ± 71	280 ± 5	218 ± 23
1,3 Butadiene	157 ± 27	373 ± 13	NR
Ethylbenzene	130 ± 10	79.6 ± 1.4	59.3 ± 5.3
Formaldehyde	1310 ± 348	1333 ± 34	1065 ± 75
Pyridine	428 ± 122	218 ± 5	210 ± 15
Styrene	147 ± 24	94.3 ± 1.8	60 ± 5
Toluene	656 ± 107	498 ± 11	518 ± 45
3-Ethenylpyridine	662 ± 155	334 ± 8	188 ± 23
m-p Xylene	299 ± 52	238 ± 5	200 ± 18
o-Xylene	67 ± 16	58.6 ± 1.6	41.3 ± 3.8
Nicotine	919 ± 240	1585 ± 42	405 ± 105
PM$_{2.5}$ or RSP	8100 ± 2,000	13,674 ± 411	10,935 ± 480

[a]Daisey et al. 1998
[b]Martin et al. 1997
[c]Nelson et al. 1998
NR=not reported.

findings into field observations is another matter. Heavner et al. (1995, 1996) conducted extensive apportionment studies in homes and workplaces in two geographic locations within the U.S. The findings of these studies are discussed in more detail in Chap. 11. It is clear that a number of sources in field environments contribute to the levels of VOCs that are not specific to tobacco. Given the complexity of the field environment, it may be more useful to think of "real-world" ETS as having a range of potential relative compositions, depending on both the source input strength, ventilation rates, aging, and the presence of confounding sources.

With the exception of studies directed specifically at the determination of potential ETS markers or tracers, there have been few studies of the composition of ETS particulate matter. Ramsey et al. (1990) determined concentrations of high-molecular-weight, straight-chain hydrocarbons in the RSP of highly diluted SS (54–600 μg/m^3) as a surrogate for ETS. Benner et al. (1989) reported levels of chemical classes (n-alkanes, branched alkanes, bases, sterols, fatty acids, and sterenes) in RSP of diluted SS as well as quantitative values for selected nitrogen-containing compounds and some sterols. Hydrocarbons were also identified without quantification. In the same study, annular denuders were used to collect inorganic species. Mean chloride concentrations were determined to be about 50 μmoles per gram RSP, while nitrite, nitrate, sulfate, and ammonium ions were present at levels of about 28–50, 7–111, 2–13, and 40 μmoles/g RSP, respectively. Duc and Huynh (1989) determined levels of PAH in SS RSP diluted to about 13.8 mg/m^3. Some of the data from these latter three studies are summarized in Table 4.5. Because of the different focuses of the individual studies, there is little comparative data among them. Most of the data were obtained at moderately high (ca. 600 μg/m^3), or very high (13,800 μg/m^3) RSP levels, probably to maximize sample collection.

Probably the most extensive study performed to date not directed primarily at the determination of marker compounds is that of Rogge et al. (1994). The investigatory team studied a mixture of exhaled MS and SS diluted and cooled in a wind tunnel. Using gas chromatography/mass spectrometry (GC/MS), the team was able to resolve 39% of the particulate mass of the simulated ETS. Of this fraction, more than 110 compounds, which composed about 70% of the resolved fraction, were identified and quantified. The compound class emission rates (in micrograms per cigarette) are summarized in Table 4.6. The reader is referred to the original reference for details regarding individual emission constituents. Interestingly, the most prevalent compound was nicotine (reflected in the very high levels of nitrogen-containing species), comprising nearly 12 % of the identified material. Because most of the nicotine in ETS is present in the vapor phase, the finding of Rogge et al. is likely an artifact of a short time between cigarette emission and collection (30 s) and the use of particle-collection media (quartz fibers), which have a high affinity for gas-phase nicotine. Other major class fractions

Table 4.5 Selected ETS Particle Phase Constituents

Constituent	Benner et al. 1989 30-m³ Teflon Bag	Duc and Huynh 1989 10-m³ Glass/Stainless Steel Chamber	Ramsey et al. 1990 1.4-m³ Stainless Steel Chamber
RSP $\mu g/m^3$	677	13,800	600
Nicotine-Vapor phase ($\mu g/m^3$)	722	92	NR
Nicotine-Particle phase ($\mu g/m^3$)	9.1	NR	NR
Myosmine ($\mu mole/g$ RSP)	3.5 ± 1.7		
Nicotyrine ($\mu mole/g$ RSP)	3.3 ± 0.8		
Cotinine ($\mu mole/g$ RSP)	11.3 ± 2.0		
Cholesterol ($\mu mole/g$ RSP)	1.4 ± 0.3[a]		
Stigmasterol ($\mu mole/g$ RSP)	2.9 ± 1.6		
Campesterol ($\mu mole/g$ RSP)	1.5 ± 0.6[a]		
24-Methylcholesta-3,5-diene, ($\mu mole/g$ RSP)	2.1 ± 2.0[a]		
Solanesol ($\mu mole/g$ RSP)	22.2 ± 3.3[a]		
Heptacosane (n-$C_{27}H_{56}$) ($\mu g/m^3$)			7.2 ± 1.6
Nonacosane (n-$C_{29}H_{60}$) ($\mu g/m^3$)			3.3 ± 0.3
Hentriacontane (n-$C_{31}H_{64}$) ($\mu g/m^3$)			15.6 ± 2.1
Tritriacontane (n-$C_{33}H_{68}$) ($\mu g/m^3$)			4.6 ± 0.2
Phenanthrene (ng/m³)		112	
Anthracene (ng/m³)		13.1	
Pyrene (ng/m³)		194	
Benz(a)anthracene (ng/m³)		101	
Chrysene + triphenylene (ng/m³)		120	
Benzo(a)pyrene) (ng/m³)		58.4	
Indeno(1,2,3-cd)pyrene (ng/m³)		16.5	
Benzo(ghi)perylene (ng/m³)		12.1	
Dibenz [a,c]- + dibenz[a,h]anthracene (ng/m³)		4.4	
Coronene (ng/m³)		4.7	

[a]RSP level not specified

Table 4.6 Emission Rates of Gas Chromatographically Resolved Organic Class Fractions in Simulated ETS Particulate Matter (Data Abstracted from Rogge et al. 1994)

Class Type	Emission Rate, μg/cigarette
n-Alkanes	549.0
Iso- and Anteiso-alkanes	289.2
Isoprenoid Alkanes	81.6
N-Alkanoic Acids	617.5
N-Alkenoic Acids	124.1
Dicarboxylic Acids	38.4
Other Aliphatic and Cyclic Acids	116.6
n-Alkanols	310.7
Phenols	386.6
Phytosterols	496.9
Triterpenoids	8.7
N-containing Compounds	1691.2
Polycyclic Aromatic Hydrocarbons	13.5
Others	8.2

identified include n-alkanes, n-alkanoic acids, and phytosterols. Solanesol, a C_{45} isoprenoid alcohol used as a tracer for ETS particulate matter in a large number of personal exposure studies (see Chap. 6), was not identified. This is likely to be due to the chromatographic visualization procedure used, which was not optimized for specific compounds.

The comparison between the relative polynuclear aromatic hydrocarbon (PAH) content of surrogate ETS RSP determined in three studies and other types of commonly encountered particulate matter is informative. Such a comparison is reported in Table 4.7, with the PAH content of both diesel particulate and ambient air particulates encountered on an Army base (from Griest et al. 1988). First, it is clear that there are some important relative compositional differences among the three studies of "ETS" particulate matter. This may be due to differences in generation methods or differences in the concentrations at which the samples were obtained. Higher particulate levels would likely shift the particle-vapor phase equilibrium such that a larger fraction of the more volatile PAHs would remain in the particle phase (Liang and Pankow 1996). For example, in the study reported by Duc and Huynh, with particle concentrations upwards of 10 mg/m^3, the pyrene concentration of surrogate ETS RSP was 14 μg/g. At RSP levels below 1 mg/m^3, the other studies reported pyrene levels of 3 and 1 μg/g. Chrysene levels in ETS RSP vary over quite a range. Perhaps most interestingly, benzo[a]pyrene (BaP) concentrations in ETS RSP were higher than those in diesel particulate but less than those of ambient air particulate. Interestingly, the ETS BaP levels reported in Duc and Huynh (1989) are not much different from those reported for the particulate matter of SS from a reference cigarette (8.8 ng/g) R. J. Reynolds (1988). However, the ETS particulate matter BaP concentration is

Table 4.7 Polynuclear Aromatic Hydrocarbons (PAHs) in Selected Particulate Matter Samples (μg/g or ppm in Particulates)

Constituent	NBS SRM 1650[a] Diesel Exhaust Particulates[a]	Air Particulate at U.S. Army Installation[b]	ETS Particulate Matter in 10 m³ Chamber[c]	Dilute SS as a Surrogate for ETS in 36-m³ Chamber[d]	Exhaled MS + SS Diluted in a Wind Tunnel as a Surrogate for ETS[e]
Anthracene	NR	NR	0.95	< 0.1	0.76
Pyrene	NR	NR	14.1	3.1	1.0
3,6,Dimethylphenanthrene	NR	NR	2.5	NR	NR
Benzo[b]fluorene	NR	NR	7.1	NR	0.54[f]
Benz[a]anthracene	6.5	4.9	7.3	11.6	0.27
Chrysene	22	11.5	8.7[g]	31.4	0.67[g]
benzo[a]pyrene	9.6	9.4	3.1	NR	NR
Benzo[a]pyrene	1.2	8.0	4.2	5.7	NR
Indeno[1,2,3-cd]pyrene	2.3	17	1.2	NR	NR
Benzo[g,h,i]perylene	2.4	21.8	0.87	NR	NR
Dibenz[a,a;h,c]anthracene	1.2[b]	3.0[h]	0.32	NR	NR
Coronene	NR	NR	0.34	NR	NR

[a] Data from National Bureau of Standards.
[b] Data from Oak Ridge National Laboratory Analysis of SRM 1650,Griest et al. 1988.
[c] Duc & Huynh 1989.
[d] Gundel et al. 1995.
[e] Rogge et al. 1994.
[f] Mixture of [a] and [b] isomers.
[g] Data for sum of chrysene + triphenylene.
[h] Data for dienz[a,h]anthracene only.
NR= not reported.

about a factor of 10 lower than that reported by Elliot and Rowe (1975) in field measurements of CO, RSP, and BaP in a large arena (ca. 32–45 ppm). Possible explanations for the higher levels in the arena may include contamination from other sources of RSP and use of much less sophisticated analytical methodology for separation and identification of BaP.

Environmental tobacco smoke is not a static matrix. It undergoes physical and chemical changes as it ages. A number of factors can influence the extent to which these changes occur. These include ventilation, the presence of surfaces and reactive species from other sources, and ultraviolet (UV) radiation. Probably one of the most significant factors in the aging of ETS is the adherence of nicotine to surfaces. This has the effect of reducing the amounts of nicotine relative to other components in ETS. The longer that ETS can remain within a given volume (lower ventilation rate), the more likely that nicotine will be depleted through surface adsorption (e.g., Tang et al. 1989). For example, Table 4.8 summarizes data from Thompson et al. (1989), which demonstrates the influence of air exchange rate on the nicotine:RSP ratio.

Table 4.8 Influence of Air Exchange Rate on Nicotine: Respirable Suspended Particulate (RSP) Ratio[a,b]

"Chamber" Type	Ventilation Rate Air changes per hour	Nicotine/RSP
0.4 m³ Stainless Steel	21-105	0.39 ± 0.05
1.4 m³ Stainless Steel	21-105	0.52 ± 0.07
Unoccupied Office	5.4	0.16 ± 0.06
18 m³ Stainless Steel	0.05	0.08 ± 0.04

[a] ETS and/or dilute sidestream smoke was the only significant contaminant for this study.
[b] Data from Thompson et al. 1989.

As the air exchange rate was decreased from >20 ACH to nearly static, the nicotine:RSP ratio decreased by nearly a factor of 5. Nelson et al. (1992) have reported on a more detailed ventilation study, demonstrating important changes in nicotine:RSP ratios as a function of ventilation. Nicotine has also been shown to be adsorbed on wall board, glass, fabric, carpeting, and clothing (Van Loy et al. 1998, Piadé et al. 1999). In some cases, nicotine can off-gas from surfaces onto which it has been adsorbed, and can contribute to the level of ambient nicotine (Van Loy et al. 1998, Piadé et al. 1999). For example, nicotine desorbing from surfaces and cigarette butts was considered responsible for nicotine levels of 2.8 to 8.7 $\mu g/m^3$ in the smoking section of a Boeing 767 commercial aircraft that was parked at the gate and unoccupied by either passengers or crew (Nelson et al. 1990). Clearly, the adsorption/desorption of nicotine is a very complex phenomenon, and a factor likely to impact its utility as a marker for other ETS compounds.

Chemical reactions with surfaces or airborne components may be responsible for reductions with time of other ETS components. For example, from data in Table 4.1, it is clear that formaldehyde:acrolein ratios increase substantially when SS is diluted to surrogate ETS in a large chamber. The most plausible explanation for this is the reactivity of acrolein. Studies using sulfur hexafluoride as a tracer for leaks in a controlled environmental chamber have demonstrated that nicotine, ethenylpyridine, RSP, nitric oxide, hydrocarbons, and carbon monoxide all had shorter half-lives than could be accounted for by leaks from the chamber (Eatough et al. 1989b). Of course, the "loss" of nitric oxide (NO) is to be expected due to its conversion to nitrogen dioxide (NO_2,) through reaction with atmospheric oxygen. Well known for atmospheric chemistry, the reaction has been confirmed for MS (Jenkins and Gill 1980) and has been postulated for ETS (Baker and Proctor 1990). Because NO is reactive, it is possible that species also may be formed through reaction with NO as ETS ages. Such has not been confirmed experimentally. Data generated in a study by Nelson and Conrad (1997) suggest that NO_2 and formaldehyde may interact with water films in HVAC system cooling coils.

A phenomenon that has not been thoroughly studied, but one that may play a significant role in the aging of ETS, is the influence of UV light. Eatough's group at Brigham Young University has studied some of the gas and particle phase constituents of surrogate ETS in large Teflon bags during irradiation from UV lamps (Eatough et al. 1989b, Benner et al. 1989). Nicotine concentration decreased significantly, and 3-EP decreased somewhat, as a result of the UV irradiation. There were significant increases in gas-phase 2-ethenylpyridine and particle-phase nicotine. The authors hypothesized that about half of the loss of gas-phase nicotine was due to the formation of acidic compounds by UV treatment and an increase in the number of particles, leading to acid-base condensation and formation of particle-phase nicotine. The other half of the loss of vapor-phase nicotine was believed to be due to the photochemical reactions of nicotine, which led to the formation of other products, such as particle-phase myosmine, nicotyrine, and cotinine. Ogden and Richardson (1998) have demonstrated postcollection instability of selected particulate-phase constituents when they were collected on filters mounted in clear plastic holders and stored in sunlight. Ogden and Richardson have thus recommended collection of ETS particles in opaque filter holders, and many investigators have adopted this practice (e.g., Phillips et al. 1998). Because virtually all indoor environments have some sort of artificial or natural lighting, these studies are, at a minimum, suggestive of the chemical instability of key components of ETS. Taken together, studies of chemical reactivity and/or sorption/desorption further reinforce the concept that ETS should be considered to have a range of likely compositions under field conditions.

MARKERS FOR ENVIRONMENTAL TOBACCO SMOKE

Investigators have long sought a marker for ETS. Given that ETS is not a single entity but comprises thousands of chemical components, and that many of the components may change chemically and/or physically with time, the lack of complete success should not be surprising. Some of the criteria for an ideal marker for any material are that it should be a specific, unambiguously determinable species, present at sufficiently high concentration in the target matrix so that it can be readily quantified. The marker should be unique to the material of interest, and should behave similarly to the material for which it is a marker (National Research Council 1986). No component of ETS fully meets these criteria.

Until the mid-1980s, carbon monoxide (CO) was probably the most widely cited marker of ETS (Aviado 1984). It could be easily determined in near real time with relatively inexpensive equipment. (Analytical methods for CO, as well as potential interferences, are discussed in greater detail in Chap. 8.) CO is produced in relatively substantial quantities in SS, the major precursor for ETS (see Chap. 3). However, the massive dilution that SS undergoes in the process of becoming ETS reduces the CO concentration to near-background levels. In addition, because it is produced in nearly all combustion-related processes, it is not at all unique to ETS. Indoor gas stoves, heaters, appliances, and infiltrated motor vehicle exhaust appear to be the primary sources of CO in indoor air. Because of the widespread nature of non-ETS sources of CO and their daily variations (changes in traffic volume, etc.), only one study has been able to demonstrate a statistically significant increase in CO levels of indoor air environments that could be attributed to smoking (Akland et al. 1985). Because of these reasons, CO, while an important component of indoor air, is no longer considered a suitable marker for ETS.

RSP has also been used as an indicator of the presence of ETS (Repace and Lowrey 1980). In most cases where it has been possible to assess both smoking and nonsmoking levels of RSP in the same or similar environments, RSP levels in smoking environments are usually higher (see Chap. 6). However, the extent of increase due to smoking has in most cases been found to be a fraction varying between a few percent to more than half (see the analysis in Chap. 6). Spengler et al. (1985) reached similar conclusions when discussing ETS-related RSP. Because there are so many other sources of RSP in the indoor environment, it has become increasingly clear that RSP is not a good marker for ETS, but is best used only as a general determinant of respirable particles in the indoor setting.

Better markers have been developed for the assessment of particles derived from combustion, of which ETS is an important contributor in many indoor environments. In the mid- to late 1980s, Conner et al. (1990) and Ogden et al. (1989) developed methods based on the UV absorption and/or fluorescence of many of the components of the particulate phase of ETS. So-called UV-absorbing particulate matter (UVPM) and fluorescing particulate matter (FPM) are described in more detail in Chap. 6. These markers have been shown to track the particle phase of ETS well and have been used in tens of studies, for both personal monitoring and area sampling of ETS. FPM can be determined to slightly lower limits of quantification and is increasingly becoming the preferred of the two markers. There is also a growing understanding that both of the markers are conservative tracers for ETS. That is, both tend to overestimate the fraction of particles contributed by ETS because they are measures of total particles derived from combustion. In many environments, there will be numerous other sources of these particles, including infiltrated diesel exhaust; carbon from copying machines; and soot from fireplaces, gas stoves, or other combustion appliances. Heavner et al. (1996) have described in detail the methods for calculation and interpretation of UVPM and FPM values. Nelson et al. (1997a, 1997b) have conducted a detailed examination of the UVPM and FPM emissions from cigarettes available commercially in more that 15 nations.

Probably the best marker for the particulate matter of ETS is solanesol (Ogden and Maiolo 1989, Tang et al. 1990). Solanesol is a trisesquiterpenoid alcohol found in tobacco leaf, cigarette smoke condensate, and the RSP of ETS. Because of its high molecular weight (631 g/mole), it has very low volatility and is expected to remain part of the particulate matter, even at high dilutions. Although the analytical procedure for solanesol is less simple than that for UVPM and FPM, detailed studies have indicated that it is the ETS RSP constituent in greatest abundance, averaging about 3 to 4% of the weight of the RSP collected from the ETS generated from U.S. commercial and reference cigarettes (Ogden and Maiolo 1989). Studies conducted on cigarettes from several nations (Nelson et al. 1997a, 1997b) have indicated that the fraction of solanesol in ETS RSP typically varies from 2.5 to 4%. One notable exception is ETS produced from Canadian brands, which exhibit a fraction of 1.5%. The difference is important when trying to estimate the fraction of RSP contributed by ETS. Tang et al. (1990) have demonstrated that solanesol concentration decays rapidly under intense UV light, but otherwise, its decay rates are comparable with those of particulate-phase nicotine (another proposed indicator of ETS RSP) and 3-EP. Solanesol has now been used with considerable success in a number of large personal-monitoring studies. However, it should be recognized that in many modern environments where limited smoking occurs, the solanesol concentration may be too low to quantify with existing sampling and analytical methodology.

Scopoletin (7-hydroxy-6-methoxy-2H-1-benzopyran-2-one) is a polyphenol distributed throughout the *Solanaceae* family, including, of course, tobacco. It has also been reported in oak leaves. While its molecular weight is much lower than that of solanesol, its polarity is such that it remains in the particle phase of tobacco smoke. An analytical method has been developed for it determination in ETS (Risner 1994), and it has been used in conjunction with other markers of ETS in at least two large personal-exposure studies (Heavner et al. 1996, Jenkins et al. 1996). However, its concentration in ETS particulate matter is less than 5% that of solanesol's (Martin et al. 1997), and its detection limit in ETS is somewhat higher (Jenkins et al. 1996). Currently, it appears to offer no important advantages over solanesol as a marker for the particle phase of ETS.

Eatough and co-workers (Eatough et al. 1989a, 1989b; Benner et al. 1989; Tang et al. 1989) have investigated the utility of particle-phase nicotine as a marker for ETS particle phase. In controlled atmospheres, about 5 to 10% of ETS nicotine has been shown to be associated with the particle phase. The proportion in field environments was somewhat higher (ca. 20%), perhaps due to rapid removal of vapor-phase nicotine (Eatough et al. 1989a). The determination of particle-phase nicotine is more difficult than for that of vapor-phase nicotine because the former requires the upstream removal of the vapor phase, usually with a coated diffusion denuder tube. The authors (Eatough et al.) suggest that particle-phase nicotine may be a better marker for ETS RSP than vapor-phase nicotine is, but they indicated that there would still be a fairly high degree of variability. The authors of this monograph analyzed the field data reported in Eatough et al. (1989a) for situations in which both particle-phase nicotine and RSP measurements were made. For 12 venues, in which RSP levels ranged from 19 to 801 $\mu g/m^3$, RSP:particle-phase nicotine ratios ranged from about 54 to 264, a five-fold range. This range is indeed smaller than that reported by Oldaker et al. (1989) for RSP:vapor-phase nicotine ratios and suggests that the former may be of somewhat greater, but certainly less than ideal, predictive value.

From data obtained during their examination of the elemental composition of indoor air, Lebret et al. (1987) suggest that cadmium (Cd) may be a good marker for ETS. In a survey of 20 homes, they found that homes with no smokers had undetectable levels of Cd. In homes with smokers present, there was a relatively high correlation between Cd and RSP [correlation coefficient (R) = 0.837]. Because tobacco consumption appeared to be an important factor in elevated RSP levels in the homes, the investigators interpreted their findings as a suggestion that airborne Cd may be indicative of tobacco smoking. Further laboratory studies and limited field studies conducted by a team at the University of Illinois (Wu et al. 1995) indicated that Cd shows good promise as a marker of ETS. However, a relatively robust analytical method (neutron activation analysis) was required to quantify Cd at the 10-ng/m^3 range. In general, such observations will require more investigation to support these tentative conclusions.

Iso-alkanes and ante-iso-alkanes are branched, long-chain hydrocarbons that are enriched in the leaf waxes of tobacco relative to other common vegetation. At least two research groups have studied their utility as a marker for ETS-derived particulate matter in some indoor and outdoor urban air (Rogge et al. 1994, Kavouras et al. 1998). Although the complexity of analysis is significant and the absolute magnitude of emissions from nonanthropogenic sources (mainly leaf abrasion products from natural vegetation) is much greater than that from ETS exfiltrating from indoor environments and outdoor smoking, their claimed advantage over other common ETS RSP markers is that they are more stable. They also appear to be more stable than the straight-chain hydrocarbons, fatty acids, or PAHs that compose a significant portion of ETS particles. The utility of a tracer that is more stable than the material being traced is not clear. The iso-alkanes and anteiso-alkanes have been used to estimate overall contribution of ETS to Los Angeles urban air fine particulate at 1 to 1.3%. No studies have reported the use of these species for estimating human exposure to ETS.

Because of its uniqueness to tobacco products at all but trace levels, nicotine has become quite popular as a marker for ETS. Since the mid-1980s, when improved analytical methods for airborne nicotine became available, the number of studies that has determined this species in both controlled atmosphere and field studies has increased exponentially (see Chap. 7). Although it is unique to tobacco in many (if not most) field situations, it is present at relatively modest levels. Levels typically range from below detection limits to 10 $\mu g/m^3$ with occasional levels reaching 20 to 100 $\mu g/m^3$. Nicotine has a strong affinity for surfaces, which decreases the utility of vapor-phase nicotine as a marker for ETS. Several investigators have demonstrated that vapor-phase nicotine is depleted from the ambient environment more rapidly than is the particulate portion of ETS (Eudy et al. 1986, Eatough et al. 1987). Interestingly, the same affinity for surfaces, which tends to decrease its concentration relative to many other ETS constituents as ETS ages, also makes the analysis of trace quantities of nicotine somewhat more difficult. Strong bases, such as ammonia or triethylamine, are usually added to the nicotine collection media after sampling to prevent the adsorption of the nicotine to glass or steel surfaces within the analytical equipment (Ogden 1989, Thompson et al. 1989, Hammond et al. 1987). For chamber studies, in which ETS is the only contaminant and ventilation rates are defined and constant, vapor-phase nicotine is probably an adequate marker for ETS. For example, Thompson et al. (1989) reported a correlation coefficient (R) of 0.976 for nicotine and RSP in an unoccupied office that had few other sources of RSP. Leaderer and Hammond (1991) demonstrated a lower correlation for ETS generated by humans in a chamber (R=0.505). However, and independent analysis performed by Ogden and Jenkins (1999) of two large chamber studies (Nelson et al. 1997a, 1997b) indicated considerable variation in the ratios of nicotine to other accepted ETS markers.

More important than nicotine's behavior in well-defined chamber atmospheres is its performance as a marker in field settings. Oldaker et al. (1989) examined extensive data sets for field determinations of nicotine and RSP in restaurants, offices, and aircraft cabins. The investigators determined nicotine:RSP correlation coefficients for these venues of 0.236, 0.198, and 0.781, respectively. The higher correlations in aircraft cabins were thought to be due to the longer sampling times employed (up to 5 h), which would have minimized the influence of temporal variations on the data. However, the high ventilation rates in aircraft cabins (up to 20 air changes per hour) and the lack of other sources of RSP make the commercial aircraft cabin a highly specialized environment, more like that of a test chamber. Leaderer and Hammond (1991) demonstrated a very good correlation between vapor-phase nicotine and ETS RSP in residences sampled over a period of a week. This finding may suggest that RSP source fluctuations average out over longer sampling times, and that the "excess" RSP due to ETS can be discerned more easily. Van Loy et al. (1998) studied the adsorption and desorption of nicotine on common indoor building materials and recommend that nicotine may not be useful as a tracer for short duration field measurements where smoking is sporadic, but may be much more effective for measurements of longer duration, where smoking rates are more consistent. Daisey (1999) analyzed the reports from personal-exposure measurements by Jenkins et al. (1996) and found strong correlations between nicotine and RSP exposures of large groups. Presumably, the analysis of groups dampens out individual subject-to-subject variation. However, it is clear that for prediction of individual exposures to other components of ETS, the use of vapor-phase nicotine as the sole tracer must be undertaken with a great deal of caution. Ogden and Jenkins (1999) analyzed reported correlations and ratios of nicotine to other common ETS markers from a number of the same personal-monitoring studies described in Chap. 7. The analysis indicated that there was considerable variation in the ratios and that the coefficients of determination (R^2) ranged from 0.14 to 0.97. In summary, nicotine has been used frequently as a marker for ETS because it meets three of the four primary criteria for a good marker. However, in individual situations, its direct relationship with other compounds of interest in ETS is certainly not constant, and while it is useful for providing comparisons with previous studies, reliance on it solely is not advised.

3-ethenylpyridine (3-EP), a pyrolysis product of nicotine degradation during smoking, is a far better marker for the vapor phase of ETS. It is adsorbed onto surfaces to a much smaller extent, decays much like nonreactive gases in ETS (Eatough 1989b), and is emitted preferentially into SS, the primary precursor of ETS. It has also been shown to interact less with ventilation systems than does nicotine (Nelson and Conrad 1997). An additional advantage is that 3-EP can be quantitatively collected and analyzed by the same methodology that is used for nicotine. (Note that the exception to this is active sampling for nicotine collected on filters treated with bisulfate because breakthrough volumes for 3-EP are too small.) Chamber studies have indicated that correlation with levels of other ETS gas-phase components is strong (Nelson et al. 1992). It has gained sufficient confidence among investigators to be used for source apportionment studies (Heavner et al. 1995, Hodgson et al. 1996). 3-EP has been used as a marker in more than a dozen large personal-monitoring studies comprising in total of more than 4000 individual subjects. In short, is seems to have become the tracer of choice for ETS vapor-phase components.

Myosmine has been proposed as an additional vapor-phase tracer from ETS. Like 3-EP, it is a pyrolysis product of nicotine and is found in measurable quantities in ETS, albeit at much lower concentrations than 3-EP (Eatough et al. 1989a, 1989b; Ogden and Nelson 1994; Jenkins et al. 1996). It can also be collected and analyzed simultaneously with nicotine and 3-EP. While myosmine is tobacco specific, it appears to offer few advantages over 3-EP as a marker. Higgins and Guerin (1988) reported data suggesting that isoprene might be a potential marker for the vapor phase of ETS. Isoprene arises from the pyrolytic degradation of higher terpenes in the tobacco. It is likely to be generated during the combustion of any plant material, such as firewood. However, it may be of utility in those environments where such interferences are limited.

CIGARETTE EQUIVALENTS

Beginning with Hoegg in 1972, a number of investigators has used the concept of "cigarette equivalent" (CEQ) to quantitatively describe the potential exposure of humans to ETS. The use has been thoroughly reviewed by Ogden and Martin (1997). The term has been described in detail in these manuscripts. Briefly, the use of "CEQ" is an approach to placing the exposure of an individual breathing ETS in context by comparing the amount of a component inhaled from ETS over a given time period with the amount potentially inhaled from direct smoking of the cigarette in question. The latter amount is usually assumed to be the MS delivery of the cigarette when smoked under standard conditions. For example, if an individual breathes about 5 m^3 of indoor air, with a nicotine concentration of 5 $\mu g/m^3$, in an 8-h workday, for a cigarette with an MS nicotine delivery of 1 mg, the individual would be said to have inhaled 0.025 CEQs during that period. Using this approach as estimation of CEQs will tend to produce high values for smoke components that are preferentially emitted into SS (compared with MS) and are sufficiently stable to exist in comparatively higher concentrations in true ETS. Examples may include ammonia and nitrogen-containing species such as the volatile nitrosamines and 3-EP. More modest estimates of CEQ are likely to be obtained for constituents whose emissions into SS and "survival" in ETS are more comparable with those of MS deliveries. Examples are likely to include the tobacco-specific nitrosamines and many particle-phase constituents. Still lower CEQ estimates will be obtained for species that are less stable in ETS or that interact with surfaces to a high degree, such as nicotine in environments where smoking is sporadic. This is illustrated in the results in Table 4.9. The estimates were calculated from results presented elsewhere on mainstream, sidestream, and ETS constituent levels in chambers for the Kentucky Reference IR4F cigarette (R. J. Reynolds 1988). The IR4F is comparable with many "low-tar" cigarettes currently on the American market. The data suggest that while an individual would breathe about one-eighth of a cigarette's worth of

Table 4.9 Estimated Cigarette Equivalents for Selected ETS Constitutents[a]

Constituent	Mainstream Smoke Delivery, per cigarette 1R4F cigarette	ETS Chamber Concentrations[b] ($\mu g/m^3$)	Calculated Cigarette Equivalents[c]
Carbon Monoxide	11.3 mg	2710	0.99
Particulate Matter	ca. 10 mg	420	0.17
Nicotine	0.79 mg	24.4	0.127
Pyridine	2.096 μg	30.5	60
Ammonia	18 μg	244	58

[a] Data calculated from values published in R. J. Reynolds 1988
[b] Per cigarette 2-h average concentrations.
[c] Assumes 8-h workday exposure at a breathing rate of 8.86 l/min.

nicotine in an 8-h exposure, during that same period, he/she would breathe more than 50 cigarettes' (2½ packs) deliveries of ammonia and pyridine.

Unfortunately, in many cases, if the level of the component of interest has not been quantified under realistic field conditions, investigators have estimated its concentration. The estimation was performed by measuring some other compound in ETS and then assuming that the SS ratio between the measured compound and the one of interest would remain constant in ETS. There are several problems with such a simplistic approach. First, the assumption of the maintenance of an SS ratio under field conditions, especially for components that are potentially reactive or that are present in trace quantities (e.g., Hammond 1993, Hammond et al. 1995), is not supported by existing data. For example, ammonia:CO ratios for true ETS from a 1R4F reference cigarette are half those found in SS for the 1R4F (Reynolds 1988). It seems critical that direct measurement of the compound of interest in the environment of interest be made. Secondly, except for a few tobacco-specific compounds, such as the alkaloids (nicotine, myosmine, etc), some leaf wax components (e.g., solanesol) and the tobacco-specific nitrosamines [e.g., 4-[N-methylnitrosamino]-1-[3-pyridyl]-1-butanone (NNK)], most environments have multiple sources for the compounds of interest (Heavner et al. 1995, Hodgson et al. 1996). For example, nitrosodimethylamine (NDMA) has often been used to demonstrate a high level of cigarette equivalents in smoking environments, based on its relatively high emission rate in SS. However, based on the normal background levels of NDMA in urban environments (Shah and Singh 1988) and the levels actually determined in environments where ETS is present (Klus et al. 1992, Mahanama and Daisey 1996), it is clear that the fraction of NDMA exposure from typically encountered levels of ETS is likely to be very small. Finally, the simplistic approach of relating ETS dose to MS delivery ignores the fact that the individuals likely to be exposed to the greatest levels of ETS are the smokers themselves. For example, Ogden (1996) demonstrated that because of source proximity, smokers are exposed to 10 to 20 times the levels of ETS vapor-phase components as nonsmokers in similar environments. Ogden and Martin applied this finding to data reported by Jenkins et al. (1996) in a large personal-monitoring study. Clearly, the estimation of CEQ depends on whether the benchmark is the MS emanating from the butt end of the cigarette during puffing or the overall exposure that results from "being a smoker." An additional difficulty with the use of CEQs is that comparison to a single smoke component ignores the spirit of the chemical complexity of cigarette smoke, which is composed of thousands of constituents. Depending on the choice of component, reference to merely one of these thousands may act to trivialize as opposed to "simplify" the provision of perspective on exposure to ETS. Readers are cautioned to evaluate such factors as the manner in which CEQ determination is made (actual ETS measurements vs. estimation from SS:MS ratios), potential contribution of nontobacco-specific constituents from other sources, and the benchmark exposure with which ETS exposure is compared before deciding whether an estimation of CEQs provides useful comparative information regarding tobacco smoke exposure.

SUMMARY

As SS and exhaled MS enter the indoor atmosphere, they are mixed and diluted to form ETS. As the dilution continues, an increasingly larger fraction of the more volatile constituents, including nicotine, evaporate from the particulate matter into the vapor phase. As ETS continues to age, nicotine is removed by adsorption on surfaces. Other vapor compounds are depleted as well, through reaction, adsorption, or other mechanisms, although most not as quickly. Most ETS particulate matter droplets are smaller than those of MS (0.35–0.4 μm) with a mass median diameter in the range of 0.15 to 0.25 μm.

The number of comprehensive studies on ETS composition is limited but growing, and more comparative data are beginning to emerge. ETS likely contains all of the constituents of SS and MS, albeit in different proportions. ETS particle composition is dominated by higher-molecular-weight alkanes, fatty acids, high-molecular-weight phenols, and phytosterols. Extensive studies have been conducted concerning the vapor-phase composition of ETS generated in chambers, and agreement among investigatory teams is good where comparisons can be made. Studies performed under field conditions suggest that ETS can be considered as having a range of compositions, depending on the magnitude of influences of ventilation, surfaces, source-input rates, lighting, and other factors of the specific environment.

A number of markers has been proposed for ETS. CO and RSPs have been used, but because they emanate from so many other sources, they are unspecific for ETS in most field situations. Vapor-phase nicotine, a tobacco-specific compound at all but trace levels, has gained considerable acceptance as a marker. However, its behavior is so different from most other vapor-phase ETS constituents that its utility as a marker is lower than originally anticipated.

3-EP has emerged as a marker of choice for the vapor phase of ETS, and studies are now performed that routinely report both nicotine and 3-EP concentrations. Markers for combustion-derived particles in indoor environments that have gained increasing acceptance include UVPM and FPM. However, due to the presence of other sources, direct dependence on these markers may lead to an overestimate of ETS levels in some settings. Currently, the most widely used marker for ETS particles is solanesol, a high-molecular-weight alcohol. The concept of CEQs continues to be employed by many investigators to provide perspective on human exposure to ETS. However, reliance on SS:MS ratios to calculate CEQ is clearly inappropriate, and any estimate of CEQ must be analyzed carefully to determine whether it actually provides useful perspective.

REFERENCES

Akland, G. G., Hartwell, T. D., et al. (1985) Measuring human exposure to carbon monoxide in Washington, D.C. and Denver, CO during the winter of 1982–1983. *Environ. Sci. Technol.,* 19, 911–918.

Armitage, A. K. & Turner, D. M. (1970) Absorption of nicotine in cigarette smoke through the oral mucosa. *Nature,* 226, 1231–1232.

Aviado, D. M. (1984) Carbon monoxide as an index of environmental tobacco exposure. *Eur. J. of Resp. Dis. Supp.* (1984), No. 133, 65, pp. 47–60.

Baker, R. R. & Proctor, C. J. (1990) The origins and properties of environmental tobacco smoke. *Environ. Int.,* 16, 231–245.

Bayer, C. W. & Black, M. S. (1987) Thermal desorption/gas chromatographic/mass spectrometric analysis of volatile organic compounds in the offices of smokers and nonsmokers. *Biomed. Environ. Mass Spectrom.,* 14, 363–367.

Benner, C. L., Bayona, J. M., Caka, F. M., Tang, H., Lewis, L., Crawford, J., Lamb, 1. D., Lee, M. L., Lewis, E. A., Hansen, L. D., & Eatough, D. J. (1989). Chemical composition of environmental tobacco smoke. 2. Particulate-phase compounds. *Environ. Sci. Technol.,* 23, 688–699.

Black, A., Pritchard, J. N., & Walsh, M. (1987) An exposure system to assess the human uptake of airborne pollutants by radio-tracer techniques. *J. Aerosol Sci.,* 18, 757–760.

Black, A., Pritchard, J. N., & Walsh, M. (1988) An exposure system to assess the human uptake of airborne pollutants by radio-tracer techniques, with particular reference to sidestream cigarette smoke. *J. Aerosol Sci.,* 19(3).

Conner, J. M., Oldaker, G. B., III, & Murphy, J. J. (1990) Method for assessing the contribution of environmental tobacco smoke to respirable suspended particles in indoor environments. *Environ. Technol., 11,* 189–196.

Creighton, D. E. & Lewis, P. H. (1978) The effect of smoking pattern on smoke deliveries. In: Thornton, R. E., ed., *Smoking Behavior— Physiological and Psychological Influences,* Churchill Livingstone, London and New York, pp. 301–314.

Crouse, W. E. & Carson, J. R. (1989) Surveys of environmental tobacco smoke (ETS) in Washington, D.C. offices and restaurants. *43rd Tobacco Chemists' Research Conference,* Richmond, VA.

Daisey, J. M. (1999) Tracers for assessing exposure to environmental tobacco smoke: what are they tracing? *Environ. Health. Perspect.* 107 (Suppl 2), 319–327

Daisey, J. M., Mahanama, K. R. R., Hodgson, A. T. (1998) Toxic volatile organic compounds in simulated environmental tobacco smoke: emission factors for exposure assessment. *J. Expo. Anal. Environ. Epidemiol.,* 8, 313-334.

Duc, T. V. & Huynh, C. K. (1989) Sidestream tobacco smoke constituents in indoor air modeled in an experimental chamber—polycyclic aromatic hydrocarbons. *Environ. Int.,* 15, pp. 57–64.

Duc, T. V. & Huynh, C. K. (1987) Deposition rates of sidestream tobacco smoke particles in an experimental chamber. *Toxicol. Letts.,* 15, 59–65.

Eatough, D. J., Hansen, L. D., & Lewis, E. A. (1990) The chemical characterization of environmental tobacco smoke. In: Ecobichon, D. J., & Wu, J. M., eds. *Environmental Tobacco Smoke. Proceedings of the International Symposium at McGill University, 1989,* pp. 3–39.

Eatough, D. J., Benner, C. I., Bayona, J. M., Caka, F. M., Tang, H., Lewis, L., Lamb, J. D., Lee, M. L., Lewis, E. A., & Hansen, L. D. (1987) Sampling for gas phase nicotine in environmental tobacco smoke with a diffusion denuder and a passive sampler. *EPA IAPCA Symposium on Measurement of Toxic and Related Air Pollutants, pp.* 132–139.

Eatough, D. S., Benner, C. J., Bayona, J. M., Richards, G., Lamb, J. D., Lee, M. L., Lewis, E. A., & Hansen, L. D. (1989b). Chemical composition of environmental tobacco smoke. 1. Gas–phase acids and bases. *Environ. Sci. Technol.,* 23, 679–687.

Eatough, D. J., Benner, C. L., Tang, H., Landon, V., Richards, G., Caka, F. M., Crawford, J., Lewis, E. A., Hansen, L. D., & Eatough, N. L. (1989a) The chemical composition of environmental tobacco smoke. III. Identification of conservative tracers of environmental tobacco smoke. *Environ. Int.,* 15, pp. 19–28.

Elliott, L. P. & Rowe, D. R. (1975) Air quality during public gatherings. *JAPCA,* 25(6), 635–636.

Eudy, L. W., Thorne, F. W., Heavner, D. K., Green, C. R., & Ingebrethsen, B. J. (1986) Studies on the vapor–particulate phase distribution of environmental nicotine by selective trapping and detection methods. *Proceedings, 79th Annual Meeting of the Air Pollution Control Association.* Paper 86–38.7.

Goodfellow, H. D., Eyre, S., & Wyatt, J. A. S., Assessing exposures to environmental tobacco smoke (1990) In: Ecobichon, D. J., and Wu, J. M., eds. *Environmental Tobacco Smoke. Proceedings of the International Symposium at McGill University, 1989*, pp. 53–67.

Griest, W. H., Jenkins, R. A., Tomkins, B. A., Moneyhun, J. H., Ilgner, R. H., Gayle, T. M., Higgins, C. E., & Guerin, M. R. (1988) *Sampling and Analysis of Diesel Engine Exhaust and the Motor Pool Workplace Atmosphere, Final Report.* ORNL/TM-10689.

Guerin, M. R. (1991) Environmental tobacco smoke. *Organic Chemistry of the Atmosphere*, Hansen, L. D. & Eatough, D. J., eds., CRC Press, Boca Raton, FL, pp. 79–119.

Gundel, L. A., Mahanama, K. R. R., & Daisey, J. M (1995) Semivolatile and particulate polycyclic aromatic hydrocarbons in environmental tobacco smoke: cleanup, speciation, and emission factors. *Environ. Sci. Technol.*, 29, 1607–1614.

Hammond, S. K., Leaderer, B. P., Roche, A. C., & Schenker, M. (1987) Collection and analysis of nicotine as a marker for environmental tobacco smoke. *Atmos. Environ.*, 21(2), 457–462.

Hammond, S. K., Sorenson, G., Youngstrom, R., Ockene, J. K. (1995) Occupational exposure to environmental tobacco smoke. *JAMA*, 274, 956-960.

Hammond, S. K. (1993) Evaluating exposure to environmental tobacco smoke. In: Winegar, E. D., Keith, L. H., eds., *Sampling and Analysis of Airborne Pollutants*, Lewis Publishers, Boca Raton, FL, pp. 319–338.

Heavner, D. L., Morgan, W. T., & Ogden, M. W. (1996) Determination of volatile organic compounds and respirable suspended particulate matter in New Jersey and Pennsylvania homes and workplaces. *Environ. Int.*, 22, 159-183.

Heavner, D. L., Morgan, W. T., Ogden, M. W. (1995) Determination of volatile organic compounds and ETS apportionment in 49 homes. *Environ. Int.*, 21, 3-21.

Higgins, C. E. & Guerin, M. R. (1988) Studies on the source apportionment of volatile organic indoor air contaminants due to cigarette smoke. *42nd Tobacco Chemists' Research Conference*, Lexington, KY.

Hodgson, A. T., Daisey, J. M., Mahanama, K. R. R., Brinke, J. T., Alevantis, L. E. (1996) Use of volatile tracers to determine the contribution of environmental tobacco smoke to concentrations of volatile organic compounds in smoking environments. *Environ. Int.*, 22, 295-307.

Hoegg, U. R. (1972) Cigarette smoke in enclosed spaces. *Environ. Health Perspect.*, 2, 117–128.

Hugod, C., Hawkins, L. H. & Astrup, P. (1978) Exposure of passive smokers to tobacco smoke constituents. *Int. Arch. Occup. Environ. Health*, 42(1), 21–29.

Ingebrethsen, B. J. (1989) The physical properties of mainstream cigarette smoke and their relationship to deposition in the respiratory tract. In: Crapo, J. D., Miller, F. J., Smolko, E. D., Graham, J. A., & Hayes, A. W., eds., *Extrapolation of Dosimetric Relationships for Inhaled Particles and Gases,* Academic Press, San Diego, pp. 125–141.

Ingebrethsen, B. J. & Sears, S. B. (1985) Particle size distribution measurements of sidestream cigarette smoke. *39th Tobacco Chemists' Research Conference,* Montreal, Quebec, Canada.

Jenkins, R. A. & Gill, B. E. (1980) Determination of oxides of nitrogen (NO,) in cigarette smoke by chemiluminescent analysis. *Anal. Chem.,* 52, pp. 925–928.

Jenkins R.A., Palausky A., Counts R. W., Bayne C. K., Dindal A. B., Guerin M. R. (1996) Exposure to environmental tobacco smoke in sixteen cities in the United States as determined by personal breathing zone air sampling. *J. Expo. Anal. Environ. Epidemiol.,* 6, 473-502.

Kavouras , I. G., Stratigakis, N., & Stephanou, E. G. (1998) Iso- and anteiso-alkanes: specific tracers of environmental tobacco smoke in indoor and outdoor particle size distributed urban air. *Environ. Sci. Technol.,* 32, 1369–1377.

Kelly, T. J., Smith, D. L., & Satola, J. (1999) Emission rates of formaldehyde from materials and consumer products found in California homes. *Environ. Sci. Technol.,* 33, 81-88

Klus, H., Begutter, H., Scherer, G., Tricker, A. R., & Adlkofer, F. (1992) tobacco-specific and volatile n-nitrosamines in environmental tobacco smoke of offices. *Indoor Environ.,* 1, 348-350.

Klus, H., Begutter, H., Nowak, A., Pinterits, G., Ultsch, I., & Wihlidal, H. (1985) Indoor air pollution due to tobacco smoke under real conditions. Preliminary results. *Tokai J. Exp. Clin. Med.,* 10(4), 331–340.

Leaderer, B. P. & Hammond, S. K. (1991) Evaluation of vapor-phase nicotine and respirable suspended particle mass as markers for environmental tobacco smoke. *Environ. Sci. Technol.,* 25, 770–777.

Lebret, E., McCarthy, J., Spengler, J., & Chang, B. H. (1987) Elemental composition of indoor fine particles. In: Seifert, B., Esdorn, H., Fischer, M., Ruden, H., & Wegner, J., eds., *Proceedings of the 4th International Conference on Indoor Air Quality and Climate, Vol. 1,* pp. 569–574.

Liang, C. & Pankow, J. F. (1996) Gas/particle partitioning of organic compounds to environmental tobacco smoke: partition coefficient measurements by desorption and comparison to urban particulate matter. *Environ. Sci. Technol.,* 30, 2800–2805.

Lipowicz, P. (1988) Determination of cigarette smoke particle density from mass and mobility measurements in a Millikan cell. *J. Aero. Sci.,* 19, 587–589.

Löfroth, G., Burton, R. M., Forehand, L., Hammond, S. K., Sella, R. L., Zweidinger, R. B., & Lewtas, J. (1989) Characterization of environmental tobacco smoke. *Environ. Sci. Technol.*, 23, 610–614.

Mahanama, K. R. R. & Daisey, J. M. (1996) Volatile n-nitrosamines in environmental tobacco smoke: sampling, analysis, emission factors, and indoor air exposures. *Environ. Sci. Technol.*, 30, 1477–1484.

Martin, P., Heavner, D. L., Nelson, P. R., Maiolo, K. C., Risner, C. H., Simmons, P. S., Morgan, W. T., & Ogden, M. W. (1997) Environmental tobacco smoke (ETS): A market cigarette study. *Environ. Int.*, 23, 75–90.

Morawska, L., Barron, W., & Hitchins, J. (1999b) Experimental deposition of environmental tobacco smoke submicrometer particulate matter in the human respiratory tract. *Am. Ind. Hygiene Assoc. J.*, 60, 334–339.

Morawska, L., Johnson, G., Ristovski, S. D., & Agranovski, V. (1999a) Relation between particle mass and number for sub-micrometer airborne particles. *Atmos. Environ.*, 33, 1983–1990.

Muramatsu, M., Umemura, S., Okada, T., & Tomita, H. (1984) Estimation of personal exposure to tobacco smoke with a newly developed nicotine personal monitor. *Environ. Res.*, 35, 218–227.

National Research Council. (1986) *Environmental Tobacco Smoke: Measuring Exposures and Assessing Health Effects.* National Academy Press, Washington, D.C. 1986.

Nelson, P. R., Conrad, F. W., Kelly, S. P., Maiolo, K. C., Richardson, J. D., & Ogden, M. W. (1997b) Composition of environmental tobacco smoke (ETS) from international cigarettes Part II: Nine country follow-up. *Environ. Int.*, 24, 251-257.

Nelson, P. R., Kelley, S. P, & Conrad, F. W. (1998) Studies of environmental tobacco smoke generated by different cigarettes. *J. Air & Waste Manage. Assoc.*, 48, 336–344.

Nelson, P. R., Heavner, D. L., Collie, B. B., Maiolo, K. C., & Ogden, M. W. (1992) Effect of ventilation and sampling time on environmental tobacco smoke component ratios. *Environ. Sci. Techno.*, 26, 1909-1915.

Nelson , P. R. & Conrad, F. W. (1997) Interaction of Environmental tobacco smoke components with a ventilation system. *Tob. Sci.*, 41, 45-52.

Nelson, P. R., Conrad, F. W., Kelly, S. P., Maiolo, K. C., Richardson, J. D., & Ogden, M. W. (1997a) Composition of environmental tobacco smoke (ETS) from international cigarettes and determination of ETS-RSP: particulate marker ratios. *Environ. Int.*, 23, 47-52.

Nelson, P. R., Heavner, D. L., Oldaker, & G. B., III (1990) Problems with the use of nicotine as a predictive environmental tobacco smoke marker. 1990 *EPA/A&WMA Conference on Measurement of Toxic and Related Air Pollutants,* Raleigh, NC.

Ogden, M. W. & Martin, P. (1997) The use of cigarette equivalents to assess environmental tobacco smoke exposure. *Environ. Int.*, 23, 123–138.

Ogden, M. W. & Jenkins, R. A. (1999) Nicotine in environmental tobacco smoke. In: *Analytical Determination of Nicotine*, Jacob, P. & Gorrod, J., eds., Elsevier, Amsterdam.

Ogden, M. W., & Nelson, P. R. (1994) Detection of alkaloids in environmental tobacco smoke. In: Linskens, H. F., Jackson, J. F., eds., *Alkaloids*, Berlin, Springer-Verlag, pp. 163-189.

Ogden, M. (1996) Environmental tobacco smoke exposure of smokers relative to nonsmokers. *Anal. Commun.*, 33, 197-198.

Ogden, M. W. & Richardson, J. D. (1998) Effect of lighting and storage conditions on the stability of ultraviolet particulate matter, fluorescent particulate matter, and solanesol. *Tob. Sci.*, 42, 10-15.

Ogden, M. W., & Maiolo, K. C. (1989) Collection and determination of solanesol as a tracer of environmental tobacco smoke in indoor air. *Environ. Sci.Technol.*, 23, 1148-1154.

Ogden, M. W., Maiolo, K. C., Oldaker, G. B., III, & Conrad, F. W., Jr. (1989) Evaluation of methods for estimating the contribution of ETS to respirable suspended particles. *43rd Tobacco Chemists' Research Conference*, Richmond, VA.

Ogden, M. W. & Maiolo, K. C. (1989) Collection and determination of solanesol as a tracer of environmental tobacco smoke in indoor air. *Environ. Sci. Technol.*, 23(9), 1148-1154.

Ogden, M. W. (1989) Gas chromatographic determination of nicotine in environmental tobacco smoke: collaborative study. *J. Assoc. Off. Anal. Chem.*, 72(6), 1002-1006.

Oldaker, G. B., III & Conrad, F. C., Jr. (1987) Estimation of effect of environmental tobacco smoke on air quality within passenger cabins of commercial aircraft. *Environ. Sci. Technol.*, 21(10), 994-999.

Oldaker, G. B., III, Crouse, W. E., & DePinto, R. M. (1989) On the use of environmental tobacco smoke component ratios. *Exerpta Medica Int. Congress Series, Present and future indoor air quality.* C. J. Bieva, Y. Courtois, & M. Govaerts, eds., pp. 287-290.

Piadé, J. J., D'Andres, S., & Sanders, E. B. (1999) Sorption phenomena of nicotine and ethenylpyridine vapours on different materials in a test chamber. *Environ. Sci. Technol.*, 33, 2046-2052.

Phillips, K., Howard, D. A., Bentley, M. C., and Alvan, G. (1998). Environmental tobacco smoke and respirable suspended particle exposures for non-smokers in Beijing. *Indoor Built Environ.*, 7, 254-269.

Pritchard, I. N., Black, A., & McAughey, J. J. (1988) The physical behavior of sidestream tobacco smoke under ambient conditions. In: Perry, R. & Kirk, P. W., *Indoor and Ambient Air Quality,* London, Selper, pp. 49-56.

Proctor, C. J., Warren, N. D., & Bevan, M. A. J. (1989) Measurements of environmental tobacco smoke in an air-conditioned office building. *Proceedings, Conference on the Present and Future of Indoor Air Quality,* Brussels.

Proctor, C. J., Martin, C., Bevan, J. L., & Dymond, H. F. (1988) Evaluation of an apparatus designed for the collection of sidestream tobacco smoke. *Analyst,* 113, 1509–1513.

Raabe, O. G. (1980) Physical properties of aerosols affecting inhalation toxicology, *Proceedings of the 19ᵗʰ Annual Hanford Life Sciences Symposium—Pulmonary Toxicology of Respirable Particles, Series 53,* Richland, WA, pp. 1–28.

Ramsey, R. S., Moneyhun, J. H., & Jenkins, R. A. (1990) Generation, sampling and chromatographic analysis of particulate matter in dilute sidestream tobacco smoke. *Anal. Chim. Acta,* 236, 213–220.

Rando, R. J., Menon, P. K., Poovey, H. G., & Lehrer, S. B. (1992) Assessment of multiple markers for environmental tobacco smoke (ETS) in controlled, steady-state atmospheres in a dynamic test chamber. *Am. Ind. Hygiene Assoc. J.,* 53, 699–704.

Repace, J. L. & Lowrey, A. H. (1980) Indoor air pollution, tobacco smoke, and public health. *Science,* 208, 464–472.

Repace, J. L. (1989) Workplace restrictions on passive smoking: Justification on the basis of cancer risk. *Transactions of an International Specialty Conference Combustion Processes and the Quality of the Indoor Environment,* Harper, J. P., ed., Niagara Falls, NY, pp. 333–347.

Reynolds, R. J., (1988) *Chemical and Biological Studies of New Cigarette Prototypes that Heat Instead of Burn Tobacco,* R. J. Reynolds Tobacco Company, Winston–Salem, N.C.

Risner, C. H. (1994) The determination of scopoletin in environmental tobacco smoke by high performance liquid chromatography. *J. Liq. Chromatogr.,* 17, 2723 – 2736

Rogge, W. F., Hildemann, L. M., Mazurek, M. A., Cass , G. R., & Simonelt, B. R. T. (1994) Sources of fine organic aerosol. 6. Cigarette smoke in the urban atmosphere. *Environ. Sci. Technol.,* 28, 1375-1388.

Scherer, G., Ruppert, T., Daube, H., Kossien, I., Riedel, K., Tricker, A. R., et al. (1995) Contribution of tobacco smoke to environmental benzene exposure in Germany. *Environ. Int.,* 21, 779-789.

Schievelbein, H. & Eberhardt, R. (1972) Cardiovascular action of nicotine and smoking. *J. Natl. Cancer Inst.,* 48, 1785–1794.

Shah, J. J. & Singh, H. B. (1988) Distribution of volatile organic chemicals in outdoor and indoor air, A national VOC data base. *Environ. Sci. Technol.,* 22, 1381-1388.

Spengler, J. D., Treltman, R. D., Tosteson, T. D., Mage, D. T., & Soczek, M. L. (1985) Personal exposures to respirable particulates and implications for air pollution epidemiology. *Environ. Sci. Technol.,* 19(8), 700–706.

Tang, H., Eatough, D. J., Lewis, E. A., Hansen, L. D., Gunther, K., Beinap, *D. M.*, & Crawford, J. (1989) The generation and decay of environmental tobacco smoke constituents in an indoor environment. 1989 *EPA/A&WMA International Symposium on Measurement of Toxic and Related Air Pollutants,* Raleigh, NC, pp. 596–605.

Tang, H., Richards, G., Benner, C. L., Tuominen, J. P., Lee, M. L., Lewis, E. A., Hansen, L. D., & Eatough, D. J. (1990) Solanesol: a tracer for environmental tobacco smoke particles. *Environ. Sci. Technol.,* 24, 848–852.

Thompson, C. V., Jenkins, R. A., & Higgins, C. E. (1989) A thermal desorption method for the determination of nicotine in indoor environments. *Environ. Sci. Technol.,* 23, 429–435.

Turner, S. (1989) Environmental tobacco smoke and smoking policies. *Transactions of an International Specialty Conference—Combustion Processes and the Quality of the Indoor Environment,* Harper, J. P., ed., Niagara Falls, NY, pp. 236–247.

Turner, S. & Binnie, P. W. H. (1990) An indoor air quality survey of twenty-six Swiss office buildings. *Proceedings of the 5th International Conference on Indoor Air Quality and Climate, Vol. 4,* Toronto, Canada, pp. 27–32.

Ueno, Y. & Peters, L. K. (1986) Size and generation rate of sidestream cigarette smoke particles. *Aerosol Sci. Technol.,* 5, 469–476.

Van Loy, M. D., Nazaroff, W. M., & Daisey, J. M. (1998) Nicotine as a marker for environmental tobacco smoke, implications of sorption on indoor surface materials. *J. Air & Waste Manage. Assoc.,* 48, 959 – 968.

Van Loy, M. D., Lee, V. C., Gundel, L. A., Daisey, J. M., Sextro, R. G., & Nazaroff, W. W. (1997) Dynamic behavior of semivolatile organic compounds in indoor air. 1. Nicotine in a stainless steel chamber. *Environ. Sci. Technol.,* 31, 2554-2561.

Wu, D., Landsberger, S., & Larson, S. M. (1995) Evaluation of elemental cadmium as a marker for environmental tobacco smoke. *Environ. Sci. Technol.,* 29, 2311–2316.

5 General Sampling and Analysis Considerations

INTRODUCTION

Tobacco smoke is an exceedingly complex matrix, consisting of several thousand constituents. As it is dispersed in the atmosphere, either from sidestream smoke (SS) or exhaled mainstream smoke (MS), its chemical and physical complexity can be increased through reactions among its constituents and through evaporation, condensation, coagulation, and adsorption or impaction on surfaces. ETS is often collected with conceptually simple systems, although the approach used to collect the matrix can often have a considerable impact on the nature of the results obtained. The purpose of this chapter is to provide a summary of important aspects and alternatives of sampling and chemical analyses presently in use for the determination of environmental tobacco smoke (ETS) constituents in ambient air. Dube and Green (1982) have thoroughly reviewed the use of various materials and procedures for the collection of MS generated and delivered by analytical smoking machines. The present discussion will concern only materials and methods for the collection of tobacco smoke at much lower concentrations, such as those encountered in ambient environments. The reader wishing more detail is referred to an excellent compilation of reviews on the instrumentation, sampling, and analytical technology for air sampling edited by Lioy and Lioy (1983). Wu (1997) has reviewed sampling and analysis methodology for determining ETS exposures.

GENERAL CONSIDERATIONS

Numerous characteristics of ETS are likely to have a significant impact on the sampling and analysis strategies used in its determination. Tobacco smoke is formed by a complex series of processes, including combustion, pyrolysis, evaporation, distillation, and condensation. During formation, a super-saturated cloud of organic vapors and combustion gases leaves the immediate vicinity of the firecone and begins to cool and condense. The result is an aerosol comprising liquid droplets and a surrounding cloud of associated vapors. As the smoke is generated and leaves the immediate environment of the cigarette as either SS or MS, it is not at chemical or physical equilibrium. However, as it is diluted with the surrounding air, chemical reactions, evaporation, condensation, and coagulation probably slow to the point where, for many constituents, a quasi-equilibrium is attained (Robinson and Yu 1999). Not all constituents achieve such a state, however. For example, the nitric oxide that does not react with organic gas-phase constituents in the atmosphere will require several hours for its conversion to nitrogen dioxide through reaction with atmospheric oxygen.

No single sampling and analysis system is appropriate for all of the constituents of ETS. Sampling systems must be optimized for a single compound or class of constituents. From a sampling and analysis standpoint, one of the most challenging attributes of ETS is its biphasic nature. That is, the smoke comprises permanent gases and lower-molecular-weight organic compounds, which reside predominantly in the vapor phase, and higher-molecular-weight organic species (and some inorganic species), which reside predominantly in the particulate phase. As smoke leaves the vicinity of the cigarette, it is diluted by ambient air, and the more volatile of the particulate-phase constituents are likely to evaporate from the aerosol droplets and enter the vapor phase. Because of the relatively high surface-area-to-volume ratio of the ETS droplets, a significant fraction of the mass of compounds usually envisioned as being primarily distributed in the particle phase will be found in the vapor phase (e.g., Ramsey et al. 1990, Gundel et al. 1995).

Evaporation of particulate-phase constituents collected during the collection process is a problem that must be addressed during the development of any sampling strategy, especially when large volumes of air are likely to be drawn across the filter assembly. Even low-volatility constituents such as benzo[a]pyrene can be evaporated from collected diesel particulates under certain sampling conditions (Griest et al. 1988). Cautreels and Van Cawenbergh (1978) examined the gas-phase constituents of urban air and identified more than 100 individual constituents. Some were aliphatic hydrocarbons with molecular weights exceeding 400; others were polynuclear aromatic hydrocarbons (PAHs) with molecular weights as great at 252 (benzo[a]pyrene); still others were phthalic acid esters with molecular weights as high as 390 [bis-2(ethylhexyl)phthalate]. Mumford et al. (1990) reported similar experiences when quantitating PAHs in both the vapor and particle phases for homes with and without kerosene heaters. PAHs with two and three rings, such as naphthalene and phenanthrene, appeared predominantly in the vapor phase; those with five, six, or seven rings, such as benzo[a]pyrene and coronene, appeared predominantly in the particle phase; and those with four rings, such as fluoranthene and benz[a]anthracene, were found in both phases. Foreman and Bidleman (1987) described a system for determining experimentally the vapor-particle partitioning of organic species during sample collection. Riggin and Peterson (1985) presented both vapor pressures and vapor/particle distribution ratios of selected PAHs at 25 °C, which show that low-molecular-weight PAHs exhibit greater vapor pressures and vapor-particle ratios than high-molecular-weight PAHs, as expected. Gundel et al. (1995) demonstrated distributions of PAHs among ETS particles and vapors. Given such findings, it is clear that a sampling protocol for many ETS constituents that emphasizes only the vapor phase or the particle phase could yield an incomplete understanding of the ambient levels of the component in question. In many cases, both phases must be sampled for a complete determination.

Many of the constituents of cigarette smoke are highly reactive, so the relative composition of the smoke may change during the sample-collection process. PAHs collected on filters with high-volume air samplers running for several hours are subject to both air oxidation to the corresponding carbonyl or quinone compounds and nitration to various nitrated species (Lee et al. 1980). Unstable constituents can decompose or react so as to increase the concentration of constituents already present or to form new constituents not normally found in smoke. For example, when sampling tobacco smoke for nitrosamine determination, considerable attention must be paid to the prevention of oxidative formation of nitrosamines. This has been accomplished by using a citric acid buffer in the aqueous trapping medium (Brunnemann and Hoffmann 1978), or treated solid sorbent cartridges (Rounbehler et al. 1980).

In field situations, the introduction of SS and MS into the ambient environment is highly episodic in nature. Typically, the emission of SS and MS is from a number of point sources over relatively short periods of time (approximately 8 min per cigarette). These sources can be dispersed to widely varying degrees within a given environment. For example, in some restaurants, there can be a single large room segregated into smoking and nonsmoking sections. Clearly, unless the ventilation systems for the two areas are separate, it is likely that ETS will be present to some extent in the nonsmoking area. Mainstream cigarette smoke is generated by individuals, all of whom employ unique smoke-generation patterns (Creighton and Lewis 1978). The smoke is retained for varying periods of time prior to exhalation of a portion of it. The amount and composition of the exhaled smoke is dependent on the smoking parameters used by the smoker (degree of inhalation, retention time of smoke in oral cavity and lungs, etc.). There are known compositional differences between inhaled and exhaled MS, and there may be some important compositional differences among the exhaled MS contributed by individual smokers to the ETS in a room. The potential compositional differences in the point sources of smokes, coupled with their varying distribution throughout a smoking environment, suggests the likelihood of encountering significant atmospheric inhomogeneities of many ETS components within a given field environment. Inhomogeneities have been documented for nicotine levels under field conditions, in which even closely spaced stationary samplers yielded considerably different ambient nicotine concentrations on several occasions (Jenkins et al. 1991). On a broader scale, Ogden et al. (1993) documented substantial differences in ETS concentrations between heavy-use and light-use areas in private residences.

The highly absorptive nature of some ETS components suggests that certain compounds will off-gas and contribute to the indoor air atmosphere long after active smoking has ceased within a given environment. There are anecdotal accounts of many people being able to detect the odor of stale tobacco smoke in rooms long after smoking has ceased. Measurable levels of nicotine have been determined in nonsmoking environments. For example, nicotine levels measured aboard a Boeing 767 aircraft several hours after it had landed and was sitting unused have been reported to range from 1.1 to 8.7 $\mu g/m^3$ (Nelson et al. 1990). Presumably, this was due to the volatilization of nicotine from fabric and cigarette butts. Nicotine was determined at a level of 0.09 $\mu g/m^3$ in the den of a nonsmoker's home 2 days after smoking had occurred in the room (Nelson et al. 1990). Piadé et al. (1999) demonstrated adsorption and re-emission of nicotine from clothing fabric, and Van Loy et al. (1998) demonstrated similar phenomena for common indoor building materials.

Of all the individual constituents found in ETS-containing atmospheres, only nicotine and other nicotine alkaloids, their derivatives (e.g., tobacco-specific nitrosamines), a few nitriles, and some leaf wax derivatives might arise solely from tobacco smoke. Virtually all of the other constituents can arise from other processes associated with human activities. These processes include off-gassing of products typically found in enclosed environments, such as wood products, carpeting and construction materials, cosmetics, furniture and polishes, household cleaners, and gasoline. Office copying, cooking, wood-burning stoves and fireplaces, space heating, and other combustion processes can contribute to the levels of chemical pollutants inside homes and offices. Outside sources, such as vehicular traffic, can also contribute to indoor pollution. Probably the best example of this is the relatively high indoor levels of carbon monoxide in high-traffic areas. It is critical to the accurate interpretation of data that the investigator assess the relative contribution of other sources to the individual concentrations observed. Background sources and concentrations are treated in Chap. 2.

SAMPLING METHODS AND MATERIALS

Active vs. Passive Sampling

Active sampling involves drawing air into collection media by means of a pump. Passive sampling involves placing a material that irreversibly adsorbs, absorbs, or chemically fixes the chemical of interest in the environment being sampled. Active sampling is used to sample large volumes of air or to sample small volumes of air over short durations. Passive sampling is used primarily to sample small volumes of air over long periods of time.

Passive samplers differ from active samplers in that their chief transport mechanism to the collection point is molecular diffusion. In contrast, active samplers use convection, usually with the aid of pumps, as the most important form of mass transport. Passive samplers have been used for numerous years in industrial-hygiene monitoring situations. Passive systems offer the advantages of being relatively simple and inexpensive; they require no elaborate pumping apparatus to obtain quantitative results. They are also quiet, which increases their utility for unobtrusive sampling. Their effective sampling rates are typically low (tens of milliliters per minute), but they can be used unattended for days (e.g., Hammond et al. 1995). A number of passive-sampling badges are available commercially, although commercial badges have typically been employed where ambient concentrations of pollutants are greater than those observed in field ETS atmospheres (McKee et al. 1982, Kring et al. 1982). Some of the commercially available card-type passive monitors feature a chemical reaction with a color change that can be examined visually for an estimated exposure or instrumentally to provide a more quantitative evaluation. One manufacturer (Advanced Optochemical Research Inc., Raleigh, NC) offers these types of monitors for a variety of ETS constituents (including formaldehyde, ammonia, nitrogen dioxide, and carbon dioxide). McConnaughey et al. (1982) described the operating principles of these systems.

The operational theory of passive samplers has been reviewed in detail elsewhere (Fowler 1982). Briefly, the two types of passive samplers, diffusion and permeation, consist of two main components: the barrier to diffusion and the trapping or adsorbent medium. The latter may consist of any material that has a strong affinity for the analyte, including water and adsorbent paper or resin. In a diffusion sampler, the barrier to mass transport is distance; one end of the sampler (a tube) is open. The opening at one end permits the analyte to diffuse toward the adsorbent medium at some rate, which depends in part on the size of the opening, the distance to the adsorbent, and the size of the molecule. The actual sampling rate, in cubic centimeters per minute, is calculated from Fick's first law of diffusion (Palmes 1980, Mulik et al. 1989). In the permeation sampler, the primary barrier to mass transport is a permeable membrane. Many permeation samplers now employ polydimethylsiloxane membranes, whose permeability is not dependent on analyte concentrations (Jenkins and Guerin 1984). The membrane has the additional effect of improving the selectivity of the sampler. The chief driving force in both of these systems is the depletion of the constituent in question at the surface of the adsorbent. Each system has some minor disadvantages. Diffusion samplers are more affected by ambient humidity, whereas membrane samplers require calibration in standard atmospheres of the analyte in question.

Several investigators have evaluated different passive sampler designs and determined both typical behavior and optimized design. Coutant et al. (1986) and Mulik et al. (1989) described such work for passive samplers designed to trap general volatile organic compounds (VOCs) on Tenax, nitrogen dioxide, or formaldehyde on specific sorbents. The sampling badge described in Coutant et al. (1986), which was tested for benzene and toluene as well as a number of perchlorinated species, would usually be placed into a thermal desorption unit for analysis and heated directly, without removing the Tenax sorbent. Cohen et al. (1990) demonstrated that a commercially available passive sampler designed by 3M Corporation for occupational exposures worked best under conditions of high concentration and low relative humidity, or low concentrations with high relative humidity. The samplers in this study were exposed to five different VOCs for three weeks at a sampling rate of approximately 30 mL/min. Several reports of field measurements of nicotine with passive systems are described in more detail in Chap. 7; they include sodium-bisulfate-treated Teflon-coated glass-fiber filters (e.g., Hammond and Leaderer 1987, Lambert et al. 1993, Ogden 1996), chilled Petri dishes (Williams et al. 1984), and XAD-4 resin beds (Ogden et al. 1989b). Problems have been reported with nicotine adsorbing on the frames of such badges and not reaching the targeted adsorbing material (Ogden et al. 1989b, Caka et al. 1989). Ogden and Maiolo (1992) performed comparative evaluation studies on diffusion-based and active samplers and found them to be comparable for nicotine collection but not so for collection of 3-ethenylpyridine (3-EP).

Active sampling requires the convective mass transport of material to the collection device. It thus requires some sort of sampling pump, and is thus more complex. A review of sampling pumps is provided by Furtado (1983). Pumps that sample at rates approximately equal to or less than 28 L/min (1 ft^3/min) can be classified as "low-volume" pumps, and those sampling substantially faster as "high-volume" pumps. Small personal-sampling pumps, such as peristaltic pumps, small oil-free diaphragm vacuum pumps, and small rotary vacuum pumps for industrial-hygiene monitoring, could be included in the "low-volume" category. Most of the high-volume air-sampling pumps are of the rotary vacuum type. Their air handling capacity seems to be limited only by the size of the motor employed. However, the size limit imposed by the need for portability probably limits their air-sampling rate to approximately 10 m^3/min.

There are practical problems associated with the use of any type of pumping system. First, unless flow rates are controlled by mechanical or electronic means, they may change during the course of an experiment as particulate loading on the filter increases. This can reduce the accuracy of the final reported concentrations of airborne constituents. Many of the smaller industrial-hygiene monitoring pumps now have electronic flow controllers or microprocessors to increase the number of piston strokes per second as the flow resistance increases. If no flow control system is used, care should be taken to keep particulate loading to a minimum. In addition, only oil-free pumps should be used in ambient-air sampling. Otherwise, contamination of the atmosphere being sampled may occur. The use of high-volume sampling pumps does have some inherent disadvantages. First, the noise is considerable. Operators should wear ear protection if they are required to be in the vicinity of the pump for any significant length of time. Also, heat produced by the motor could potentially alter the vapor-particle distribution of the more volatile constituents in the neighborhood of the sampler. Because of the large flow rates used in high-volume samplers, the amounts of the more-volatile constituents in the particulate matter collected on the filters are probably not an accurate representation of the amounts in the ambient particulate matter. The air flow tends to evaporate these species out of the collected particles. Probably the easiest way to minimize this difficulty is to maintain the ratio of filter area to volumetric flow rate as high as is practical. Active sampling systems are described in more detail in each of the chapters on specific constituents of ETS and are the most commonly used type of systems.

Collection of Particulate Matter

The choice of sampling system is determined largely by the nature of the particulate matter to be sampled, the intended subsequent treatment (if any) of the sample, and the particle size range of interest. Systems suitable for collecting solid particles (e.g., mineral matter) may not be suitable for collecting liquid (e.g., fuel smokes, MS) particles. Collection media suitable for the subsequent determination of organic constituents may not be suitable for the subsequent determination of trace metals. An especially important consideration is whether the investigator seeks to collect respirable suspended particulate (RSP) matter or total suspended particulate matter (TSP). The mass of TSP collected almost always exceeds the mass of RSP under any given situation, and RSP is the more-relevant material to study if the concern is inhalation hazard. Many investigators have employed small cyclone separators to remove larger particles from the sampled stream when RSP is the material of interest. These issues are discussed in greater detail in Chap. 6.

Particulate matter is most commonly collected by filtration. In brief, filters remove particles from the airstream through processes of direct interception, diffusional and inertial deposition, and electrical and gravitational attraction (e.g., Lippmann 1983). Popular filter types include fibrous materials, amorphous gel membranes (tortuous channels), and nucleopore membranes (straight channels). A filter medium commonly employed is the Teflon-coated glass-fiber filter pad, available commercially in a number of shapes and sizes. The glass-fiber pad has a very high retention efficiency for particulate matter of the size range normally encountered with ETS (up to a few tenths of a micron). Coating the glass fibers with Teflon decreases the opportunity for formation of chemical artifacts on the filter surface. In addition, the filters can be treated with benzenesulfonic acid solution so that they can retain vapor-phase nicotine (Caka et al. 1990). Probably the most popular filter for ETS particle collection (in terms of the number of samples reported in the literature) is the 37-mm diam, 1.0-μm-pore-size Fluoropore membrane filter (Catalog No.

FALP 03700, Millipore Corp. Bedford MA). Despite having a pore size larger than most ETS particles, it has been demonstrated to be highly efficient for collection of ETS RSP. Ogden and Heavner (Ogden et al. 1996) have patented a now-commercially available sampling system for ETS components that collects particulates on the Fluoropore membranes and the nicotine alkaloids on XAD-4 cartridges.

A number of filter media have been used to collect indoor air samples, including uncoated glass and quartz fibers, Teflon and polycarbonate membranes, and cellulose esters. An important consideration during field sampling is the affinity of vapor-phase constituents for surfaces, including particulate filters. Some investigators have demonstrated that significant quantities of organovolatiles can be adsorbed onto the filter during the sampling for particulates (Hammond et al. 1987).

In addition to filtration, impactors and impingers have been used to collect indoor air particulate matter. Impactors are often used in cascade fashion to determine particle-size distributions (see Chap. 6) and function by routing particulates through jets of increasing nozzle velocity in order to impact particles of a selected size range on a given stage. Stober (1980) and Harrison and Perry (1986) reviewed the use of impactors for size classification of inhalable aerosols. MSP Corporation (Minneapolis, MN) markets single-stage impactors for personal-exposure particulate collection. Impingers typically collect particles by bubbling the air sample through a liquid that traps all or some of the material. Impingers tend to be more effective for vapor-phase species; the diffusion coefficients for particles are insufficiently large to permit all of the particles to contact the liquid-air interface as the bubbles travel through the liquid trapping solution.

Collection of Vapor-Phase Constituents

The primary means of collection of samples of VOCs from indoor air involves the use of solid sorbent media (e.g., Pellizzari 1991) (see also Chap. 11). A solid sorbent stabilizes the species so that it can be returned to the laboratory for analysis, and it can concentrate an air sample by factors of as much as hundreds or thousands, effectively lowering the detection limit of the target analyte. This is especially effective when the solid sorbent can be analyzed by thermally desorbing its collected material directly into the analytical measurement instrument. Nicotine, most of which is present in the vapor phase, is probably the most common ETS-specific compound sampled and has been collected on a number of solid sorbents (see Chap. 7). These have included Tenax (Thompson et al. 1989), XAD-4 (Ogden et al. 1989), OV1-coated Uniport-S (Muramatsu et al. 1984), sodium bisulfate-treated Teflon-coated glass-fiber filters (Hammond et al. 1995), benzenesulfonic-acid-coated diffusion denuder tubes (Eatough et al. 1989a), and XAD-4–coated denuder tubes (Van Loy et al. 1997). The denuder systems function by collecting gas-phase species as they diffuse toward the wall of a acid-coated tube. Particles are not effectively retained in the tube because their rate of diffusion is too slow to permit them to reach the walls of the tube during their time of passage through it. Denuder technology is especially useful for determining the distribution of vapor phase and particle phase of a constituent.

For non-ETS-specific indoor air studies, Tenax appears to be the most popular sorbent (Wallace 1987). Tenax is a porous polymer resin originally developed for use as a gas-solid chromatographic stationary phase. It received wide use in air-sampling applications, probably because it has very good retention characteristics for the C7 through C12 compounds, does not retain water, and can be reused repeatedly. However, it must undergo extensive cleanup prior to initial use and is subject to decomposition through reaction with atmospheric ozone or oxides of nitrogen (Pellizzari et al. 1984). It can be used in small traps for experiments of limited duration, and quantities as large as 2.5 g can be packed into large tubes for 24-h air-sampling experiments (Jenkins et al. 1982). The chief disadvantages of Tenax are its poorer retention of more-volatile species and its propensity to release self-decomposition products during higher-temperature thermal desorption analysis. At long sampling times, breakthrough of the more volatile constituents can occur. In order to compensate for this deficiency, Tenax has been combined in sequence with other more retentive sorbents in order to determine the more-volatile ETS constituents quantitatively (e.g., Higgins et al. 1988). Breakthrough volumes under various sampling conditions for many constituents have been compiled by the U.S. Environmental Protection Agency (EPA) (Gallant et al. 1978).

The use of carbon-based sorbents (e.g., sintered carbon blacks and high-purity carbon molecular sieves), especially packed parfait-style in sorbent traps, is increasing in popularity and threatens to displace Tenax as the prevalent general-use vapor-collection media. These materials, which use synthetic polymer starting materials, possess a very uniform carbon surface with excellent adsorptive/desorptive and hydrophobic characteristics (Betz and Firth 1988, Supelco 1990). Because of their inherent thermal stability, they are excellent candidates for thermal desorption analyses, unlike the BAD resins and polyurethane foam (PUF), and they are probably superior to Tenax GC. Application of one type of multisorbent trap to the sampling and analysis of complex atmospheres has been described by Ma et al. (1997) and Dindal et al. (1998). Baek et al. (1997) reported the use of single-stage traps filled with CarboTrap for collection of ETS-related VOCs. Multistage traps, some of which include Tenax, have been used for ETS VOC characterization in several studies (Heavner et al. 1995, Hodgson et al. 1996, Heavner et al. 1996).

The Amberlite XAD resins (XAD-2 and -4, the latter having a larger surface area per unit mass) are another class of widely used porous resins and have been employed for collecting both volatile and semivolatile constituents from air samples (Adams et al. 1977). XAD has the advantage of a somewhat greater capacity than Tenax for the more-volatile constituents. However, it has lower temperature stability, precluding desorption by thermal means. Instead, the resin bed is eluted with solvent, and then the eluate is usually concentrated prior to analysis. During such a procedure, there is the potential for loss of the more-volatile analytes. In addition, XAD exhibits somewhat higher background levels than does Tenax when used in dry atmospheres. However, XAD-4 is the most widely used sorbent media for collection of nicotine and nicotine-derived species in ETS and is specified in an American Society for Testing and Materials (ASTM) method (ASTM 1997).

Charcoal has also been used as an adsorbent for vapor-phase constituents. It is inexpensive and has good retention for organic species up through di-substituted benzenes. However, it retains water easily, and its retention characteristics may suffer from batch-to-batch variation. It, too, must be eluted with solvent to desorb organic species, and often desorption is not quantitative. The utility of charcoal for broad-spectrum sampling and analysis of ambient tobacco smoke constituents is probably limited.

Plugs of PUF have been used as collection media for vapor-phase organics (Jackson and Lewis 1980). They have the advantages of low restriction to flow (making them very suitable for high-volume sampling applications), ease of purification and handling, and low cost compared with the cost of sorbents described in this chapter (Riggin and Petersen 1985). These advantages are partially offset by a poorer retention for more-volatile species compared with the XAD or Tenax resins and by the apparent formation of mutagenic artifacts during sampling.

Collection of Gases

In the case of constituents that exist solely in the gaseous state at ambient temperatures, concentration prior to analysis is often not a viable procedure. (Exceptions are those gases for which chemical reactions are employed to "fix" the gas). These constituents have too high a vapor pressure to be efficiently and quantitatively adsorbed. Examples include the oxides of carbon and nitrogen and low-molecular-weight hydrocarbon and sulfur compounds. Nonreactive gases can be pumped into gas-sampling bags, made of Tedlar or Teflon, for later analysis by chromatography or other means. However, an increasing number of investigators are using on-line instrumental analysis of ambient samples without prior concentration (Jenkins and Guerin 1984). Usually, these procedures are both highly sensitive and specific for the individual compounds of interest. Various techniques have been employed, and are discussed in greater detail in the following sections. Low-molecular-weight carbonyls, such as formaldehyde, acetaldehyde, and acrolein, have been collected somewhat differently. Their sampling and analysis are described in detail in Jenkins (1987). Briefly, the air sample is drawn through a micro-impinger containing dinitrophenylhydrazine (DNPH), a derivatizing reagent specific for carbonyl. Alternatively, the atmosphere can be passed through a bed of DNPH-coated silica. The resulting dinitrophenylhydrazones can then be concentrated and determined quantitatively by high-performance liquid chromatography. In this case, the sample collection/concentration procedure also acts as the first step of the analysis.

MEASUREMENT METHODOLOGIES

Because they are so specific to individual constituents of indoor air and/or ETS, specific analytical technologies are described in the individual chapters describing the results of field studies. In most of the reported investigations, samples are collected in the field and are returned to the laboratory for detailed chemical analysis. In many cases, careful studies have been conducted to verify that samples can be returned from the field intact and stored for some period of time prior to analysis. However, for some of the more reactive indoor air constituents, such as the oxides of nitrogen or the volatile aldehydes, it is more difficult to return an intact sample to the laboratory for analysis. The material must be collected and analyzed immediately on site with real-time monitoring instrumentation. In other cases, because of the increasing sophistication, sensitivity, and selectivity of portable analytical instrumentation, it may be relatively convenient to perform analytical measurements in the field. These two approaches are described briefly in the following sections.

Real-Time Monitoring

The number of instrumental systems commercially available for real-time, or near-real-time determination of indoor air pollutants is considerable. For example, there are monitoring systems commercially available for carbon monoxide and carbon dioxide, oxides of nitrogen, ozone, ammonia, methane, selected low-molecular-weight hydrocarbons (aromatic and otherwise), sulfur oxides, and hydrogen sulfide. Individual monitoring systems have been discussed in more detail in the chapters describing field studies of individual constituents or compound classes. Those used most commonly for indoor air studies have included nondispersive infrared (NDIR) systems, photoacoustic infrared systems, potentiometric and coulometric electrochemical analyzers, chemiluminescent systems, flame photometric and ionization detectors, photoionization monitors, and dispersive infrared spectrometers. In many cases, a given constituent can be determined by a variety of instrumental techniques. The choice of the best approach can be a complex decision and is influenced by a number of factors, including cost, system portability, selectivity and sensitivity, and the quantity and concentration of potential interferences. There are also several types of systems available for the real-time determination of particulate concentrations, including continuous-measurement microbalances and photometers, which are discussed in greater detail in Chap. 6.

An approach often overlooked by many investigators is the use of direct-reading colorimetric indicator tubes. The tubes function by drawing a known amount of ambient air through a tube packed with a sorbent that has been treated with reactants specific for the compound of interest. If the compound is present, the sorbent changes color, and the length of the color stain is proportional to the quantity of target analyte present in the air. These are produced by a number of manufacturers, including Draeger, Kitagawa, and Gastec, and although they lack the accuracy of more expensive and sophisticated instrumental systems (typically precision and accuracy of the former are ± 25%), they can be of utility for doing screens in areas suspected of contamination with a known material. When used for their intended application (monitoring in industrial workplaces, typically drawing a few hundred milliliters of ambient air through the tube), most tubes are not likely to have the sensitivity to be effective for most indoor air settings. However, laboratory experiences have suggested that their range can be extended one to two orders of magnitude by drawing larger quantities of air through the tubes (Jenkins et al. 1990). Greater range extensions seem possible. Specific constituents for which colorimetric tubes are available that may be of interest in indoor air investigations include those that test for carbon monoxide, carbon dioxide, nitrogen dioxide, ammonia, acetone, benzene, and toluene. Saltzman (1983) has reviewed the use of these tubes in substantial detail.

Portable Analytical Instrumentation

Another approach to the field determination of indoor air components involves the use of field-portable versions of laboratory instruments. The use of such systems has increased dramatically in recent years, the major driving force being environmental investigation and remediation activities. Field analytical instrumentation for environmental studies can range from small, high-speed gas chromatographs (GCs) to triple quadrupole mass spectrometers transported in mobile homes. Portable gas chromatographs are probably of greatest interest for indoor air studies, and have been used in a number of investigations. Many of these systems are easily portable, offer a variety of detector systems, and have shown greatest utility in the determination of volatile species. One of the limitations of such systems is that those that operate under battery power can lack heated injection systems. This limitation makes them more susceptible to contamination by higher-molecular-weight, less-volatile airborne species. Without a heated injector port, such materials may condense in the injector and bleed slowly into the chromatographic column. This can have the effect of increasing the apparent background or giving rise to spurious peaks. There is at least one manufacturer that produces a portable GC designed specifically for routine, unattended air monitoring (Sentex Sensing Technology, Inc., Ridgefield, NJ). It possesses a built-in air-sampling system and sorbent cartridge. Organovolatiles are collected on the sorbent cartridge, thermally desorbed, and quantified chromatographically in a manner similar to that performed with laboratory-based systems.

Many ion-mobility spectrometers are becoming more portable and offer potential for real-time monitoring of target species. In contrast, mobile mass spectrometers, weighing only a few hundred pounds, have not been used routinely in indoor air quality studies, probably because of the difficulty of taking such systems indoors. However, these systems offer the promise of making real-time measurements of volatile organics in complex atmospheres with the degree of analytical confidence usually reserved for more laborious, laboratory-based measurements, such as GC/MS. Research under way at several institutions involves the downsizing of such instruments to the point at which they could be of utility in indoor air studies. However, in many situations, there is a need for collecting and/or analyzing the sample in as unobtrusive a manner as possible. Such criteria will probably preclude the use of portable, albeit sophisticated, analytical instrumentation in many field studies for the foreseeable future.

Interferences

With the exception of some of the major constituents (e.g., CO_2, CO, suspended particles), many of the volatile organic components of ETS are present at concentrations of 0.1 to 100 $\mu g/m^3$ (Higgins et al. 1988, Löfroth et al. 1989, Heavner et al. 1995). Many of the trace constituents of the particulate phase may be present at levels one to three orders of magnitude lower. At these levels (parts per billion or parts per trillion), the contribution to the overall ambient levels of an individual constituent from external, non-cigarette related sources can be substantial (Chap. 2) relative to the contribution from tobacco smoke. For example, Lioy et al. (1990) reported that indoor benzene levels in a nonsmoking home in Bayonne, New Jersey, reached peak levels of up to 8 $\mu g/m^3$. Higgins et al. (1988) reported 16 -$\mu g/m^3$ benzene levels in an office that was heavily polluted with tobacco smoke (particulate level: 330 $\mu g/m^3$) and for which there was no other source of benzene. Löfroth et al. (1989) reported levels of formaldehyde in smoky taverns (nicotine: ca. 65 $\mu g/m^3$) of about 100 $\mu g/m^3$. This is somewhat less than the levels reported in a one-year old, nonsmoking home (Hawthorne et al. 1984), which would periodically reach 165 $\mu g/m^3$ formaldehyde. These data highlight the need to carefully determine the concentration of the target constituents under true nonsmoking conditions if the contribution from ETS is to be accurately determined. Heavner et al. (1996) reported an extensive study of personal exposure to VOCs in smoking and nonsmoking homes and workplaces.

Another confounding factor in the accurate determination of constituent levels in indoor air is the extreme complexity of the matrix. This is especially true for trace substances, such as PAHs, for which there can be a number of isomers of a given molecular weight. For example, the separation of benzo[a]pyrene, a common, but trace, combustion-derived carcinogen, from benzo[e]pyrene, a noncarcinogenic isomer, can be difficult. In such cases, multiple chromatographic steps, such as sequential chromatography (Jenkins et al. 1982, Tomkins et al. 1985) or multidimensional chromatography (Giddings 1987, Duinker et al. 1988, Kopczynski 1989) coupled with selective detectors, may be required to confidently identify and quantify the target species. Often in these situations, fairly large samples may be required in order to meet absolute instrumental detection limit requirements. For example, at typical RSP levels of 100 $\mu g/m^3$, a constituent whose concentration in the particulate matter of ETS is 1 ppm will be present at ambient concentrations of 100 pg/cm^3. It may be necessary to sample tens or hundreds of cubic meters of indoor air to acquire an adequate amount of sample for confident determination. This may be difficult to accomplish in relatively small enclosed environments, and at these high sample volumes, the chances of loss of the target species through evaporation or chemical transformation is considerable (Griest et al. 1988).

AREA SAMPLING VS. PERSONAL MONITORING

Area sampling typically involves sampling the air at a stationary location in the area of interest; personal sampling involves the collection of air in the breathing zone of an individual by means of a device attached to the individual. Industrial-hygiene workers have struggled for years over the fundamental question of whether a stationary sampler collecting air in an environment in which a worker is performing his or her duties can adequately represent that individual's actual exposure. The inadequacy of a stationary area sampler is likely to be exacerbated if the worker is highly mobile within his or her work environment or if the worker moves in and out of the exposure situation. In general, it has been believed that a personal monitor will more accurately represent an individual's exposure (First 1983). Indeed, this is likely to have been an important factor in the marked increase of ETS-related personal exposure studies.

Indoor air studies seem to confirm that, for many constituents, an individual's personal exposure will be greater than that which would be predicted by a direct measurement of contaminant levels in the surrounding air (Wallace et al. 1987). The causes of this situation have been described as "the Pigpen Effect," named after the *Peanuts* cartoon strip character who perpetually walks around in a cloud of his own airborne debris (Spengler 1989). The activities in which an individual engages are likely to emit contaminants into that space which is closest to the individual to an extent greater than the same activities would emit into the surrounding micro-environment. Such exposures can come from the wearing of dry-cleaned or permanent-press clothing (Kelly et al. 1999), pumping gasoline, using office copying products or felt tip pens, showering (exposure to airborne chlorinated hydrocarbons) (Wilder et al. 1987), stoking of wood-burning heating appliances, smoking, etc. Ogden has demonstrated that, in residential settings, smokers are exposed to ~20 times more nicotine from ETS than nonsmokers occupying the same residences (Ogden 1996).

Studies of personal sampling vs. area sampling have yielded somewhat mixed conclusions. Avol et al. (1989) reported good agreement between personal-exposure monitors and stationary area monitors for ozone in urban conditions. Sullivan and Koines (1990) reported that measured personal exposures to benzene were lower than measured indoor area levels in the Baltimore Total Exposure Assessment Methodology (TEAM) study. There have been four studies comparing results from area monitors with results from personal monitors reported for the ETS-specific constituent nicotine. One set of investigators (Crouse and Oldaker 1990) compared levels of nicotine determined by both area samplers and personal samplers in restaurants. They found that the levels determined by a stationary sampler yielded somewhat higher levels of nicotine than those determined by sampling the breathing zone of stationary individuals using personal samplers. The investigators speculated that this discrepancy may be due to adsorption or desorption phenomena occurring with clothing or other adsorptive surfaces, or to dilution of nicotine levels in breathing-zone air with exhaled breath. In contrast, other investigators (Jenkins et al. 1991) compared indoor air nicotine levels determined by mobile breathing-zone samplers worn by individuals with those determined by paired stationary area samplers. The investigators determined that the magnitude of the differences between individual samples was sufficiently large to preclude the conclusion that a statistically significant difference existed between mobile personal sampling and stationary area sampling. In more recent studies, Sterling et al. (1996) reported no statistical differences between area sampling and personal sampling for subjects working in unrestricted smoking environments. Jenkins and Counts (1999) reported good agreement between statistical groupings of area-sampling and personal-sampling data for subjects employed in the hospitality industry. However, their results also demonstrated that, on an individual basis, area samples are useful for estimating individual exposures to ETS to within a factor of 5 to10. These mixed results indicate that investigators attempting to determine personal exposure to indoor contaminant levels should be aware that there may be great differences in spatial homogeneity of constituent concentrations, and that area samples are probably useful for providing order-of-magnitude estimates of personal exposure for subjects remaining in the same environment as that being monitored.

EXPERIMENTAL DESIGNS AND SUPPORTING OBSERVATIONS

For any scientific investigation, the experimental design should be based on the objective of the study. If the objective is to determine the concentrations of an ETS-related constituent in a given environment at a given time, it is adequate to sample that environment for the time period of interest. If the objective is to determine the breathing-zone exposure of a given individual in a given environment for a given period of time, it is adequate to employ a personal breathing-zone sampler affixed to the individual of interest. To be of value for comparison with other studies, the study must document the conditions of the sampled environment in as much detail as is possible and practical. Important parameters include the volume and air-exchange rate of the environment, a measure of the smoking density (cigarettes smoked per unit time per cubic meter of indoor air volume), and the presence of possibly confounding factors (e.g., urban or rural environment, indoor activities, and appliances). Supporting measurements of ETS-related constituents (e.g., nicotine, 3-EP, solanesol) can be especially helpful.

Experimental designs become more complex when the objective of the study is to apportion the amount of particular constituent due solely to ETS. In this case, it is necessary to establish both the contribution of non-ETS sources to the concentration of the constituent of interest and the amount of the target analyte emitted into ETS in controlled-atmosphere (chamber) experiments. Experimental designs are especially complex when the objective of the study is to determine general population exposures. Large numbers of statistically selected areas and/or individuals must be sampled with particular attention to confounding sources of exposures and to detailed measures of background exposure levels. Monsour and Waite (1981) reviewed in detail the important steps in a survey of establishments. Painstaking determination and documentation of the accuracy and precision of sampling and analysis methods are required to establish statistically significant differences due to ETS-exposures under the most-common long-duration exposure situations (e.g., in the home). Studies performed as part of the U.S. EPA TEAM program (e.g., Pellizzari et al. 1987) as well as the series of multi-nation ETS personal-exposure studies conducted by Covance Laboratories (e.g., Phillips et al. 1997) have taken this approach.

The range of sophistication in experimental design of field studies reported in the open literature thus far is considerable. In many cases, it would appear that the selection of a given environment for ETS or related sampling is a matter of convenience. Often, description of experimental protocols beyond those specifically required to collect the sample of interest has been minimal. Little supporting information about the area being sampled (air exchange, number of smokers, room volume, etc.) appears to have been acquired, and the inferences or implications of the data are only minimally relatable to other situations. Furthermore, it is important to recognize that in studies of ETS-contaminated environments in which non-tobacco-specific constituents are to be surveyed, the confounding influences of other sources can be considerable. For example, newly constructed or renovated structures can out-gas disproportionate amounts of solvents, formaldehyde, or carpet-glue components. The conditions in rooms that exhibit extremes of ventilation, temperature, or relative humidity may shift equilibria between particle phases and vapor phases of ETS components. Ventilation can introduce materials either from the outside or from other parts of buildings that might not have been present under other circumstances. It is important for investigators to document the existence of sources and confounding influences in order to maximize the utility of their data to other investigators. Careful determination of background concentrations of the target species is critical if determination of the contribution of ETS to the species concentration is the objective. It may also be important to avoid subconscious or inadvertent consideration of the locations that are most highly contaminated with ETS as being "typical."

Many of these concerns have been addressed in some studies. For example, several investigators who have wanted to ensure that the restaurants in which they collected nicotine data were representative of restaurant distribution in the area in which they were sampling have used electronic Yellow Pages, Chamber of Commerce data, and Gallup Survey results (Crouse and Carson 1989). Others have used random selection from the local Yellow Pages listing of restaurants, assuming that a sufficiently large sample of restaurants will be representative of those in the general area (Thompson et al. 1989). An increasing trend in large personal-exposure studies has been the reporting of the demographic characteristics (age, gender, household income, education, etc.) of the study population (e.g., Jenkins et al. 1996, Phillips et al. 1996). In such cases, it seems best to cite the criteria used for sampling venues or selecting subjects, so that other investigators can independently judge the representativeness of the sample suite. In such studies, the acquisition of ancillary data may be useful in extending the conclusions that can be drawn. For example, Thompson et al. (1989) reported data concerning the number of occupants and smokers of the sampled location, the number of cigarettes smoked, the distance to the nearest smoker, and estimated room size. Oldaker and Conrad (1987) reported on the exact seat location of each sampler, the number of cigarettes smoked, and the exact sampling duration to determine ETS nicotine levels in aircraft cabins. A list of ancillary factors measured in some selected studies is presented in Table 5.1. The presentation is not meant to be all inclusive, but rather to provide examples of both general and specific factors measured in some studies.

Ventilation and air-exchange data are useful for modeling studies; however, they are difficult to obtain in many true field situations or in studies of personal exposure in which subjects are selected at random. A determination of air-exchange rate usually involves the release of a tracer pollutant into the environment (e.g., sulfur hexafluoride) and repeated measurement of the tracer concentration over several half-lives (usually several hours in low-ventilation environments). This is difficult to perform without the consent of occupants and owners and may require one or more additional visits to the sampling location.

Because of the social implications of exposure to ETS, unobtrusive sampling in field environments has been used for area sampling and short-duration personal sampling. There have been anecdotal reports of smokers deliberately contaminating samplers with exhaled MS when they were aware that ETS sampling was being conducted (Moneyhun 1988). It is reasonable to assume that many investigators have used unobtrusive sampling as a way of collecting samples, although few studies have explicitly reported that they did. One group (Thompson et al. 1989) used personal samplers concealed under jackets, with the open end of the sampler extending into the breathing zone, to collect ETS nicotine samples. Others have reported concealing sampling pumps in a small briefcase (Jenkins et al. 1991). One group of investigators has developed a very sophisticated unobtrusive sampling system, which has been used in a number of studies. The Portable Air Sampling System (PASS) has been described in detail in McConnell et al. (1987). It consists of a briefcase that has been outfitted with an electrochemical CO monitor and sampling systems for ambient nicotine and particulate matter. In addition, the CO analyzer was attached to a data logger for recording the real-time data. The unit was sufficiently sound insulated so that it can be employed in even very quiet restaurants without notice by the general public. The advantage of the PASS unit was that it could obtain information about a number of ETS-related constituents simultaneously. Another group has developed a somewhat different system (Eatough, et al. 1989b) that used a combination of sorbent tubes and on-line monitors for both integrated and real-time monitoring. It contained a series of 12 Tenax-filled micro-tubes and annular denuders for nicotine collection and could measure temperature, relative humidity, oxides of nitrogen, and carbon.

SUMMARY

The chemical and physical complexity of ETS, coupled with its episodic introduction into the environment, make the development of accurate sampling and analysis strategies challenging. Chemical and physical transformation of the smoke in the air can occur both with time and as the samples are being collected. Any sampling strategy that seeks to perform source apportionment must take into account the wide variety of sources for non-tobacco-specific compounds, especially in indoor environments. Atmospheric temporal and spatial inhomogeneity must also be addressed when considering placement of stationary samplers. A variety of sampling and analytical technologies exist, and the choice of the most appropriate must consider both the target constituent and the question for which an answer is sought. Studies are most useful when the criteria for site selection or the demographic characteristics of the study population are reported.

Table 5.1 Supporting Information Collected in Selected Studies of Indoor Air Contaminants

Akland et al. 1985	Oldaker and Conrad 1987	Crouse and Carson 1989	Thompson et al. 1989	Healthy Buildings International 1990	Jenkins et al. 1996
Human Exposure to CO	ETS Levels in Jet Aircraft	ETS Levels in Restaurants and Offices	ETS Levels in Restaurants	ETS Levels in Public Buildings	Personal Exposure to ETS
Geographic Area of Interest	No. of Occupants in Smoking Section	No. of Smokers Present	No. of Smokers Present	No. of Smokers Present	Sample Population in 16 Urban Areas
Target Population	No. of Cigarettes Smoked	No. of Cigarettes Smoked	No. of Cigarettes Smoked	No. of Cigarettes Smoked	Hourly Recording of Cigarettes Being Smoked
Confounding Sources Present	Aircraft Type		Presence of Cigars or Pipes		Confounding Sources Present
		Estimated Room Volume	Estimated Room Volume	Estimated Room Area	Size and Volume of Home and Workplace
	Seat Location	Average No. of Occupants	Distance to Closest Smoker		Complete Demographic Composition of Study Population
Sampling Duration	Sampling Duration	Sampling Duration	Sampling Duration	Sampling Duration	Sampling Duration
Randomized Individual Selection		Representative Distribution of Sampling Sites	Randomized Site Selection		Randomized Subject Selection
Constituents Measured:	Constituents Measured:	Constituents Measured:	Constituents Measured:	Constituents Measured:	Constituents Measured:
CO	Nicotine	Nicotine	Nicotine	Nicotine	Nicotine
		CO		CO	3-Ethenylpyridine
		CO_2		CO_2	Myosmine
		RSP		RSP	RSP
		UVPM			UVPM
					FPM
					Solanesol
					Scopoletin

REFERENCES

Adams, J., Menzies, K., & Levins, P. (1977) *Selection and Evaluation of Sorbent Resins for the Collection of Organic Compounds.* U.S. Environmental Protection Agency Report (EPA-600-7-77-044), Springfield, VA, National Technical Information Service.

Akland, G. G., Hartwell, T. D. et al. (1985) Measuring human exposure to carbon monoxide in Washington, D.C. and Denver, CO during the winter of 1982–1983. *Environ. Sci. Technol.,* 19, 911–918.

American Society for Testing and Materials. (1997) D5075-Standard Test Method for Nicotine and 3-Ethenylpyridine in Indoor Air. In: *Annual Book of ASTM Standards.* West Conshohocken, PA: American Society for Testing and Materials, 400–407.

Avol, E., Anderson, K., Whynot, J., Daube, B., Linn, W., & Hackney, J. (1989) Progress in personal and area ozone sampling using solid reagents. *Proceedings, 1989 EPA 1A WMA Specialty Conference on Total Exposure Assessment Methodology,* Las Vegas, NV, pp. 408–416.

Baek, S. O., Kim, U. S., & Perry, R. (1997) Indoor Air Quality in Homes, Offices, and Restaurants in Korean Urban Areas—Indoor/Outdoor Relationships. *Atmos. Environ.,* 31, 529–544.

Betz, W. R. & Firth, M. C. (1988) Utilization of carbon-based adsorbents for monitoring adsorbates in various sampling modes of operation. *Proceedings, 1988 EPA/APCA International Symposium on Measurement of Toxic and Related Air Pollutants,* Research Triangle Park, NC, pp. 670–678.

Brunnemann, K. D. & Hoffmann, D. (1978) Chemical studies on tobacco smoke. LIX. Analysis of volatile nitrosamines in tobacco smoke and polluted indoor environments. In: Walker, E. A., Castegnaro, M., Griciute, L., & Lyle, R. E., eds., *Environmental Aspects of N-Nitroso Compounds (IARC Scientific Publications* No. 19), Lyon, International Agency for Research on Cancer, pp. 343–356.

Caka, F. M, Eatough, D. J., Lewis, E. A., Tang, H., Hammond, S. K., Leaderer, B., P., Pierce, J. B., Koutrakis, P., Spengler, J. D., Fasano, A., McCarthy, J., Ogden, M. W., & Lewtas, J. (1989) A comparison study of sampling techniques for nicotine in indoor environments. 1989 *EPA/AWMA Conference on Measurement of Toxic and Related Air Pollutants,* Raleigh, NC, pp. 525–541.

Caka, F. M., Eatough, D. J., Lewis, E. A., Tang, H., Hammond, S. K., Leaderer, B. P., Koutrakis, P., Spengler, J. D., Fasano, A., McCarthy, J., Ogden, M. W., & Lewtas, J. (1990) Intercomparison of sampling techniques for nicotine in indoor environments. *Environ. Sci. Technol.,* 24, 1196–1203.

Cautreels, W. & Van Cawenbergh, K. (1978) Experiments on the distribution of organic pollutants between airborne particulate matter and the corresponding gas phase. *Atmos. Environ.,* 12, 1134–1141.

Cohen, M. A., Ryan, P. B., Yanagisawa, Y., & Hammond, S. K. (1990) The validation of a passive sampler for indoor and outdoor concentrations of volatile organic compounds. *J. Air Waste Manage.,* 40, 993–997.

Coutant, R. W., Lewis, R. G., & Mulik, J. D. (1986) Modification and evaluation of a thermally desorbable passive sampler for volatile organic compounds in air. *Anal. Chem.,* 58, 445–448.

Creighton, D. E. & Lewis, P. H. (1978) The effect of smoking pattern on smoke deliveries. In: Thornton, R. E., ed., *Smoking Behavior—Physiological and Psychological Influences,* Churchill Livingstone, London and New York, pp. 301–314.

Crouse, W. E. & Carson, J. R. (1989) Surveys of environmental tobacco smoke (ETS) in Washington, D.C. offices and restaurants. *43rd Tobacco Chemists' Research Conference,* Richmond, VA.

Crouse, W. E. & Oldaker, G. B., 111 (1990) Comparison of area and personal sampling methods for determining nicotine in environmental tobacco smoke. 1990 *EPA/AWMA Conference on Toxic and Related Air Pollutants,* Raleigh, NC.

Dindal, A. B., Ma, C. Y., Skeen , J. T., & Jenkins, R.A. (1998) Novel calibration technique for headspace analysis of semi-volatile compounds. *J. Chromatog. A,* 793, 397–402.

Dube, M. F. & Green, C. R. (1982) Methods of collection of smoke for analytical purposes. *Recent Adv. Tob. Sci.,* 8, 42–102.

Duinker, J. C., Schulz, D. E., & Petrick, G. (1988) Multidimensional gas chromatography with electron capture detection for the determination of toxic congeners in polychlorinated biphenyl mixtures. *Anal. Chem., 60,* 478–482.

Eatough, D. I., Benner, C. L., Bayona, J. M., Richards, G., Lamb, J. D., Lee, M. L., Lewis, E. A. & Hansen, L. D. (1989a) Chemical composition of environmental tobacco smoke. 1. Gas-phase acids and bases. *Environ. Sci. Technol.,* 23(6), 679–687.

Eatough, D J., Caka, F. M., Wall, K., Crawford, J., Hansen, L. D., & Lewis, E. A. (1989b) An automated sampling system for the collection of environmental tobacco smoke constituents in commercial aircraft. *Proceedings, 1989 EPA/AWMA International Symposium on Measurement of Toxic and Related Air Pollutants,* Raleigh, NC, pp. 564–575.

First, M. W. (1983) Air sampling and analysis for contaminants in workplaces. In: Lioy, P. J. & Lioy, M. J. Y., eds., *Air Sampling Instruments for Evaluation of Atmospheric Contaminants, 6th Edition,* Cincinnati, OH, pp. A2–A13.

Foreman, W. T. & Bidleman, T. F. (1987) An experimental system for investigating vapor-particle partitioning of trace organic pollutants. *Environ. Sci. Technol.,* 21(9), 869–875.

Fowler, W. K. (1982). Fundamental of passive vapor sampling. *Am. Lab.,* 14, 80–87.

Furtado, V. C. (1983) Air movers and samplers. *Proceedings, 1983 Air Sampling Instruments for Evaluation of Atmospheric* Contaminants, *6ᵗʰ Edition,* Lioy, P. J. & Lioy, M. J. Y., eds., Cincinnati, OH, pp. M2–M22.

Gallant, R. F., King, J. W., Levins, P. L., & Piecewice, J. F. (1978) *Characterization of Sorbent Resins for Use in Environmental Sampling,* U.S. Environmental Protection Agency Report, EPA-600/7-78-054, Springfield, VA.

Giddings, J. C. (1987) Concepts and comparisons in multidimensional separation. *JHRC&CC,* 319–323.

Griest, W. H., Jenkins, R. A., Tomkins, B. A., Moneyhun, J. H., Ilgner, R. H., Gayle, T. M., Higgins, C. E., & Guerin, M. R. (1988) *Sampling and Analysis of Diesel Engine Exhaust and the Motor Pool Workplace Atmosphere, Final Report.* ORNL/TM-10689.

Gundel, L.A., Mahanama, K. R. R., & Daisey, J. M (1995) Semivolatile and particulate polycyclic aromatic hydrocarbons in environmental tobacco smoke: cleanup, speciation, and emission factors. *Environ. Sci. Technol.,* 29, 1607–1614.

Hammond, S. K., Sorenson, G., Youngstrom, R., & Ockene, J. K. (1995) Occupational exposure to environmental tobacco smoke. *JAMA,* 274, 956–960.

Hammond, S. K., Coghlin, J., & Leaderer, B. P. (1987) Field study of passive smoking exposure with passive sampler. *Proceedings of the 4ᵗʰ International Conference on Indoor Air Quality and Climate, Vol. 2,* West Berlin, pp. 131–136.

Hammond, S. K. & Leaderer, B. P. (1987) A diffusion monitor to measure exposure to passive smoking. *Environ. Sci. Technol.,* 21, 494–497.

Harrison, R. M. & Perry, R., eds. (1986) *Handbook of Air Pollution Analysis, 2nd Edition,* Chapman and Hall, New York, NY, 634 pp.

Hawthorne, A. R., Gammage, R. B., Dudney, C. S., Hingerty, B. E., Schuresko, D. D., Parzyck, D. C., Womack, D. R., Morris, S. A., Westley, R. R., White, D. A., & Schrimsher, J. R. (1984) *An Indoor Air Quality Study of Forty East Tennessee Homes.* ORNL-5965.

Healthy Buildings International (1990) *Measurements of Environmental Tobacco Smoke in General Office Areas,* Fairfax, VA.

Heavner, D. L., Morgan, W. T., & Ogden, M. W. (1995) Determination of volatile organic compounds and ETS apportionment in 49 homes. *Environ. Int.,* 21, 3–21.

Heavner, D. L., Morgan, W. T., & Ogden, M. W. (1996) Determination of volatile organic compounds and respirable suspended particulate matter in New Jersey and Pennsylvania homes and workplaces. *Environ. Int.*, 22, 159–183.

Higgins, C. E., Jenkins, R. A., & Guerin, M. R. (1988) Organic vapor phase composition of sidestream and environmental tobacco smoke from cigarettes. Procs., 1987 *EP41APCA Symposium on Measurement of Toxic and Related Air Pollutants*, pp. 140–151.

Hodgson, A. T., Daisey, J. M., Mahanama, K. R. R., Brinke, J. T., & Alevantis, L. E. (1996) Use of volatile tracers to determine the contribution of environmental tobacco smoke to concentrations of volatile organic compounds in smoking environments. *Environ. Int.*, 22, 295–307.

Jackson, M. D. & Lewis, R. G. (1980) Polyurethane foam and selected sorbents as collection media for airborne pesticides and polychlorinated biphenyls. *Sampling and Analysis of Toxic Organics in the Atmosphere, ASTM STP 721*, American Society for Testing and Materials, pp. 36–47.

Jenkins, R. A. & Counts, R. W. (1999) Occupational exposure to environmental tobacco smoke: Results of two personal exposure studies. *Environ. Health Perspect.*, 107 (Suppl. 2), 341–348.

Jenkins, R. A., Thompson, C. V., Gayle, T. M., Ma, C. Y., & Tomkins, B. A. (1990) *Characterization of Rocket Propellant Combustion Products —Description of Sampling and Analysis Methods for Rocket Exhaust Characterization Studies, Interim Report*. ORNL/TM-11643.

Jenkins, R. A. & Guerin, M. R. (1984) Analytical chemical methods for the detection of environmental tobacco smoke constituents. *Eur. J. Resp. Dis.*, 65, 33–46.

Jenkins, R. A., Moody, R. L., Higgins, C. E., & Moneyhun, J. H. (1991) Nicotine in environmental tobacco smoke (ETS): comparison of mobile personal and stationary area sampling. *Proceedings of the EPA/AWMA Conference on Measurement of Toxic and Related Air Pollutants*, Durham, NC.

Jenkins, R. A., Gayle, T. M., Wike, J. S., & Manning, D. L. (1982) Sampling and chemical characterization of concentrated smokes. *Toxic Materials in the Atmosphere, ASTM STP 786*, American Society for Testing and Materials, pp. 153–166.

Jenkins, R. A., Griest, W. H., Moody, R. L., Buchanan, M. V., Maskarinec, M. P., Dyer, F. F., & Ho, C.-h. (1988) *Technology Assessment of Field Portable Instrumentation for Use at Rocky Mountain Arsenal, Final Report*. ORNL/TM-10542.

Jenkins, R. A. (1987) Method 5—volatile aldehydes in tobacco smoke-polluted indoor air. In: O'Neill, 1. K., Brunnemann, K. D., Dodet, B., & Hoffmann, D., eds., *Environmental Carcinogens Methods of Analysis and Exposure Measurement, Vol. 9—Passive Smoking (IARC Publications No. 81)*, Lyon, International Agency for Research on Cancer, pp. 213–220.

Jenkins, R. A., Palausky, A., Counts, R. W., Bayne, C. K., Dindal, A. B., & Guerin, M. R. (1996) Exposure to environmental tobacco smoke in sixteen cities in the United States as determined by personal breathing zone air sampling. *J. Expo. Anal. Environ. Epidemiol.*, 6, 473–502.

Kelly ,T. J., Smith, D. L., & Staola, J. (1999) Emission rates of formaldehyde from materials and consumer products found in California homes. *Environ. Sci. Technol.*, 33, 81–88.

Kopczynski, S. L. (1989) Multidimensional gas chromatographic determination of cotinine as a marker compound for particulate-phase environmental tobacco smoke. *J. Chromalogr.*, 463, 252–260.

Kring, E. V., Graybill, M. W., Morello, J. A., Ansul, G. R., Adkins, J., E., Jr., & Lautenberger, W. J. (1982) PRO-TEK organic vapor air monitoring badges. *Toxic Materials in the Atmosphere, A STM STP 786,* American Society for Testing and Materials, pp. 85–103.

Lambert, W. E., Sarnet, J. M., and Spengler, J. D. (1993). Environmental tobacco smoke concentrations in no-smoking and smoking sections of restaurants. *Am. J. of Public Health*, 83(9), 1339–1341.

Lee, F. S.-C., Harvey, T. M., Prater, T. J., Paputa, M. C., & Schuetzle, D. (1980) Chemical analysis of diesel particulate matter and an evaluation of artifact formation. *Sampling and Analysis of Toxic Organics in the Atmosphere, ASTM STP 721,* American Society for Testing and Materials, pp. 92–110.

Lioy, P. J., Wallace, L., & Pellizzari, E. D. (1990) Indoor/outdoor and personal monitor relationships for selected volatile organic compounds measured at three homes during New Jersey Team— 1987. *Proceedings, 1989 EPA/AWMA Specialty Conference on Total Exposure Assessment Methodology,* Las Vegas, NV, pp. 17–37.

Lioy, P. J. & Lioy, M. J. Y. (1983) *Air sampling instruments for evaluation of atmospheric contaminants.* American Conference of Governmental Industrial Hygienists, Cincinnati, OH.

Lippmann, M. (1983) Sampling aerosols by filtration. In: Lioy, P. J. & Lioy, M. J. Y., eds., *Air Sampling Instruments for Evaluation of Atmospheric Contaminants, 6th Edition,* Cincinnati, OH, pp. PI–P30.

Löfroth, G., Burton, R. M., Forehand, L., Hammond, K. S., Seila, R. L., Zweidinger, R. B., & Lewtas, J. (1989) Characterization of environmental tobacco smoke. *Environ. Sci. Technol.*, 23, 610–614.

Ma, C. Y., Skeen, J. T., Dindal, A. B., Bayne, C. K., & Jenkins R.A. (1997) Performance evaluation of a thermal desorption /gas chromatography/mass spectrometric method for the characterization of waste tank headspace samples, *Environ. Sci. Technol.*, 31, 853–859.

McConnaughey, P. W., McKee, E. S., & Pritts, I. M. (1982) Passive colorimetric dosimeter tubes for ammonia, carbon monoxide, carbon dioxide, hydrogen sulfide, nitrogen dioxide, and sulfur dioxide. Toxic *Materials in the Atmosphere, ASTM STP 786,* American Society for Testing and Materials, pp. 113–132.

McConnell, B. C., Perfetti, P. F., Walsh, R. F., Oldaker, G. B., III, Conrad, F. W., Jr., Heavner, D. L., Conngr, J, M., Ingebrethsen, B. J., Eudy, L. W., Ogden, M. W., & Stancill, M. W. (1987) Development and evaluation of portable air sampling system (PASS) for environmental tobacco smoke (ETS). *41ˢᵗ Tobacco Chemists' Research Conference,* Greensboro, NC.

McKee, E. S., McConnaughey, P. W., & Pritts, I. M. (1982) Organic vapor dosimeter. *Toxic Materials in the Atmosphere, ASTM STP 786,* American Society for Testing and Materials, pp. 104–112.

Moneyhun, J. H. (1988) Personal communication to R. A. Jenkins, Oak Ridge National Laboratory.

Monsour, N. J., & Waite, P. J., (1981) Principal steps in an establishment sample survey. *Proceedings, 1981 DOE Statistical Symposium, November 4–6, Brookhaven National Laboratory,* Brookhaven, NY, pp. 1– 13.

Mulik, J. D., Lewis, R. G., & McClenny, W. A. (1989) Modification of a high efficiency passive sampler to determine nitrogen dioxide or formaldehyde in air. *Anal. Chem.,* 61, 187–189.

Mumford, J. L., Lewtas, J., Burton, R. M., Svendsgaard, D. B., Houk, V. S., Williams, R. W., Walsh, D. B., & Chuang, J. C. (1990) Unvented kerosene heater emissions in mobile homes: studies on indoor air particles, sernivolatile organics carbon monoxide, and mutagenicity. *Proceedings of the 5ᵗʰ International Conference on Indoor Air Quality and Climate, Vol. 2,* Toronto, Canada, pp. 257–262.

Muramatsu, M., Umemura, S., Okada, T., & Tomita, H. (1984) Estimation of personal exposure to tobacco smoke with a newly developed nicotine personal monitor. *Environ. Res.,* 35, 218–227.

Nader, J. S. & Lauderdale, J. F. (1983) Direct reading instruments for analyzing airborne gases and vapors. In: Lioy, P. J. & Lioy, M. J. Y., eds., *Air Sampling Instruments for Evaluation of Atmospheric Contaminants, 6ᵗʰ Edition,* Cincinnati, OH, pp. V5–V118.

Nelson, P. R., Heavner, D. L., & Oldaker, G. B., III (1990) Problems with the use of nicotine as a predictive environmental tobacco smoke marker. *1990 EPA/AWMA Conference on Toxic and Related Air Pollutants,* Raleigh, NC.

Ogden, M. W., Heavner, D. L., Foster, T. L., Maiolo, K. C. Cash, S. L., Richardson, J. D., Martin, P., Simmons, P. S., Conrad, F. W., & Nelson, P. R. (1996) Personal monitoring system for measuring environmental tobacco smoke exposure. *Environ. Technol.*, 17, 239–250.

Ogden, M. (1996) Environmental tobacco smoke exposure of smokers relative to nonsmokers. *Anal. Commun.*, 33, 197–198.

Ogden, M. W. and Maiolo, K. C. (1992). Comparative evaluation of diffusive and active sampling systems for determining airborne nicotine and 3-ethenylpyridine. *Environ. Sci. Technol.*, 26, 1226–1234.

Ogden, M. W., Maiolo, K. C., Oldaker, G. B., III, & Conrad, F. W. Jr. (1989b) Evaluation of methods for estimating the contribution of ETS to respirable suspended particles. *43rd Tobacco Chemists' Research Conference,* Richmond, VA.

Ogden, M. W., Eudy, L. W., Heavner, D.L., Conrad, F. W. Jr., & Green, C. R. (1989) Improved gas chromatographic determination of nicotine in environmental tobacco smoke. *Analyst,* 114, 1005–1008.

Ogden, M. W., Davis, R. A., Maiolo, K. C., Stiles, M. F., Heavner, D. L., Hege, R B., et al. (1993) Multiple measures of personal ETS exposure in a population-based survey of nonsmoking women in Columbus, Ohio. In: Jaakkola, J. J. K., Ilmarinen, R., & Seppänen, O., eds., *Indoor Air '93. Proceedings of the 6th International Conference on Indoor Air Quality and Climate. Helsinki, Finland: Indoor Air '93,* pp. 523–528.

Oldaker, G. B., III, Ogden, M. W., Maiolo, K. C., Conner, J. M., Conrad, F. W., & DeLuca, P.O. (1990) Results from surveys of environmental tobacco smoke in restaurants in Winston-Salem, North Carolina. *Proceedings of the 5th International Conference on Indoor Air Quality and Climate, Vol. 2,* Toronto, Canada, pp. 281–285.

Oldaker, G. B., III & Conrad, F. W., Jr. (1987) Estimation of effect of environmental tobacco smoke on air quality within passenger cabins of commercial aircraft. *Environ. Sci. Technol.,* 21, 994–999.

Palmes, E. D. (1980) Sampling rates and other considerations for selected passive air sampling techniques. *Sampling and Analysis of Toxic Organics in the Atmosphere,* ASTM STP 721, American Society for Testing and Materials, pp. 178–183.

Pellizzari, E. D. (1991) Sampling of gas-phase compounds with sorbent beds. In: Hansen, L. D. & Eatough, D. J., eds., *Organic Chemistry of the Atmosphere,* CRC Press, Boca Raton, FL, pp. 1–52.

Pellizzari, E. D., Perritt, K., Hartwell, T. D., Michael, L. C., Sparacino, C. M., Sheldon, L. S:, Whitmore, R., Leninger, C., Zelon, H., Hardy, R. W., & Smith, D. (1987). *Total Exposure Assessment Methodology (TEAM) Study. Elizabeth and Bayonne, New Jersey, Devils Lake, North Dakota and Greensboro, North Carolina, Volume 11, Part 11, Final Report,* EPA/600/6-87/002b, p. 847.

Pellizzari, E. D., Sheldon, L. S., Sparacino, C. M., Bursey, J. T., Wallace, L., & Bromberg, S. (1984) *Proceedings of the 3ʳᵈ International Conference on Indoor Air Quality and Climate, Vol. 4,* Stockholm, pp. 303–308.

Phillips, K., Bentley, M. C., Howard, D., & Alvan, G. (1998) Assessment of environmental tobacco smoke and respirable suspended particle exposures for nonsmokers in Kuala Lumpur using personal monitoring. *J. Expo. Anal. Environ. Epidemiol.,* 8, 519–541.

Phillips, K., Howard, D. A., Bentley, M. C., & Alvan, G. (1998) Assessment of environmental tobacco smoke and respirable suspended particle exposures for nonsmokers in Hong Kong using personal monitoring. *Environ. Int.,* 24, 851–870.

Phillips, K., Howard, D. A., & Bentley, M. C. (1997) Assessment of air quality in Turin by personal monitoring of nonsmokers for respirable suspended particles and environmental tobacco smoke. *Environ. Int.,* 23, 851–871.

Phillips, K., Howard, D. A., Bentley, M., & Alvan, G. (1998) Measured exposures by personal monitoring for respirable suspended particles and environmental tobacco smoke of housewives and office workers resident in Bremen, Germany. *Int. Arch. Occup. Environ. Health,* 71, 201–212.

Phillips, K., Howard, D. A., Bentley, M. C., & Alvan, G. (1998) Assessment of environmental tobacco smoke and respirable suspended particle exposure of nonsmokers in Lisbon by personal monitoring. *Environ. Int.,* 24, 301–324.

Phillips, K., Howard, D. A., Bentley, M., & Alvan, G. (1999) Assessment of environmental tobacco smoke and respirable suspended particle exposures for nonsmokers in Basel by personal monitoring. *Atmos. Env.,* 33, 1889–1904.

Phillips, K., Bentley, M., Howard, D., & Alvan, G. (1998) Assessment of environmental tobacco smoke and respirable suspended particle exposures for nonsmokers in Prague using personal monitoring. *Int. Arch. Occup. Environ. Health,* 71, 379–390.

Phillips, K., Bentley, M. C., Howard, D. A., & Alvan, G. (1998) Assessment of air quality in Paris by personal monitoring of nonsmokers for respirable suspended particles and environmental tobacco smoke. *Environ. Int.,* 24, 405–425.

Phillips, K., Bentley, M. C., Howard, D. A., & Alvan, G. (1996) Assessment of air quality in Stockholm by personal monitoring of nonsmokers for respirable suspended particles and environmental tobacco smoke. *Scan. J. Work. Environ. Health,* 22, 1–24.

Phillips, K., Howard, D. A., Bentley, M. C., & Alvan, G. (1998) Assessment by personal monitoring of respirable suspended particles and environmental tobacco smoke exposure for non-smokers in Sydney, Australia. *Indoor Built Environ.,* 7, 188–203.

Phillips, K., Bentley, M. C., Howard, D. A., Alvan, G., & Huici, A. (1997) Assessment of air quality in Barcelona by personal monitoring of nonsmokers for respirable suspended particles and environmental tobacco smoke. *Environ. Int.*, 23, 173–196.

Phillips, K., Howard, D. A., Browne, D., & Lewsley, J. M. (1994) Assessment of personal exposures to environmental tobacco smoke in British nonsmokers. *Environ. Int.*, 20, 693–712.

Piadé, J. J., D'Andres, S., Sanders, E. B. (1999) Sorption phenomena of nicotine and ethenylpyridine vapours on different materials in a test chamber. *Environ. Sci. Technol.*, 33, 2046–2052.

Ramsey, R. S., Moneyhun, J. H., & Jenkins, R. A. (1990) Generation, sampling and chromatographic analysis of particulate matter in dilute sidestream tobacco smoke. *Anal. Chim. Acta,* 236, 213–220.

Riggin, R. M. & Peterson, B. A. (1985) Sampling and analysis methodology for semivolatile and nonvolatile organic compounds in air. In: Gammage, R. B. & Kaye, S. V., eds., *Indoor Air and Human Health,* Lewis Publishers, Boca Raton, FL, pp. 351–359.

Robinson, R. J. & Yu, C. P. (1999) Coagulation of cigarette smoke particles. *J. Aerosol Sci.*, 30, 533–548.

Rounbehler, D. P., Reisch, J. W, & Find, D. H. (1980) Nitrosamine air sampling using a new artifact-resistant solid sorbent system. *Sampling and Analysis of Toxic Organics in the Atmosphere, ASTM STP 721,* American Society for Testing and Materials, pp. 80–91.

Saltzman, B. E. (1983) Direct reading colorimetric indicators. In: Lioy, P. J. & Lioy, M. J. Y., eds., *Air Sampling Instruments for Evaluation of Atmospheric Contaminants, 6th Edition,* Cincinnati, OH, pp. T2–T29.

Sheldon, L. S., Sparacino, C. M.9 & Pellizzari, E. D. (1985) Review of analytical methods for volatile organic compounds in the indoor environment. In: Gammage, R. B., Kaye, S. V., & Jacobs, V. A., eds., *Indoor Air and Human Health,* pp. 335–402.

Spengler, J. (1989) Personal communication to R. A. Jenkins, Oak Ridge National Laboratory.

Sterling, E. M., Collett, C. W., & Ross, J. A. (1996) Assessment of non-smokers' exposure to environmental tobacco smoke using personal-exposure and fixed-location monitoring. *Indoor Built Environ.*, 5, 112–125.

Stober, W. (1980) Aerosol sampling and characterization for inhalation exposure studies with experimental animals. In: Willeke, K., ed., *Generation of Aerosols and Facilities for Exposure Experiments,* Ann Arbor Science Publishers, Ann Arbor, MI, pp. 31–63.

Sullivan, D. A. & Koines, A. (1990) An evaluation of the effectiveness of dispersion modeling to estimate human exposure: a case study based on the Baltimore total exposure assessment methodology (TEAM) study. *Proceedings, 1989 EPA/AWMA Specialty Conference on Total Exposure Assessment Methodology,* Las Vegas, NV, pp. 499–513.

Supelco, Inc. (1990) High-purity carbon molecular sieves for monitoring many volatile airborne contaminants. *The Supelco Reporter, 9,* pp. 13, 16.

Swift, D. L. (1983) Direct reading instruments for analyzing airborne particles. *Proceedings, 1983 Air Sampling Instruments for Evaluation of Atmospheric Contaminants, 6th Edition,* Lioy, P. J. & Lioy, M. J. Y., eds., Cincinnati, OH, pp. U2–U27.

Thompson, C. V., Jenkins, R. A., & Higgins, C. E. (1989) A thermal desorption method for the determination of nicotine in indoor environments. *Environ. Sci. Technol., 23,* 429–435.

Tomkins, B. A., Jenkins, R. A., Griest, W. H., Reagan, R. R., & Holladay, S. K. (1985) Liquid chromatographic determination of benzo[a]pyrene in total particulate matter of cigarette smoke. *J. Assoc. Off. Anal. Chem., 68*(5), 935–940.

Van Loy, M. D., Nazaroff, W. M., and Daisey, J. M. (1998). Nicotine as a marker for environmental tobacco smoke, implications of sorption on indoor surface materials. *J. Air & Waste Manage. Assoc., 48,* 959–968.

Van Loy, M. D., Lee, V. C. Gundel, L. A. Daisey, J. M. Sextro, R. G., and Nazaroff, W. M. (1997). Dynamic behavior of semivolatile organic compounds in indoor air. 1. Nicotine in a stainless steel chamber. *Environ. Sci. Technol., 31,* 2554–2561.

Wallace, L. A., Pellizzari, E. D., Hartwell, T. D., Sparacino, C., Whitmore, R., Sheldon, L., Zelon, H., & Perritt, R. (1987) The TEAM study: personal exposures to toxic substances in air, drinking water, and breath of 400 residents of New Jersey, North Carolina, and North Dakota. *Environ. Res., 43,* 290–307.

Wallace, L. A. (1987) *The Total Exposure Assessment Methodology (TEAM) Study: Summary and Analysis: Volume 1,* EPA/600/6-87/002a.

Wilder, L., Heston, G. T., Zickler, M., & Habrukowich, R. (1987) Inhalation exposure from volatile organic contaminants in drinking water. *Proceedings, 1987 EPA/APCA Symposium on Measurement of Toxic and Related Air Pollutants,* pp. 123–131.

Williams, D. C., Whitaker, J. R., & Jennings, W. *(1984)* Air monitoring for nicotine contamination. *J. Chromatogr. Sci., 22,* 259–261.

Wu, J. (1997) Methods for measuring exposure to environmental tobacco smoke (ETS). *J. Korean Soc. Tob. Sci., 19,* 162–169.

INTRODUCTION

Particulate matter is one of the most commonly measured indoor air pollutants. Particulate matter is the most visible form of air pollution, and it is reasonably easy to determine quantitatively with minimal instrumentation. The term "total suspended particulates" (TSP) refers to all solid- and liquid-phase matter suspended in the air and that is collectable. The TSP must also be sufficiently stable (nonvolatile) to be maintained on the filter for mass determination and other analyses. Respirable suspended particulates (RSP) refers to only those particulates that are small enough to reach the lower airways of the human lung. There are many interpretations as to the maximum particle size that such deposition occurs (Raabe 1980, Heyder et al. 1980). Some investigators use a very conservative value of 3 μm, others use values of 10 or 15 μm. For particulate matter arising solely from environmental tobacco smoke (ETS), the distinction is probably not significant, since virtually all ETS particulate matter is in the submicron size range (Benner et al. 1989, McAughey et al. 1990, Morawska et al. 1999). However, many indoor air particulates are much larger than this. Some investigators use the term "respirable" to refer to only those particles less than 2.5 μm in diameter (the so-called fine fraction), reserving the term "inhalable" to denote those particles 2.5–10 μ (the "coarse" fraction). The Occupational Safety and Health Administration (OSHA) defines RSP as particles less than 4.0 μm in diameter. For the purposes of this chapter and to minimize confusion, discussion will be focused predominantly on RSP. RSP is taken here to include particles reported by the investigator as RSP, up to a mass median diameter of 10 μm.

Particulate matter found in indoor air can be composed of both solid- and liquid-phase material. Solids can be either symmetrically or irregularly shaped. Indoor air particulates can be present in many forms, including mold spores, pollen, human, insect, or animal dander, solid particulates such as dusts and infiltrated diesel exhausts, inorganic aerosols (sulfates), and consumer product sprays (antiperspirants, furniture polishes). Most probably, ETS RSP is comprised of liquid or waxy droplets, although there are some data that suggest that the droplets may contain very small amounts of cigarette ash, which act as condensation nuclei (Stober 1982). ETS RSP originates as both sidestream and exhaled mainstream cigarette smoke particulate matter. As the particles are diluted and mixed with the ambient air, the more volatile components of the smoke particles evaporate, and the droplets shrink somewhat. As a result, the RSP of ETS is composed of the higher molecular weight and/or the less volatile constituents of cigarette smoke. Described below are the studies that have reported suspended particulates in a number of field situations. In these cases, field is distinguished from test environments, chamber studies, and studies that may have been conducted in realistic settings (e.g., test homes) by the absence of manipulation of the environment. The data described in this chapter suggest that while RSP levels tend to be higher in smoking environments than in comparable nonsmoking environments, not all the RSP in ETS-contaminated environments results from ETS. In most cases, the fraction of RSP resulting from ETS is substantially less than 50%.

SAMPLING AND ANALYSIS METHODS

Introduction

Ambient aerosols are distributed over a range of sizes. Typically, that range is expressed in terms of average diameter. The mean or median diameter is typically a number median or a mass median. The number distribution refers to the number of particles in any given size range. The mass distribution refers to the total mass of particles in any given size range. The distinction is important, because a few particles of larger diameter represent a much larger fraction of the total mass than the same number of particles of a smaller diameter. The mass distribution can be calculated from the number distribution by multiplying the latter by the density (mass per unit volume) of the particles. However, care must be exercised when this approach is taken, as the density of the particles can vary with particle size. In some cases, the term "aerodynamic diameter" is used. It is associated with those particulates that are irregularly shaped and have a settling velocity *equivalent* to a spherical particle at unit density in normal air of a certain size. Typically, distributions of particle sizes are near log-normal. That is, the greater number or mass of particles is skewed to smaller values. Often, ambient aerosols have more than one contributing source, and thus have complex, overlapping distributions. Marple and Rubow (1980) have reviewed particle size distribution characteristics.

When sampling predominantly liquid-based aerosols, such as ETS particulate matter or other volatilization/combustion-derived "smokes," an important consideration is to avoid artificially diluting the material. This is especially important for matrices containing constituents of differing volatilities. As the aerosol is diluted, the more volatile species evaporate from the droplet, and the mass of the individual droplets, and thus the collectable mass, decreases. However, small changes in the mass of the droplet will not have an important effect on the droplet size, because the droplet's mass is proportional to the cube of the radius and depends on the density. For example, a 50% reduction in particle volume or mass results in a 21% decrease in the particulate diameter. Evaporation of collected particulate-phase constituents during the collection process can be a serious problem, especially when large volumes of air are drawn across the filter assembly. Even constituents as low in volatility as benzo[a]pyrene can be evaporated from collected airborne diesel particulates under long-term, large-volume sampling conditions (Griest et al. 1988). In addition, there is the potential for chemical reactions between ambient gases drawn through the filter and collected particles that can increase the volatility of the resulting compounds.

Mass Concentration Measurements

While a number of technologies are available for making real-time measurements of particle concentrations, including condensation nuclei counters, single particle analyzers, integrating nephelometers, forward scattering optical devices and β-attenuation systems, the discussion here will be limited to that reported to have been used in the published indoor air quality studies described below. Certainly the most popular systems for real-time determination of particle mass concentrations are optically based particle monitors. Most of these types of instruments are based on the Tyndall effect, in which particles attenuate a light beam due to the scattering of incident radiation (Moore 1962). The actual scattering efficiency is a complex function of the angle of incident vs. scattered radiation, the volume concentration, particle size and shape, and refractive index of the droplet (Harrison and Perry 1986). This latter factor is important and underscores the need to calibrate such instrumentation with either the material of interest or one with a chemical composition similar to that being monitored (Ingebrethsen et al. 1988). Virtually all indoor air quality studies, both ETS- and non-ETS-related, that report the use of true real-time instrumentation for particulate monitoring have relied on some form of forward-scattering photometer. Such a device examines light scattered 45–90° off the axis of incident radiation (Swift 1983). At these angles, scattering is still fairly intense, but sophisticated electronics are not required to sort the incident beam from the scattered radiation, the former being quite weak at such an angle. The amount of light scattered is proportional to the aerosol concentration. Typically, such a system employs a pump operating at a few liters per minute to draw an ambient air sample through a sensing chamber. As stated above, these instruments provide the most accurate result if they have been previously calibrated against a gravimetrically measured sample of the material being monitored. Some of the more popular versions of such a system include the DataRAM and Personal DataRAM, manufactured by MIE Corp., the DUSTTRAK airborne dust monitor, sold by TSI, Inc., and the Haz-DUST II and EPAM-5000, sold by SKC, Inc. Swift (1983) provides an excellent review of direct reading instrumentation for particle analysis. In-depth discussions of particle measurement technology can be found in a number of air pollution monitoring handbooks (e.g., Harrison and Perry 1986).

In contrast to light-scattering units, instrumental systems for *gravimetrically* determining airborne particulate matter are typically based on the near real-time determination of deposition on a specific surface. The two most popular systems have been the automated piezobalance (Quant et al. 1982) and the tapered element oscillating microbalance (TEOM) (Patashnick and Rupprecht 1986). The piezobalance measures aerosol-mass concentration through its relationship to the frequency change of an oscillating quartz crystal. Ambient air is drawn through the system at a fixed rate. The quartz crystal acts as the collecting surface for an electrostaticlprecipitator capable of efficiently collecting particles from $0.01–10$ μm mass median diameter (MMD). The rate of change of the output frequency is proportional to the mass of particles deposited. In the 1980s and early 1990s, a number of instruments were available using this principle. The more sophisticated versions had sensitivities down to about 10 μg/m^3 and data-logging and automated crystal cleaning capability. Typical minimum sampling and reporting periods are 1–3 min. One of the most popular models of this genre was the Model 5000 Automated Respirable Aerosol Mass Monitor, manufactured by TSI, Inc., of St. Paul, Minnesota; however, the device is no longer sold by TSI.

The TEOM (Rupprecht & Patashnick, Inc., Albany, NY) measures aerosol mass directly. In this case, a pump draws ambient air through a small filter. The filter is attached to a vibrating reed, the vibrational frequency of which is related to the mass of the filter. As the filter loads with particulates, its mass increases, changing the frequency at which the reed vibrates. One advantage of such a system over an automated piezobalance is that a sample of the particulate matter is collected on the filter, and can be chemically analyzed after the measurement has been completed. Another is that the system can be calibrated directly, by adding "microdots" of known weight to the filter. The limit of sensitivity is in the range of 5–15 μg/m^3, and the minimum sampling and reporting period is about 15 s. EPA recently ruled that TEOM methodology is equivalent to gravimetric determinations of airborne particulate matter <10 μm in diameter (EPA 1990). Nelson et al. (1997, 1998) has performed a multi-nation evaluation of ETS RSP generated from popular brand styles, and compared TEOM response with gravimetric RSP collected from chamber atmospheres. The studies have indicated that TEOM response and gravimetric response were usually within 10% of each other.

Collecting airborne particulates on filter media has been by far the most popular method for particle mass measurements. This is probably because the relationship between the actual component measured and aerosol mass concentrations is very straightforward. Also, the equipment required to make the measurement tends to be relatively inexpensive and easy to operate. However, it is critical to distinguish between TSP and RSP. The former is determined by an absolute filtration of all airborne material within a given volume of air. RSP can only be determined if particles larger than those thought to be respirable are removed from the air upstream of the filter. This is typically accomplished with small cyclone separators (e.g., SKC, Inc., Eighty Four, PA), or impactors that act by allowing those particles with higher kinetic energy (and larger size) to impact against the walls of the separator, while the smaller particles pass through the system. An excellent review of the technology of filtration and filter media has been prepared by Lippmann (1983). Briefly, filters remove particles from the airstream through processes of direct interception, diffusional and inertial deposition, and electrical and gravitational attraction. Popular filter types include fibrous materials, amorphous gel membranes (tortuous channels) and nucleopore membranes (straight channels). Most investigators using gravimetric technology sample for periods up to a few hours, collecting air particulates on filters, and weighing the individual filters before and after collection. Sampling rates vary from tens of milliliters per minute up to a few liters per minute, although some investigators have reported sampling in a high-volume mode, using rates of 1 m^3/minute, or more (e.g., Griest et al. 1988) in selected indoor air situations. (Caution should be exercised when using high-volume sampling in small enclosed areas. In some cases, it is possible to over-sample the environment—essentially scrub the air in the room of its particulate matter—and yield lower apparent overall concentrations.) A number of types of filter media have been used, including glass fiber, Teflon-coated glass fiber, Teflon membrane, and Fluoropore membranes. Teflon and Teflon-coated glass fibers are popular because of their relative inertness. In terms of sheer numbers of samples acquired in ETS-related studies, the Fluoropore membrane is probably the most widely used for ETS-related studies. The filter has been employed as one component of the Portable Air Sampling System (PASS—see Chap. 7) and comprises a polytetrafluoroethylene membrane backed with a polyethylene membrane. Sampling rates up to 2.2 L/min have been used. These filters are typically used in conjunction with small cyclone separators mounted upstream of the filter. These perform a continuous size fractionation, such that only those particles smaller than 2.5 – 4.0 μm reach the filter, depending on the air flow rate. The larger particles inertially impact on the walls of the separator, or drop into a receiving well. A now-patented dual pump sampling system, the Double Take sampler from SKC Inc., developed by Ogden et al. (1996), is now commercially available for studies where both particle and vapor phase samples must be collected simultaneously.

While the theory and equipment required to make a gravimetric mass measurement may be straightforward, great care must be taken to weigh small quantities of collected particles accurately. Typically, if filter loading is expected to be only a few tens of micrograms, filters must be equilibrated at a constant temperature and humidity for several hours (24 to 48 h in some procedures) prior to their initial weighing. Also, beta-ray emitting foils must be placed both in and near the microgram balance to discharge static electricity, several independent weighings of the used filter must be made [typically five or more (Conner et al. 1990)], and a mean value reported. Humidity control helps reduce problems with static charge effects. Frequent calibration checks of the balance are essential. Another potential source of uncertainty in the measurement is the calibration of sampling pumps. Without the use of electrically or mechanically controlled flow rates, sampling flow rates can change as filter loadings increase. Most modern sampling systems provide for some degree of flow control. However, it is not unusual for flow rates to change over a 5- to 15-h sampling period. Also, calibration errors can occur due to changes in temperature and/or ambient pressure. Many investigators report the use of pre- and post-sampling flow rate calibration and normalization of data to standard temperature and pressure.

A generation ago, it appeared that many investigators assumed that all or most of the RSP present in smoking environments was derived from ETS (e.g., Repace and Lowrey 1980). However, since then, there has been an increasing realization that there are many sources of RSP in enclosed spaces, even where smoking is occurring (e.g., Leaderer and Hammond 1991). As a result, there has been considerable experimental effort made towards developing quantitative measures of RSP derived solely or predominantly from ETS. At this writing, there are three methods in routine use to accomplish this objective. Two of the methods [determining ultraviolet absorbing particulate matter (UVPM) and fluorescing particulate matter (FPM)] target the determination of combustion-derived particles, while a third method, measuring solanesol, targets direct determination of tobacco-specific RSP.

Because ETS is a combustion-derived material, it contains a number of chemical species, including single- and multi-ring aromatic compounds, which absorb UV light. Of course, because other combustion-derived aerosols may be present in the environment, such as those from wood-burning stoves, kerosene heaters, and diesel exhaust, if all UV-absorbing particulate matter is taken as ETS-derived, its measurement would overestimate the contribution of ETS. Nevertheless, most investigators employing such an approach have justified it on the basis that UVPM provides a *more accurate* assessment of ETS-derived particulate matter than the use of simple RSP levels.

The analytical method developed (Conner et al. 1990) to determine UVPM and employed in a number of studies was designed to "piggy back" on a standard gravimetric method for RSP. Particulates are collected on a Fluoropore membrane filter and extracted with methanol. An aliquot of the extract is analyzed on a "columneas" high-performance liquid chromatographs (HPLC) with a UV detector. 2,2',4,4'-tetrahydroxybenzophenone is used as a surrogate standard for quantitation. In most of the studies reporting both the use of this and a conventional gravimetric method for total RSP, the fraction of RSP reported as UVPM ranges up to a maximum of 50% (see Table 6.1), although for certain individual measurements, especially at high particulate concentrations, UVPM may be as large as 80% of the RSP value. Ogden et al.

(1989) have developed a variation of this method, in which fluorescing particulate matter (FPM) is determined simultaneously. The measurement is made by adding a fluorescent detector in series with the UV detector on the HPLC system used to determine UVPM. Excitation is at 300 nm; emission is measured at 420 nm. Scopoletin is used as a surrogate standard. The method is reported to be less subject to interferences than the UVPM method and would be likely to overestimate ETS contribution to RSP to a lesser extent.

The method by which comparative responses of the filter extracts to the standards are converted to a particulate matter value bears explanation. Essentially, cigarettes are smoked by humans in a controlled atmosphere chamber to generate true ETS. In the chamber, considerable effort is made to preclean the atmosphere, so that the only source of RSP is ETS. Then, as both RSP and the response to the UV and fluorescing standards are determined, a factor is obtained that equalizes the UVPM and FPM to be identical to the RSP reading. That same factor is used in field studies, when the sources of RSP may be considerable. Because tobaccos may vary in composition among nations, it is important when conducting field studies that appropriate conversion factors specific to the country in question are used. Nelson et al. (1997 and 1998) have reported an extensive study of nation-specific conversion factors for UVPM, FPM, and Sol-PM (see below).

Solanesol is a C_{45} isoprenoid alcohol present in tobacco leaves (Ogden and Maiolo 1989, Tang et al. 1990). It is the chemical species that provides the basis of the name of the Solanaceae family, which includes peppers, eggplant, tomatoes, and of course, tobacco, and thus is not unique to tobacco. However, no sources of airborne solanesol have been reported, and thus solanesol, which is distributed nearly 100% into the particle phase of tobacco smoke, is taken to be specific to tobacco combustion. Ogden and Maiolo (1992) have developed a high-performance liquid chromatography (HPLC) method for solanesol, which has been used by a number of investigators (e.g., Jenkins et al. 1996, Phillips et al. 1997) to determine ETS-specific particulate matter. Note that both solanesol and solanesol-specific particulate matter (Sol-PM) have been reported in several cases. Factors that convert solanesol concentrations to Sol-PM concentrations are developed similarly to those for UVPM and FPM, and have been reported by Nelson et al. (1997a, 1997b).

Method Intercomparisons

Few systematic studies have been reported comparing RSP determinations made with different methodologies under field conditions. Because the techniques are sensitive to various physical properties (mass, particle diameter, light-scattering efficiency, etc.), differences among techniques, especially when the instrumentation is not calibrated specifically for typical indoor air particulates, should not be unexpected. Löfroth et al. (1989) compared TSP determined gravimetrically with TSP determined on a TSI Model 3500 piezobalance in two field locations. The piezobalance yielded readings that averaged about 85% of those determined directly by weight. During preliminary chamber studies, the authors reported that an Electric Aerosol Analyzer (EAA—Thermo-System, Inc., Model 3030) gave TSP values that ranged from 0–60% greater than those obtained gravimetrically. Traynor et al. (1987) estimated that submicron RSP measured by an electrical mobility analyzer was probably a factor of 2 greater than TSP determined gravimetrically due to assumptions concerning particle shape and density. Benner et al. (1989) reported comparisons between an automated piezobalance (TSI Model 5000) and gravimetrically determined RSP for a series of chamber studies. Agreement was reported to be good, with the gravimetric levels about 94% of those determined instrumentally. The slightly lower manual method response was believed to be due to volatilization of materials from the filter during collection. Miesner et al. (1989) reported on the comparison of two optical monitors [the Handheld Aerosol Monitor (HAM) and the MINIRAM, a miniaturized version of the RAM-I, formerly produced by MIE Inc, but no longer commercially available] in both chamber and field studies. Only after the two systems were calibrated against a test aerosol [di(2-ethylhexyl)phthalate] did they yield comparable results in both test and field settings. Finally, Ingebrethsen et al. (1988) have reported an extensive intercomparison study conducted in a chamber with diluted sidestream cigarette smoke and ETS as the aerosol of interest. A piezobalance and an optical particle counter-condensation nuclei counter were shown to agree reasonably well with gravimetric measurements. The MINIRAMs also showed good agreement, but only after correction of readings for particle density. The response of the HAM gave consistently lower readings than those determined gravimetrically and showed a strong dependence on the particular method used to generate ETS. One can conclude from these studies that automated gravimetric measurements (such as a piezobalance) can yield comparable results to those of manual methods. However, optically based instrumentation probably must be calibrated with the aerosol of interest before meaningful results can be obtained

From the published data, it is difficult to tightly specify limits of detection (LODs) for various instrumental systems. For example, one group of investigators (Miesner et al. 1989) reports values obtained with the HAM optical unit down to 1 $\mu g/m^3$ RSP. In that same study, simultaneous gravimetric and optical measurements were performed with the MINIRAM. For several of these experiments, gravimetric RSP ranged up to 20 $\mu g/m^3$, while the simultaneous MINIRAM measurement was listed as below the LOD. However, the same set of experiments listed some MINIRAM-determined levels as low as 10 $\mu g/m^3$. Sterling and Mueller (1988) report sensitivity of an optical scattering device as low as 4 $\mu g/m^3$. In an extensive study, Ingebrethsen et al. (1988) reported piezobalance readings in the range of 20–30 $\mu g/m^3$ RSP for essentially clean air. The lowest readings for this chamber study for the HAM, piezobalance, and the MINIRAM appeared to be in the range of 25

μg/m3. A summation of the literature reports would seem to suggest that while individual manufacturers may provide optimistic estimates of LODs in the range of 10 μg/m^3 RSP, practical LODs for such instruments would seem to be in the range of 20–25 μg/m^3. Note that technological advances over the last decade have resulted in improvements in size and response linearity for the predecessors of the instruments described above. However, most manufacturers, if instrumentation is likely to be used in indoor air quality studies, provide estimated LODs on the order of 10 μg/m^3. Instrumentation is available that can detect lower concentrations, but the chances of encountering such low particle concentrations in indoor environments other than semiconductor fabrication facilities seem small.

For the gravimetric determination of RSP, it appears that the LOD is governed by the practical weighing limit of about 2 μg (Crouse et al. 1988). Others have suggested that the limit is 15 μg. Clearly, increasing the sampling duration would serve to lower the effective LOD relative to ambient air concentration. However, it is important to realize that the longer the sampling duration, the greater the likelihood that some semivolatile species will be evaporated from the filter during collection. If a 2 μg absolute weighing detection limit is assumed, a 2 L/min sample for 1 h would yield an LOD of about 16 μg/m^3 RSP. Indeed, this is in the range in which many investigators of earlier studies (e.g., Sterling and Mueller 1988, Ogden et al. 1990, Crouse et al. 1988, Healthy Buildings International 1990) chose to establish an LOD. Interestingly, in modern studies with large numbers of samples, limits of detection or quantification do not appear appreciably better. For example, Phillips et al. (1998, Kuala Lumpur) reported LOQs of 8.2, 13.4, and 21.7 μg/m^3 for 24-h, 15-h, and 9-h RSP determinations, respectively. Jenkins et al. (1996) reported a mean LOD for 24-h RSP determinations of 15.4 μg/m^3. Higher detection limits for these larger studies may be due to the sheer volume of samples, which necessitates compromises on the amount of time allotted to repeated weighings of filters to achieve the lowest possible detection limits.

UVPM, FPM, and Sol-PM measurements are not intended to be comparable with RSP, since the former two measure only those particulates containing some species that absorb radiation in the ultraviolet spectrum (325 nm). (FPM includes only those species that absorb radiation of 300 nm and emit at 420 nm.) Usually, such species are generated from combustion processes. Sol-PM refers only to tobacco-combustion-derived particles. In Table 6.1, results are compared from selected field studies where *area* samples were acquired in which both UVPM and gravimetric RSP were determined simultaneously. The data indicate that UVPM typically comprises from about one-sixth up to about half of the gravimetrically determined RSP. The exception to this is an environment in which a relatively high proportion of particulates are expected to be combustion derived (mainly from ETS), that in which bartenders work. In that case, UVPM comprised 70% of the RSP.

Table 6.1 Comparisons of UV-Absorbing Particulate Matter (UVPM) with Gravimetrically Determined Respirable Suspended Particulate Matter (RSP) for Area Samples Mean Concentrations ($\mu g/m^3$)

Location	Reference	UVPM	RSP
42 restaurants	Crouse et al. 1988	26.1[a]	62.0[a]
10 trains, smoking compartments	Proctor 1989	59.8	216
10 trains, nonsmoking compartments	Proctor 1989	33	186
5 betting shops	Proctor et al. 1989b	164	333
125 offices in 4 cities	Oldaker et al. 1990	27[a]	126[a]
82 restaurants in 3 cities	Oldaker et al. 1990	36[a]	126[a]
53 bartender work areas	Jenkins and Counts 1999	96	136
32 wait staff work areas	Jenkins and Counts 1999	27	71

[a] Geometric mean.

Table 6.2 summarizes the mean and median data from studies of *personal exposure* in smoking workplaces, where RSP, FPM, and solanesol (here reported as Sol-PM) were all measured on the same sample. RSP provides a measure of respirable particles in general, FPM, the combustion-derived fraction, and Sol-PM, the tobacco-specific fraction. The studies cover a wide range of societies and work situations, but the conclusions are not radically different from the area comparisons of UVPM and RSP. That is, the fraction of RSP composed of combustion-derived particles ranged from 16–72% (means), while the fraction of RSP composed of ETS-specific particles ranged from 7–57% (means). Typical values were on the order of 20–30%. Jenkins et al. (1996) performed a detailed analysis of the ETS-specific fraction of RSP to which a large number of subjects were exposed. For those subjects expected to be most highly exposed (living and working in smoking environments), the extremes of the distribution fraction of Sol-PM:RSP (80th and 95th percentiles) were 37% and 61%, respectively. These studies together indicate that in typically encountered environments, ETS may account for a third of the overall particulate burden. However, in some environments, this fraction may be as much as two-thirds.

RESULTS OF FIELD STUDIES

There are numerous published reports of levels of RSP in indoor air. Appendix 1 cites and summarizes major studies. The range of the data is considerable. There are a number of studies reporting levels below the LOD, and several reports of levels greater than 1000 $\mu g/m^3$. As noted above, there exist considerable differences among the various techniques for the determination of RSP. Because of these differences, absolute comparisons among various studies are meaningful only in a semiquantitative sense. Optical measurement systems must be calibrated with the matrix of interest if the results are to be accurate (Ingebrethsen et al. 1988). Usually, that calibration is performed gravimetrically. Even with the "primary standard method," gravimetry, differences in filter media, sampling rates and durations are considerable. This affords the opportunity for varying fractions of RSP to evaporate from the filter prior to weighing. Thus, there are likely to be differences if the same atmosphere was monitored with different filter systems. Probably the most accurate relative comparisons are within a given day in which the identical sampling and analysis method has been used for all samples. Nearly identical protocols for collection and gravimetric analysis of RSP (and combustion-derived and ETS-specific particles) appear to be emerging for personal exposure studies, and this may act to standardize such. These problems notwithstanding, where possible, results from field studies are discussed here as though they were directly comparable. Comments regarding those factors that might impact on comparability are included.

Table 6.2 Personal Exposure to ETS-associated Particulate Matter in Smoking Workplaces

Condition	Location	Reference	Number of Observations	RSP Median	RSP Mean	FPM Median	FPM Mean	Sol-PM Median	Sol-PM Mean
Workers[a]	U.S.	Jenkins and Counts 1999	134	40.4	62.0	7.7	25.5	0.96	16.4
Workers[b]	U.S.	Jenkins and Counts 1999	52	14.2	29.5	1.7	9.4	Below LOD	7.0
Office workers	Germany	Philips et al. 1998	63–65	37	49	3.7	12	0.59	7.9
Workers[c]	Australia	Phillips et al. 1998	20	34	34	2.1	5.7	1.5	9.5
Workers	France	Phillips et al. 1998	104	63	71	12	18	3.8	16
Workers	Spain	Phillips et al. 1997	60	94	112	30	48	37	64
Workers	Italy	Phillips et al. 1997	71	90	92	16	31	9.1	30
Workers	Czech Republic	Phillips et al. 1998	109–110	61	79	35	57	15	44
Workers	Switzerland	Phillips et al. 1999	62–64	28	40	4.0	11	1.3	8.6
Workers	Sweden	Phillips et al. 1996	53	16	24	NR	NR	1.1	3.6
Workers	Hong Kong	Phillips et al. 1998	47–48	51	66	5.9	22	5.2	18
Workers	Malaysia	Philips et al. 1998	73–74	44	52	3.7	6.2	0.71	3.6
Workers	U.S.	Heavner et al. 1996	28	53	67	17	26.5	12.6	21.2
Workers	U.S.	Sterling et al. 1996	25	29	30	7.4	9.6	8.1	9.3

[a]Unrestricted smoking workplaces.
[b]Smoking restricted to designated areas.
[c]24 h over approx. 3 work shifts.

Establishing background levels of RSP in indoor air quality studies of ETS is a complex issue, and establishing criteria for true nonsmoking environments is difficult. Consequently, those criteria vary considerably among studies. A number of investigators have used outdoor air levels of RSP as an indicator of true background. Results reported in the literature from investigators performing indoor air quality studies and summarized in Appendix 1 indicate those levels generally range from about 1–80 μg/m^3. One exception to this observation is the study reported by Zhao and YuFeng (1990) of coal-fired stoves in residences in rural China. In that study, outdoor RSP levels as high as 990 μg/m^3 were measured. For urban areas in developed nations, it is generally accepted that many factors can contribute to RSP, such as traffic, meteorologic events, and combustion-related sources, and that typical indoor ventilation systems remove some of the RSP load as air is drawn into the building. While many private residences do not have a forced ventilation system that provides for intake of outdoor air, the existence of passive infiltration can result in some filtering of particulates as the air passes through cracks and through window screens. The effect of this passive infiltration is often a penetration rate of less than unity. Thus, it seems reasonable that the RSP load in nearby outdoor air should not be considered as the lowest level attainable within a given indoor environment. A good example of this phenomenon is the study by Proctor et al. (1989b). In this study, urban outside air ranged from 108–624 μg/m^3 RSP, while measurements inside a nonsmoking betting shop ranged from 33-63 μg/m^3. Other examples are plentiful, (e.g., Spengler et al. 1981 1985, Quackenboss et al. 1989).

Perhaps more relevant to assessing the excess exposure to RSP resulting from ETS is consideration of the "background" as the RSP level present in nonsmoking situations. However, the criteria used to determine a nonsmoking condition vary considerably from study to study. For example, many investigators use the absence of observable smoking during the sampling period, while others (Kirk et al. 1988) have used the criteria of no smoking for 2 h prior to the start of sample acquisition. Depending on ambient ventilation and particle settling rates, it is conceivable that some ETS-derived RSP could still be airborne after a period of 2 h. Turk et al.(1987) used the criteria of no smoking within 30 ft of the sampling system. Jenkins et al. (1996) have specified more detailed criteria. In this study of personal exposure, nonsmokers had to report themselves as living or working in nonsmoking environments, and then confirm the absence of smoking through lack of sight or smell of tobacco smoke on a diary maintained hourly during their time at work or away from work. However, given the degree of air current movement within large open areas in public buildings, and mixing of nonfiltered air from smoking zones into nonsmoking zones, it may be that such a criterion is not sufficiently stringent to exclude ETS RSP from the sampling systems.

In most, but not all, of the studies in which nonsmoking, or control, RSP levels have been reported, values less than 100 μg/m^3 are the rule. For example, Repace and Lowrey (1980) reported levels of RSP all less than 60 μg/m^3 in a variety of situations. Leaderer and Hammond (1991) reported area RSP levels in nonsmoking residences ranging up to 30 μg/m^3. Heavner et al. (1996) reported personal exposure of workers in nonsmoking workplaces to RSP up to 98 μg/m^3, with a mean of 30 μg/m^3. However, there are a few exceptions to this general rule. Proctor et al. (1989) reported means for 5 replicate gravimetric measurements for RSP in 2 offices in which no smokers were housed as 16 and 118 μg/m^3, with individual levels as high as 208 μg/m^3. Phillips et al. (1998, Beijing) reported personal exposure to RSP (mean and 90th percentile) at 84 and 161 μg/m^3, respectively, for housewives living in nonsmoking homes in Beijing. Zhao and YuFeng (1990) reported levels of RSP in rural homes heated with coal stoves to be as high as 4550 μg/m^3. In an earlier study of ETS levels in subway trains before and after a ban on smoking (Proctor 1987), measurements of RSP levels in nonsmoking compartments gave RSP concentrations of 180 μg/m^3. This finding was supported in a later study (Proctor 1989) of aboveground trains, in which the mean level of RSP in nonsmoking compartments was shown to be 186 μg/m3. Probably the major exception to the general trend of < 100 μg/m^3 RSP for nonsmoking situations is the very extensive study performed by Kirk et al. (1988). In this study, they examined more than 2800 locations in the U.K., including work, home, travel, and leisure (see Table 6.3). More than half of these situations were nonsmoking. All of the reported means are greater than 250 μg/m^3. The measurements were made with an optical device (the MINIRAM), and while no detection limits are reported, the lowest non-zero level reported in the study was 70 μg/m^3 RSP. This, plus the very high levels reported throughout the study, suggests that either the areas sampled were unusually high in the levels of RSP observed, or that the instrumental monitor was not properly calibrated against the ambient aerosols present. However, in a follow-up sub-study, comparisons between gravimetrically and optically determined RSP levels were generally within a factor of two, the gravimetric measurements always being lower. This suggests that the data in Table 6.3 may be artificially high, by at least a factor of two, but likely greater. Thus, the data should be evaluated in a relative, rather than an absolute, manner.

The relative results of the study, and that of the other studies reported in Appendix 1, illustrate probably the most important observation regarding RSP levels determined for smoking and nonsmoking situations. That is, regardless of the particular method employed, levels of RSP determined in the presence of ETS are virtually always higher than those determined for nonsmoking situations. Typically, the differences range from small (i.e., <10% relative) up to a factor of 3, larger. The results of Miesner et al. (1989) are representative. Hospital public areas in which no smoking was allowed averaged 17 μg/m^3 RSP, whereas smoking areas in the same hospitals averaged 36 μg/m^3. Spengler et al. (1981) found that 35 nonsmoking residences averaged 24 μg/m^3 RSP, while homes with 1 smoker averaged 36 μg/m^3, and homes with

Table 6.3 RSP Levels in the United Kingdom[a,b] (μg/m^3)

Activity	No. Data Points	Mean	Standard Deviation	Minimum	Maximum
Travel—smoking	297	790	750	0	4980
Travel—nonsmoking	241	420	360	70	1830
Work—smoking	224	610	590	70	5780
Work—nonsmoking	480	310	260	0	2200
Home—smoking	156	700	520	70	3150
Home—nonsmoking	592	270	230	0	2050
Leisure—smoking	703	910	850	70	6220
Leisure—nonsmoking	108	330	260	70	1240
Total—smoking	1380	810	770	0	6220
Total—nonsmoking	1421	310	270	0	2200

[a]Data from Kirk et al. 1988. Levels determined using a MiniRam optical particle monitor.
[b]Note that levels are much higher than those reported by other investigators in comparable situations. Thus, absolute levels should be considered suspect.

2 smokers averaged 70 $\mu g/m^3$. Ott et al. (1996) reported that over the course of 2 years of measurements, average RSP levels in one tavern decreased from 83 $\mu g/m^3$ to 26 $\mu g/m^3$ when smoking was banned in the facility. Heavner et al. (1996), in a study of personal exposures to RSP in the workplace, reported that levels for 28 workers in smoking environments averaged 67 $\mu g/m^3$, while levels for 52 workers in nonsmoking environments average 30 $\mu g/m^3$.

Based on the existing data sets, the extent to which the differences in RSP for smoking vs. nonsmoking environments can be completely ascribed to ETS is unclear. When the differences are small, a large fraction of the difference can be accounted for in the amount of ETS-specific particulate matter (Sol-PM). For example, Phillips et al. (1997) reported on personal exposures to Sol-PM and RSP in smoking and nonsmoking workplaces in Turin, Italy. Geometric mean values of RSP that subjects encountered were 72 $\mu g/m^3$ and 60 $\mu g/m^3$, respectively. The geometric mean values for Sol-PM were 7.8 $\mu g/m^3$ and 1.3 $\mu g/m^3$, respectively. The difference in Sol-PM levels, 6.5 $\mu g/m^3$, accounts for slightly more than half the RSP difference of 12 $\mu g/m^3$. (That there was quantifiable Sol-PM levels in supposedly nonsmoking workplaces speaks to the challenges in establishing and maintaining criteria for defining the differences in environments.) In other cases, a much smaller fraction of the RSP difference can be ascribed to ETS. For example, the same research group has reported similar data for workers in Bremen, Germany (Phillips et al. 1998). Personal exposure, geometric-mean levels of RSP in smoking and nonsmoking workplaces were 36 $\mu g/m^3$ and 24 $\mu g/m^3$, respectively. However, the Sol-PM levels in these environments were low: 1.3 $\mu g/m^3$ and 0.55 $\mu g/m^3$, respectively. The difference in Sol-PM levels, 0.75 $\mu g/m^3$, accounts for only 6% of the RSP difference between the two environments. There are likely to be several explanations for this phenomenon. In many cases, especially in many developed societies, overall RSP and ETS-derived RSP (Sol-PM) are low and approach the limits of quantification. The challenge of accurately comparing results from two totally different analytical procedures (RSP: gravimetric; Sol-PM: chromatographic), when both species are at or near detection limits is considerable. Most importantly, there are differences between the two types of environments that can transcend the mere presence or absence of tobacco smoking. These factors can include differences in activities within a facility, the amount of carpeting and office equipment, differences in ventilation systems, the location relative to traffic sources and industrial emissions, etc. All of these can contribute to differences in particle levels in the indoor environment. For subjects for which personal exposure is measured, demographic differences impact behaviors in all environments, which in turn contribute to differences in RSP exposure. In summary, it is clear that important differences in RSP concentrations exist between smoking and nonsmoking environments. However, the extent to which ETS contributes to those differences probably ranges from completely to a small fraction.

Given that there are many other sources of particulate matter in addition to ETS, and that removal and infiltration rates, as well as ventilation patterns, vary from room to room, it is not surprising that there may be little direct relationship between the number of cigarettes smoked in many public unmanipulated field environments and RSP level. For example, for a large multi-city study of ETS levels in restaurants and offices (Oldaker et al. 1990, Oldaker et al. 1990a, Oldaker et al. 1991, Crouse and Oldaker 1990, Crouse and Carson 1989, Oldaker and McBride 1988), a regression of RSP level on smoking intensity (number of cigarettes smoked per cubic meter per hour), yielded coefficients of determination $(R^2) = 0.07447$ and 0.0009325 for the restaurants and offices, respectively.

Table 6.4 summarizes comparisons of geometric and arithmetic means for those studies for which such data were available. In all the reported studies, the geometric mean is less than the arithmetic mean. In studies for which individual data points were available, it was clear that the values were distributed approximately log-normally, rather than in a Gaussian fashion.

Figure 6.1 illustrates a typical distribution pattern. This summarizes RSP data, all of which were acquired by the same method (gravimetric), collected in restaurants in six cities (Oldaker et al. 1990, Oldaker et al. 1990a, Oldaker et al. 1991, Crouse and Oldaker 1990, Crouse and Carson 1989, Oldaker and McBride 1988). (Additional, unpublished data from these studies were obtained from the investigators through the Center for Indoor Air Research, and evaluated independently by the authors of this monograph.) While the RSP levels are reported up to 840 $\mu g/m^3$, the bulk of the data is represented by values less than 240 $\mu g/m^3$. The geometric mean for the data set is about 122 $\mu g/m^3$, while the arithmetic mean is 157 $\mu g/m^3$. Because the arithmetic means (and to a lesser extent, the geometric means) can be markedly affected by a few extreme values, investigators are increasingly reporting median (50th percentile) and graphically portraying data in cumulative distribution plots [e.g., Hammond et al. 1995, Jenkins et al. 1996, Phillips et al. 1997 (Turin)].

Figure 6.2 presents such a plot for 8-h TWA RSP concentrations determined from personal exposure measurements in smoking and nonsmoking workplaces (data extracted from Jenkins et al. 1996). (Note that smoking workplaces included those for which some restrictions on smoking were reported.) The plots indicate that the overall 2- to 3-fold difference between smoking and nonsmoking workplace RSP levels is maintained across most of the distribution. However, it is also clear that there is considerable overlap in the levels encountered. For example, 12.5% of the subjects in nonsmoking workplaces encountered RSP levels higher than 50% of the subjects in smoking workplaces.

Because of the differences in sampling and analysis methodologies, as well as sampling intervals and societies in which the studies were conducted, direct quantitative comparisons among various studies are difficult. Nevertheless, in Table 6.5 are summarized the findings of the major field

Table 6.4 Respirable Suspended Particulate Matter ($\mu g/m^3$)
Comparison of Arithmetic and Geometric Means for Selected Studies

Location/Conditions	Reference	Reported Range	Arithmetic Mean	Geometric Mean
165 offices in 5 cities	[a]	0–115	157	122.3
264 restaurants in 6 cities	[b]	0–843	145	113.7
37 restaurants	Crouse et al. 1988	16–221	80.8	62.0
224 homes in Onandaga City	Perritt et al. 1990	0.72–172	36.7	25.7
209 homes in Suffolk City	Perritt et al. 1990	2.18–284	46.4	35.9
70 smoking locations in 40 buildings	Turk et al. 1987	< 5–308	70	44
73 workers in smoking workplaces (personal exposure)	Phillips et al.1998 (Kuala Lumpur)	23–88[c]	52	43
63 workers in smoking workplaces (personal exposure)	Phillips et al. 1998 (Bremen)	15–105[c]	46	36
60 workers in smoking workplaces (personal exposure)	Phillips et al. 1997 (Barcelona)	47–197[c]	112	94
71 workers in smoking workplaces (personal exposure)	Phillips et al. 1997 (Turin)	20–172[c]	92	72
104 workers in smoking workplaces (personal exposure)	Phillips et al. 1998 (Paris)	16–127[c]	71	56

[a]Data compiled from the following references: Oldaker et al. 1990; Oldaker & McBride, 1988; Crouse & Carson, 1989.
[b]Data compiled from the following references: Oldaker et al. 1990, 1990a, 1991; Crouse & Oldaker, 1990; Crouse & Carson 1989.
[c]Reported range is actually the interdecile interval, from the 10th to the 90th percentile.

Table 6.5 Major Field Studies Indoor Particulate Matter Concentrations ($\mu g/m^3$).

			Smoking		Controls			Method	
	Location/Condition	References	Mean	Range	Mean	Range	Measure	Type	Interval
Residences Area Sampling									
19	Residences/nonsmoking	Nitschke et al. 1985			26	6–88	RSP	G	168 h
15	Residences/1 smoking occup.	Spengler et al. 1981	36				RSP	G	24 h
35	Residences/nonsmoking	Spengler et al. 1981			24		RSP	G	24 h
28	Residences/smoking	Spengler et al. 1981	74				RSP	G	24 h
73	Residences/smoking	Spengler et al. 1981			28		RSP	G	24 h
23	Homes	Mumford et al. 1989	74.1		28.4		RSP	G	24 h
224	Homes—Onandaga City	Perritt et al. 1990	61.3	0.7–172	18.1		RSP	G	NS
209	Homes—Suffolk City	Perritt et al. 1990	70.1	2.2–284	24.9		RSP	G	NS
15	Residences/Zigui area	Zhao and YuFeng 1990			1120	110–2230	RSP	G	8 h
16	Residences/WuShan area	Zhao and YuFeng 1990			1810	610–4550	RSP	G	8 h
98	Homes Netherlands	Lebret et al. 1990	191[c]		41[c]		RSP	G	1 week
54	Homes Netherlands	Lebret et al. 1990					RSP	G	1 week
43	Residences/nonsmoking	Quackenboss et al. 1989			30.3		RSP (PM₁₀)	G	NS
27	Residences/smoking ≤20 cig./day	Quackenboss et al. 1989		46.2			RSP (PM₁₀)	G	NS
18	Residences/smoking ≤20 cig./day	Quackenboss et al. 1989		75			RSP (PM₁₀)	G	NS
93	Residences western India	Ramakrishna et al. 1989			3900		TSP	G	39–51 min
56	Residences northern India	Ramakrishna et al. 1989			3100		TSP	G	46–53 min
60	Residences southern India	Ramakrishna et al. 1989			2600		TSP	G	58–68 min
19	Residences/nonsmoking	Nitschke et al. 1985			26	6–88	RSP	G	168 h
15	Residences/1 smoker	Spengler et al. 1981	36	–			RSP	G	24 h
35	Residences/nonsmoking	Spengler et al. 1981			24	–	RSP	G	24 h
28	Residences/smoking	Spengler et al. 1985	74	–			RSP	G	24 h
73	Residences/nonsmoking	Spengler et al. 1985			28	–	RSP	G	24 h

Table 6.5 continued

	Location/Condition	References	Smoking		Controls			Method	
			Mean	Range	Mean	Range	Measure	Type	Interval
23	Homes	Mumford et al. 1989	74.1		28.4		RSP	G	14 h
15	Residences/ Zigui area	Zhao & YuFeng 1990			1120	110–2230	RSP	G	8 h
16	Residences/ WuShan area	Zhao & YuFeng 1990			1810	610–4550	RSP	G	8 h
98	Homes, Netherlands	Lebret et al. 1990			41[a]		RSP	G	1 w
54	Homes, Netherlands	Lebret et al. 1990	191[a]				RSP	G	1 w
47	Smoking homes	Leaderer and Hammond 1991	44	10–150			RSP	G	1 w
49	Nonsmoking homes	Leaderer and Hammond 1991			15	5–30	RSP	G	1 w
	Activity rooms of 580 children in smoking homes	Neas et al. 1994	48.5	5–115			PM_{25}	G	4 w
	Activity rooms of 470 children in nonsmoking homes	Neas et al. 1994			17.3	5–65	PM_{25}	G	4 w
224	Homes/Onandaga City	Perritt et al. 1990	61.3	0.7–172	18.1		RSP	G	NS
209	Homes/ Suffolk City	Perritt et al. 1990	70.1	2.2–284	24.9		RSP	G	NS
43	Residences/nonsmoking	Quackenboss et al. 1989			30.3		RSP[b]	G	NS
27	Residences/smoking ≤20 cigarettes/day	Quackenboss et al. 1989	46.2				RSP[b]	G	NS
18	Residences/smoking >20 cigarettes/day	Quackenboss et al. 1989	75.0				RSP[b]	G	NS

Residences – Personal Exposure

n	Group	Reference	Mean	Range / 95th	Type	Method	Duration
306	Workers in smoking homes	Jenkins et al. 1996	44.1	125[c]	RSP	G	15 h
899	Workers in nonsmoking homes	Jenkins et al. 1996	19.7	46.8[c]	RSP	G	15 h
8	Workers in smoking homes	Phillips et al. 1998 (France)	66	112[d]	RSP	G	15 h
67	Workers in nonsmoking homes	Phillips et al. 1998 (France)	36	66[d]	RSP	G	15 h
51	Housepersons in smoking homes	Phillips et al. 1998 (France)	71	130[d]	RSP	G	24 h
30	Workers in smoking homes	Phillips et al. 1997 (Italy)	76	135[d]	RSP	G	15 h
75	Workers in nonsmoking homes	Phillips et al. 1997 (Italy)	52	81[d]	RSP	G	24 h
43	Housepersons in nonsmoking homes	Phillips et al. 1998 (France)	44	84[d]	RSP	G	24 h
36	Housepersons in smoking homes	Phillips et al. 1997 (Italy)	83	140[d]	RSP	G	24 h
47	Housepersons in nonsmoking homes	Phillips et al. 1997 (Italy)	55	81[d]	RSP	G	24 h
28	Workers in smoking homes	Phillips et al. 1997 (Spain)	95	160[d]	RSP	G	15 h
40	Workers in nonsmoking homes	Phillips et al. 1997 (Spain)	59	105[d]	RSP	G	15 h
43	Housepersons in smoking homes	Phillips et al. 1997 (Spain)	82	155[d]	RSP	G	24 h
40	Housepersons in nonsmoking homes	Phillips et al. 1997 (Spain)	63	90[d]	RSP	G	24 h
9	Housepersons in smoking homes	Phillips et al. 1996 (Sweden)	39	15 – 154	RSP	G	24 h
31	Housepersons in nonsmoking homes	Phillips et al. 1996 (Sweden)	18	8.2 – 58	RSP	G	24 h
38	Workers in smoking homes	Phillips et al. 1999 (Switzerland)	46	85[d]	RSP	G	15 h
66	Workers in nonsmoking homes	Phillips et al. 1999 (Switzerland)	48	55[d]	RSP	G	15 h
25	Housepersons in smoking homes	Phillips et al. 1999 (Switzerland)	60	88[d]	RSP	G	24 h
58	Housepersons in nonsmoking homes	Phillips et al. 1999 (Switzerland)	31	49[d]	RSP	G	24 h
30	Workers in smoking homes	Phillips et al. 1998 (Australia)	37	70[d]	RSP	G	24 h
43	Workers in nonsmoking homes	Phillips et al. 1998 (Australia)	31	41[d]	RSP	G	24 h
23	Workers in smoking homes	Phillips et al. 1998 (Germany)	44	89[d]	RSP	G	15 h

Table 6.5 continued

	Location/Condition	References	Smoking Mean	Smoking Range	Controls Mean	Controls Range	Measure	Method Type	Method Interval
81	Workers in nonsmoking homes	Phillips et al. 1998 (Germany)			25	39[d]	RSP	G	15 h
21	Housepersons in smoking homes	Phillips et al. 1998 (Germany)	39	63[d]			RSP	G	24 h
58	Housepersons in nonsmoking homes	Phillips et al. 1998 (Germany)			27	37[d]	RSP	G	24 h
70	Workers in smoking homes	Phillips et al. 1998 (Czech Republic)	61	105[d]			RSP	G	15 h
67	Workers in nonsmoking homes	Phillips et al. 1998 (Czech Republic)			32	55[d]	RSP	G	15 h
50	Housepersons in smoking homes	Phillips et al. 1998 (Czech Republic)	61	112[d]			RSP	G	24 h
36	Housepersons in nonsmoking homes	Phillips et al. 1998 (Czech Republic)			35	55[d]	RSP	G	24 h
56	Housepersons in smoking homes	Phillips et al. 1998 (China)	123	221[d]			RSP	G	24 h
45	Housepersons in nonsmoking homes	Phillips et al. 1998 (China)			84	161[d]	RSP	G	24 h
59	Workers in smoking homes	Phillips et al. 1998 (Malaysia)	53	83[d]			RSP	G	15 h
88	Workers in nonsmoking homes	Phillips et al. 1998 (Malaysia)			59	87[d]	RSP	G	15 h
40	Housepersons in smoking homes	Phillips et al. 1998 (Malaysia)	83	89[d]			RSP	G	24 h
51	Housepersons in nonsmoking homes	Phillips et al. 1998 (Malaysia)			53	89[d]	RSP	G	24 h
50	Workers in smoking homes	Phillips et al. 1998 (Hong Kong)	61	96[d]			RSP	G	15h
74	Workers in nonsmoking homes	Phillips et al. 1998 (Hong Kong)			54	96[d]	RSP	G	15h
35	Housepersons in smoking homes	Phillips et al. 1998 (Hong Kong)	49	77[d]			RSP	G	15h
34	Housepersons in nonsmoking homes	Phillips et al. 1998 (Hong Kong)			47	70[d]	RSP	G	24h
22	Housepersons in smoking homes	Phillips et al. 1998 (Portugal)	40	67[d]			RSP	G	24h
56	Housepersons in nonsmoking homes	Phillips et al. 1998 (Portugal)			38	54[d]	RSP	G	24h
Residences—Personal Exposure									
255	Nonsmokers	Phillips et al. 1994 (UK)	179	89–420			RSP	G	24h
29	Workers in smoking homes	Heavner et al. 1996	89	11.8–825			RSP	G	15h
58	Workers in nonsmoking homes	Heavner et a. 1996			28	8 – 100	RSP	G	15h

Offices									
31	Offices/smoking	Carson & Erikson 1988	44	6–426			RSP	UV	59–84 min
30	Offices/smoking	Crouse & Carson 1989	61	11–279			RSP	G	1h
17	Offices/mechanically ventilated	Turner & Binnie 1990	63	15–357			RSP	P	NS
131	Offices	Oldaker et al. 1990	126	0–1088			RSP	G	1h
22	Offices/smoking occupants	Sterling & Sterling 1984		32			TSP		
44	Offices/smoking occupants	Weber & Fischer 1980	133[e]				RSP	P	2 min
194	Offices	Healthy Buildings Intl. 1990	28	6–180			RSP	G	1h
70	Offices in four buildings	Oldaker et al. 1995	5[f]–34[f]	<7–74			RSP	G	7 h
Work Situations Personal Exposure									
28	Workers in smoking workplaces	Heavner et al. 1996	67	18–217			RSP	G	9 h
52	Workers in nonsmoking workplaces	Heavner et al. 1996			30	0–98	RSP	G	9 h
15	Workers in smoking workplaces	Coultas et al. 1990	64	4–146			RSP	G	2–8 h
82	Waiters	Jenkins and Counts 1999	109	386[c]			RSP	G	4–8 h
80	Bartenders	Jenkins and Counts 1999	151	428[c]			RSP	G	4–8 h
Public Buildings									
27	Public buildings/smoking occupants	Repace & Lowrey 1980	278	86–1140			RSP	P	2 min
		Repace & Lowrey 1982		<50–308					
40	Public buildings	Grimsrud et al. 1990	44[a]	<5–308			RSP	G	NS
70	Smoking locations in 40 buildings	Turk et al. 1987			24[a]	5–63	RSP	G	75–100 h
106	Nonsmoking venues in 40 buildings	Turk et al. 1987			15[a]	5–63	RSP	G	75–100 h
30	Outside sites near 40 buildings	Turk et al. 1987			14[a]	<5–68	RSP	G	75–100 h
45	Public buildings	Health Buildings Intl. 1990	36	6–284			RSP	G	1h

Table 6.5 continued

Location/Condition	References	Smoking Mean	Smoking Range	Controls Mean	Controls Range	Measure	Method Type	Method Interval
Restaurants								
83 Restaurants	Oldaker et al. 1990	126	0–685			RSP	G	1 h
30 Restaurants/smoking	Crouse & Carson 1989	111[a]	16–366			RSP	G	1 h
37 Restaurants	Crouse et al. 1988	62.0[a]	16–231			RSP	G	1 h
41 Restaurants	Ogden et al. 1990	106				RSP	G	1 h
62 Restaurants	Healthy Buildings Intl. 1990	53	10–228			RSP	G	1 h
Transportation								
28 Boeing 747smoking/nonsmoking	Oldaker et al. 1990	39	3–185	15	3–98	RSP	G	3–12 h
29 Boeing 747smoking/smoking	Oldaker et al. 1990	75.8	?–883	34.8		RSP	G	3–12 h
92 Aircraft	Nagda et al. 1990					RSP	O,G	NS
Other Work Situations								
286 Work situations	Healthy Buildings Intl. 1990	36	6–519			RSP	G	1 h

[a] Geometric Mean.
[b] PM₁₀.

O: Optical
P: Piezoelectric balance.
RSP: Respirable suspended particulates.
TSP: Total suspended particulates.
UV: Ultraviolet absorbing particulate matter
w: week/s

NS: Not Specified.

[c] 95th percentile.
[d] 90th percentile.
e Mean of means.
[f] Included in public buildings as well.
G: Gravimetric methods.

studies (those with 15 or more data points) in which RSP was determined in conjunction with ETS levels. These data are excerpted from Appendix 1 (with the exception of the study by Kirk et al. (1988) due to the aforementioned issues), and permit some general conclusions to be drawn. (Note: to avoid repetition, data reported in Table 6.2 are not repeated in Table 6.5.) For private residences, while some concentrations of RSP may be as high as 200– 300 μg/m^3, average levels of RSP in smoking households appear to be 40–70 μg/m^3, with levels in nonsmoking households about half or less of this value. In those studies that report such, medians or geometric means are almost always less than the arithmetic means, indicating that the distribution of those data is somewhat skewed. For personal exposure in residential settings, the average RSP concentrations encountered by subjects are not importantly different from the area measurements. Only one study (Phillips et al. 1998, Beijing) reported a mean personal exposure level in private residences greater than 100 μg/m^3.

In offices, the range of RSP concentrations obtained from area samples is greater, with some studies exhibiting means of greater than 100 μg/m^3. Levels as high as several hundred μg/m^3 were noted. In contrast, for personal exposure monitoring studies in nonhospitality smoking workplaces (see both Tables 6.2 and 6.5), only one study in Spain (Phillips et al. 1997, Barcelona) reported average RSP concentrations above 100 μg/m^3. Lower concentrations in personal exposure studies suggest that workers may limit—deliberately or otherwise—their time or exposure in smoking environments.

In areas sampled in public buildings, the data sets are more limited, but in 2 of the 3 major studies, RSP levels averaged less than 50 μg/m^3 where smoking was observed. One contributor to this may be the large volumes in public areas (bank lobbies, post offices, etc.) in which the ETS can be dispersed. Another contributor may be more sophisticated ventilation/filtration systems. RSP levels in restaurants and taverns are higher; several of the studies reported levels above 100 μg/m^3. The higher concentrations may be due in part to the cooking plus the high level of activity in such locations. In aircraft, average RSP levels ranged from 35–76 μg/m^3. As with other studies, nonsmoking RSP levels in the same studies were about half those of the smoking levels. It is important to note that in many of the smaller studies listed in Appendix 1, RSP levels seem to be higher than those of the studies that include a larger number of locations. Many factors may enter into this finding. First, the smaller studies are more susceptible to the impact of higher levels of RSP in 2 or 3 locations than the larger studies. Also, in selecting only a few sampling locales to represent a given venue, there may be a subconscious trend on the part of the selection team to sample only the "smokier" places. In terms of typical human exposures, again, it is difficult to draw firm conclusions because of the differences in sampling and analysis methods which can often yield different results for the same atmosphere. However, it is probably safe to conclude that typical concentrations encountered where ETS is present

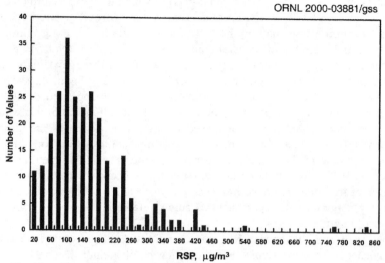

Fig. 6.1. Distribution of respirable suspended particulate matter concentrations in restaurants in six cities. (Data combined from several studies; see discussion on Fig. 6.1 on previous page.)

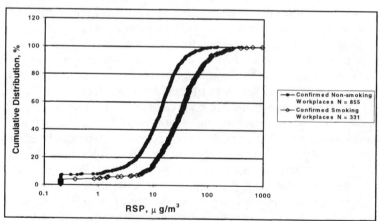

Fig. 6.2. Cumulative distribution of personal exposure to RSP for workers in confirmed smoking and nonsmoking environments. Data extracted from original data of 16 U.S. Cities Study (Jenkins et al. 1996) and processed for this monograph. Note that 59% of subjects in smoking workplaces reported some degree of smoking restriction. (Note that to be included in this compilation, subjects had to report in their diaries observation of smoking practices consistent with their responses regarding the smoking status of their workplace environment.)

would be up to about 100 $\mu g/m^3$, measured by gravimetric means. From 100–300 $\mu g/m^3$ should be considered high, and probably above 300 $\mu g/m^3$ constitutes an extreme. Indoor air levels of RSP, where no smoking is occurring, are about half the level found in smoking situations.

SUMMARY

Indoor air RSP can exist in many forms; that resulting from ETS seems likely to be present in the form of liquid or waxy droplets, most of which are much smaller than 1 μm mass median diameter. There are both instrumental and manual means for determining RSP levels in indoor air. However, the results from different systems can only be considered comparable if the instrumental systems have been carefully calibrated with the airborne material of interest. Chemical means for estimating the contribution of ETS to RSP have been evaluated, and suggest that ETS RSP may comprise from 6–70% of indoor air RSP, in the field scenarios to which the methods have been applied.

RSP levels have been reported from >1000 $\mu g/m^3$ down to below detection limits. The latter are usually about 20–25 $\mu g/m^3$ for instrumental systems, and about 10–20 $\mu g/m^3$ for manual methods, depending on the sampling duration. Background levels of RSP depend on numerous factors, including local vehicular traffic patterns, quality of ventilation systems, and the presence of other sources, such as cooking or wood-burning stoves. Usually these levels are typically in the range of a few tens of $\mu g/m^3$. Because of differences in sampling and analysis methods, direct absolute comparisons between studies using different measurement techniques are difficult. Study comparisons between smoking and nonsmoking locations inevitably reveal higher levels of RSP in smoking areas. The differences can range from small to quite large, but typically, smoking RSP levels are a factor of 1.5–3 greater. The extent to which ETS-derived RSP contributes to the difference is highly variable. The published literature suggests that large, public buildings, due to volume and ventilation, are likely to have lower levels of RSP, while hospitality venues have the highest. Residential levels above 100 $\mu g/m^3$ are observed, but more typical levels, especially averaged over several hours to days, are in the range of 30 – 60 $\mu g/m^3$. Extreme concentrations of RSP, above 300 $\mu g/m^3$, are only rarely encountered. Thus, it would appear that all but the most uncommon exposures would not exceed government standards for RSP levels in the workplace. OSHA has established a personal exposure limit of 5000 $\mu g/m^3$ (OSHA 1974). However, it does appear that EPA-proposed guidelines for outdoor ambient air fine particles (EPA 1997) are frequently exceeded in indoor situations. The EPA limit is 65 $\mu g/m^3$ $PM_{2.5}$ for 24 h. Thus, it would appear that such a level is frequently exceeded in indoor situations, with or without the presence of ETS. Nevertheless, levels > 65 $\mu g/m^3$ are more likely to be attained when smoking is occurring.

REFERENCES

Benner, C. L., Bayona, J. M., Caka, F. M., Tang, H., Lewis, L., Crawford, J., Lamb, J. D., Lee, M. L., Lewis, E. A., Hansen, L. D., & Eatough, D. J. (1989) Chemical composition of environmental tobacco smoke. 2. Particulate-phase compounds. *Environ. Sci. Technol.,* 23(6), 688–699.

Benton, G., Miller, D. P., Reimold, M., & Sisson, R. (1981) A study of occupant exposure to particulates and gases from woodstoves in homes. *Proceedings, 1981 International Conference on Residential Solid Fuels—Environmental Impacts and Solutions,* Cooper, J. A. & Malek, D., eds., Portland, OR, June 1–4, 1981.

Brunekreef, B. & Boleij, J. S. M. (1982) Long-term average suspended particulate concentrations in smokers' homes. *Int. Arch. Occup. Environ. Health,* 50, 299–302.

Burton, R. M., Scila, R. A, Wilson, W. E., Pahl, D. A., Mumford, J. L. & Koutrakis, P. (1990) Characterization of kerosene heater emissions inside two mobile homes. *Proceedings of the 5th International Conference on Indoor Air Quality and Climate, Vol. 2,* Toronto, Canada, pp. 337–342.

Carson, J. R. & Erickson, C. A. (1988) Results from survey of environmental tobacco smoke in offices in Ottawa, Ontario. *Environ. Toxicol. Letts.,* 9, 501–508.

Colome, S. D. & Spengler, J. D. (1981) Residential indoor and matched outdoor pollutant measurements with special consideration of wood-burning homes. *Proceedings, 1981 International Conference on Residential Solid Fuels—Environmental Impacts and Solutions,* Cooper, J. A. & Malek, D., eds., Portland, OR, pp. 435–455.

Colome, S. D., Kado, N. Y., Jacques, P., & Kleinman, M. (1990) Indoor-outdoor relationship of particles less than 10 ~m in aerodynamic diameter (PM 10) in homes of asthmatics, *Proceedings of the 5th International Conference on Indoor Air Quality and Climate, Vol. 2,* Toronto, Canada, pp. 275–280.

Conner, J. M., Oldaker, G. B., III, & Murphy, J. J. (1990) Method for assessing the contribution of environmental tobacco smoke to respirable suspended particles in indoor environments. *Environ. Technol.,* 11, 189–196.

Coultas, D. B., Samet, J. M., McCarthy, J. F., & Spengler, J. D. (1990) A personal monitoring study to assess workplace exposure to environmental tobacco smoke. *Am. J. Public Health,* 80, 988–990.

Crouse, W. E., Ireland, M. S., Johnson, J. M., Striegel, R. M., Jr., Williard, C. S., DePinto, R. M., Oldaker, G. B., III, & McBride, R. L. (1988) Results from a survey of environmental tobacco smoke (ETS) in restaurants. *Transactions of an International Specialty Conference—Combustion Processes and the Quality of the Indoor Environment,* Harper, J. P., ed., Niagara Falls, NY, pp. 214–222.

Crouse, W. E. & Oldaker, G. B. (1990) Comparison of area and personal sampling methods for determining nicotine in environmental tobacco smoke. 1990 *EPA/ AWMA Conference on Toxic and Related Air Pollutants,* Raleigh, NC.

Crouse, W. E. & Carson, J. R. (1989) Surveys of environmental tobacco smoke (ETS) in Washington, D.C. offices and restaurants. *43rd Tobacco Chemists' Research Conference,* Richmond, VA.

Cuddeback, J. E., Donovan, J. R., & Burg, W. R. (1976) Occupational aspects of passive smoking. *Am. Ind. Hyg. Assoc. J.,* 263–267.

Elliott, L. P. & Rowe, D. R. (1975) Air quality during public gatherings. *JAPCA,* 25(6), 635–636.

(EPA) Environmental Protection Agency (1990) *Federal Register,* 55 (209), 43406–43407.

Environmental Protection Agency (1997) EPA's Revised particulate Matter Standards http://ttnwww.rtpnc.epa.gov/naaqsfin.pmfact.htm

First, M. W. (1984) Environmental tobacco smoke measurement: retrospect and prospect. *Eur. J. Respir. Dis.,* 65 (Suppl.), 9–16.

Georghiou, P. E., Blagden, P. A., Snow, D. A., Winson, L., & Williams, D. T. (198~) Air levels and mutagenicity of PM-10 in an indoor ice arena. *JAPCA,* 39(12), 1583–1585.

Griest, W. H., Jenkins, R. A., Tomkins, B. A., Moneyhun, J. H., Ilgner, R. H. Gzy e. T. M., Higgins, C. E., & Guerin, M. R. (1988) *Sampling and Analysis of Diesel Engine Exhaust and the Motor Pool Workplace Atmosphere, Final Report.* ORNL/TM-10689.

Grimsrud, D. T., Turk, B. H., Prill, R. J., & Geisling-Sobotka, K. L. (1990) Pollutant concentrations in commercial buildings in the U.S. pacific northwest. *Proceedings of the 5th International Conference on Indoor Air Quality and Climate, Vol. 2* Toronto, Canada, pp. 483–488.

Harrison, R. M., & Perry, R., eds., (1986) *Handbook of Air Pollution Analysis, 2nd Edition,* Chapman and Hall, New York, NY, 634 pp.

Hawthorne, A. R., Gammage, R. B., Dudney, C. S., Hingerty, B. E., Schuresko. D. D., Parzyck, D. C., Womack, D. R., Morris, S. A., Westley, R. R., White. D. A . & Schrimsher, J. M. (1984) *An indoor air quality study of forty east Tennessee homes.* ORNL-5965.

Healthy Buildings International (1990) *Measurement of Environmental Tobacco Smoke in General Office Areas.*

Heavner, D. L., Morgan, W. T., & Ogden, M. W. (1996) Determination of volatile organic compounds and respirable suspended particulate matter in New Jersey and Pennsylvania homes and workplaces. *Environ. Int.,* 22, 159–183.

Hedge, A., Erickson, W. A., & Rubin G. (1994) The Effects of alternative smoking policies on indoor air quality in 27 smoking office buildings. *Ann. Occup. Hyg.,* 38, 265–278.

Heyder, J., Gebhard, J., & Stahlhofen, W. (1980) Inhalation of aerosols: particle deposition and retention. In: Willeke, K., ed., *Generation of Aerosols and Facilities for Exposure Experiments,* Ann Arbor Science Publishers, Ann Arbor, MI, pp. 65–103.

Husgafvel-Pursiainen, K., Sorsa, M., Moller, M., & Benestad, C. (1986) Genotoxicity and polynuclear aromatic hydrocarbon analysis of environmental tobacco smoke samples from restaurants. *Mutagenesis,* 1, 287–291.

Ingebrethsen, B. J., Heavner, D. L., Angel, A. L., Conner, J. M., Steichen, T. J., & Green, C. R. (1988) A comparative study of environmental tobacco smoke particulate mass measurements in an environmental chamber. *JAPCA,* 8(4), 413–417.

Jenkins, R. A., Palausky, A., Counts, R. W., Bayne, C. K., Dindal, A. B., & Guerin, M. R. (1996) Exposure to environmental tobacco smoke in sixteen cities in the United States as determined by personal breathing zone air sampling. *J. Expo. Anal. Environ. Epidemiol.,* 6, 473–502.

Kirk, P. W. W., Hunter, M., Baek, S. O., Lester, J. N., & Perry, R. (1988) Environmental tobacco smoke in indoor air. *Proceedings, Indoor Ambient Air Quality Conference,* London, 99–112.

Klepeis, N. E., Ott, W. R., & Switzer, P. (1996) A multiple-smoker model for predicting indoor air quality in public lounges. *Environ. Sci Technol.,* 30, 2813–2820.

Lambert, W. E., Sarnet, J. M., & Spengler, J. D. (1993) Environmental tobacco smoke concentrations in no-smoking and smoking sections of restaurants. *Am. J. Public Health,* 83(9), 1339–1341.

Leaderer, B. P. & Hammond, S. K. (1991) Evaluation of vapor-phase nicotine and respirable suspended particle mass as markers for environmental tobacco smoke. *Environ. Sci. Technol.,* 25, 770–777.

Lebret, E., Boleij, J., & Brunekreef, B. (1990) Environmental tobacco smoke in Dutch homes. *Proceedings of the 5th International Conference on Indoor Air Quality and Climate, Vol. 2,* Toronto, Canada, pp. 263–268.

Lippmann, M. (1983) Sampling aerosols by filtration. *Air Sampling Instruments for evaluation of atmospheric contaminants, 6th edition,* Lioy, P. J. & Lioy, M. J. Y., eds., Cincinnati, OH, pp. Pl–P30.

Löfroth, G., Burton, R. M., Forehand, L., Hammond, K. S., Seila, R. L., Zweidinger, R. B., & Lewtas, J. (1989). Characterization of environmental tobacco smoke. *Environ. Sci. Technol.,* 23, 610–614.

Marple, V. A. & Rubow, K. L. (1980) Aerosol generation concepts and parameters. In Willeke, K., ed., *Generation of Aerosols and Facilities for Exposure Experiments.* Ann Arbor Science, Publishers, pp. 3–29.

McAughey, J. J., Pritchard, J. N., & Strong, J. C. (1990) Respiratory deposition of environmental tobacco smoke. *Proceedings of the 5th International Conference on Indoor Air Quality and Climate, Vol. 2,* Toronto, Canada, pp. 361–366.

Miesner, E. A., Rudnick, S. N., Hu, F.-C., Spengler, J. D., Preller, L., Ozkaynak, H., & Nelson, W. (1989) Particulate and nicotine sampling in public facilities and offices. *JAPCA,* 39(12), 1577–1582.

Moore, W. J., ed. (1962), *Physical Chemistry Third Edition,* Prentice-Hall Inc., Englewood Cliffs, NJ, p. 761.

Morawska, L., Barron, W., & Hitchins, J. (1999) Experimental deposition of environmental tobacco smoke submicrometer particulate matter in the human respiratory tract. *Am. Ind. Hyg. Assoc. J.,* 60, 334–339.

Mumford, J. L., Lewtas, J., Burton, R. M., Henderson, F. W., Forehand, L., Allison. J. C., & Hammond, S. K. (1989) Assessing environmental tobacco smoke exposure of pre-school children in homes by monitoring air particles, mutagenicity, and nicotine. *1989 EPA/AWMA International Symposium on Measurement of Toxic and Related Air Pollutants,* Raleigh, NC, pp. 606–616.

Nagda, N., Fortmann, R., Koontz, M. & Konheim, A. (1990) Investigation of cabin air quality aboard commercial airliners. *Proceedings of the 5ᵗʰ International Conference on Indoor Air Quality and Climate, Vol. 2,* Toronto, Canada, pp. 245–250.

National Institute for Occupational Safety & Health [NIOSH] (1985) NIOSH/OSHA pocket guide to chemical hazards. *Environmental Tobacco Smoke—Measuring Exposures and Assessing Health Effects.* DHEW Publ. No. 85-14. Cincinnati, OH National Institute for Occupational Safety and Health, 241 pp.

Neas, L. M., Dockery, D. W., Ware, J. H., Spengler, J. D., Ferris, B. G., Jr., & Speizer, F. E. (1994) Concentration of indoor particulate matter as a determinant of respiratory health in children. *Am. J. Epidemiol.,* 139, 1088–1099.

Nelson, P. R., Conrad, F. W., Kelly, S. P., Maiolo, K. C., Richardson, J. D., & Ogden, M. W. (1997) Composition of environmental tobacco smoke (ETS) from international cigarettes and determination of ETS-RSP:particulate marker ratios. *Environ. Int.,* 23, 47–52.

Nelson, P. R., Conrad, F. W., Kelly, S. P., Maiolo, K. C., Richardson, J. D., & Ogden, M. W. (1998) Composition of Environmental tobacco smoke (ETS) from international cigarettes Part II: Nine country follow-up. *Environ. Int.,* 24, 251–257.

Nitschke, I. A., Clark, W. A., Clarkin, M. E., Traynor, G. W., & Wadach, J. B. (1985) *Indoor Air Quality, Infiltration, and Ventilation in Residential Buildings,* NYSERDA 85-10, Albany New York State Energy Research and Development Authority.

(OSHA) Occupational Safety and Health Administration (1974) *Code of Federal Regulations 29 CFR 1910.1000 Part Z.*

Ogden, M. W. Maiolo, K. C., Oldaker, G. B., 111, & Conrad, F. W., Jr. (1990) Evaluation of methods for estimating the contribution of ETS to respirable suspended particles. *Proceedings of the 5ᵗʰ International Conference on Indoor Air Quality and Climate, Vol. 2,* Toronto, Canada, pp. 483–488.

Ogden, M. W., Maiolo, K. C., Oldaker, G. B., III, & Conrad, F. W., Jr. (1989) Evaluation of methods for estimating the contribution of ETS to respirable suspended particles. *43rd Tobacco Chemists Research Conference,* Richmond, VA.

Ogden, M. W., Heavner, D. L., Foster, T. L., Maiolo, K. C. Cash, S. L., Richardson, J. D., Martin, P., Simmons, P. S., Conrad, F. W., & Nelson, P. R. (1996) Personal monitoring system for measuring environmental tobacco smoke exposure, *Environ. Technol.,* 17, 239–250.

Ogden, M. W. & Maiolo, K. C. (1989) Collection and determination of solanesol as a tracer of environmental tobacco smoke (ETS). *Proceedings, Indoor Ambient Air Quality Conference,* London, pp. 77–88.

Ogden, M. W. & Maiolo, K. C. (1989) Collection and determination of solanesol as a tracer of environmental tobacco smoke in indoor air. *Environ. Sci. Technol.* , 23, 1148–1154.

Ogden, M. W., Maiolo, K. C. (1992) Comparison of GC and LC for determining solanesol in environmental tobacco smoke. *LC-GC,* 10, 459–462

Oldaker, G. B., III, Perfetti, P. F., Conrad, F. W., Jr., Conner, J. M., & McBride, R. L. (1990) Results from surveys of environmental tobacco smoke in offices and restaurants. *Int. Arch. Occup. Environ. Health,* 99–104.

Oldaker, G.B., III & McBride, R. L., (1988), Portable air sampling system for surveying levels of environmental tobacco smoke in public places. *Symposium on Environment and Heritage. World Environment Day, Hong Kong, 1988,* Hong Kong University, Hong Kong, June 6, 1988.

Oldaker, G. B., III, Stancill, J. O., Conrad, F. W, Jr., Collie, B. B., Fenner, R. A. Lephardt, J. O., Baker, P. G., Lyons-Hart, J., & Parrish, M. E. (1990) Estimation of effect of environmental tobacco smoke on air quality within passenger cabins of commercial aircraft. In: Lunau, F. & Reynolds, G. L., eds., *Indoor Air Quality and Ventilation,* Selper Ltd., London, pp. 447–454.

Oldaker, G. B., Taylor, W. D., & Parrish, K. B. (1995) Investigations of ventilation rate, smoking activity and indoor air quality at four large office buildings. *Environ. Technol.,* 16, 173–180.

Oldaker, G. B., III, Stancill, J. O., Baker, P. G., Lyons-Hart, J., & Parrish, M. E. (1991) Results from a survey of environmental tobacco smoke in Hong Kong restaurants. *Unpublished work. Personal Communication.*

Oldaker, G. B., III, Ogden, M. W., Maiolo, K. C., Conner, J. M, Conrad, F. W., Jr. Stancill, M. W., & DeLuca, P. O. (1990a) Results from surveys of environmental tobacco smoke in restaurants in Winston-Salem, North Carolina. *Proceedings of the 5th International Conference on Indoor Air Quality and Climate, Vol. 2.* Toronto, Canada, pp. 281–285.

Ott, W. R., Switzer, P., & Robinson, J. (1996) Particle concentrations inside a tavern before and after prohibition of smoking: Evaluating the performance of an indoor air quality model. *J. Air & Waste Manage. Assoc.*, 46, 1120–1134.

Parker, G. B., Wilfert, G. L., & Dennis, G. W. (1984) Indoor air quality and infiltration in multifamily naval housing. *Annual PNWIS/APCA Conference, November 12–14,1984,* Portland, OR, pp. 1–14.

Patashnick, H. & Rupprecht, G. (1986) Microweighing goes on line in real time. *Research & Development,* June 1986.

Perritt, R. L., Hartwell, T. D., Sheldon, L. S., Cox, B. G., Smith, M. L., & Rizzuto, J. E. (1990) Distribution of NO2, CO, and respirable suspended particulates in New York state homes. *Proceedings of the International Conference on Indoor Air Quality and Climate, Vol. 2,* Toronto, Canada, pp. 251–256.

Phillips, K., Howard, D. A., Bentley, M. C., & Alvan, G. (1998) Assessment of environmental tobacco smoke and respirable suspended particle exposure of nonsmokers in Lisbon by personal monitoring. *Environ. Int.,* 24, 301–324.

Phillips, K., Bentley, M. C., Howard, D. A., Alvan, G., & Huici, A. (1997) Assessment of air quality in Barcelona by personal monitoring of nonsmokers for respirable suspended particles and environmental tobacco smoke. *Environ. Int.,* 23, 173–196.

Phillips, K., Howard, D. A., & Bentley, M. C. (1997) Assessment of air quality in Turin by personal monitoring of nonsmokers for respirable suspended particles and environmental tobacco smoke. *Environ. Int.,* 23, 851–871.

Phillips, K., Bentley, M. C., Howard, D. A., & Alvan, G. (1998) Assessment of air quality in Paris by personal monitoring of nonsmokers for respirable suspended particles and environmental tobacco smoke. *Environ. Int.,* 24, 405–425.

Phillips, K., Bentley, M. C., Howard, D., & Alvan, G. (1998) Assessment of environmental tobacco smoke and respirable suspended particle exposures for nonsmokers in Kuala Lumpur using personal monitoring. *J. Expo. Anal. Environ. Epidemiol.,* 8, 519–541.

Phillips, K., Howard, D. A., Bentley, M., & Alvan, G. (1999) Assessment of environmental tobacco smoke and respirable suspended particle exposures for nonsmokers in Basel by personal monitoring. *Atmos. Env.,* 33, 1889–1904.

Phillips, K., Bentley, M., Howard, D., & Alvan, G. (1998) Assessment of environmental tobacco smoke and respirable suspended particle exposures for nonsmokers in Prague using personal monitoring. *Int. Arch. Occup. Environ. Health,* 71, 379–390.

Phillips, K., Howard, D. A., Browne, D., & Lewsley, J. M. (1994) Assessment of personal exposures to environmental tobacco smoke in British nonsmokers. *Environ. Int.*, 20, 693–712.

Phillips, K., Howard, D. A., Bentley, M. C., and Alvan, G. (1998). Environmental tobacco smoke and respirable suspended particle exposures for non-smokers in Beijing. *Indoor Built Environ.*, 7, 254–269.

Phillips, K., Howard, D. A., Bentley, M. C., & Alvan, G. (1998) Assessment of environmental tobacco smoke and respirable suspended particle exposures for nonsmokers in Hong Kong using personal monitoring. *Environ. Int.*, 24, 851–870.

Phillips, K., Bentley, M. C., Howard, D. A., & Alvan, G. (1996) Assessment of air quality in Stockholm by personal monitoring of nonsmokers for respirable suspended particles and environmental tobacco smoke. *Scan. J. Work. Environ. Health*, 22, 1–24.

Phillips, K., Howard, D. A., Bentley, M., & Alvan, G. (1998) Measured exposures by personal monitoring for respirable suspended particles and environmental tobacco smoke of housewives and office workers resident in Bremen, Germany. *Int. Arch. Occup. Environ. Health*, 71, 201–212.

Phillips, K., Howard, D. A., Bentley, M. C., & Alvan, G. (1998) Assessment by personal monitoring of respirable suspended particles and environmental tobacco smoke exposure for non-smokers in Sydney, Australia. *Indoor Built Environ.*, 7, 188–203.

Proctor, C. J., Warren, N. D., & Bevan, M. A. J. (1989b) An investigation of the contribution of environmental tobacco smoke to the air in betting shops. *Environ. Technol. Letts*, 10, 333–338.

Proctor, C. (1987) A study of the atmosphere in London underground trains before and after the ban on smoking. *Toxicol. Letts.*, 35, 131–134.

Proctor, C. J., Warren, N. D., & Bevan, M. A. J. (1989) Measurements of environmental tobacco smoke in an air-conditioned office building. *Procs., Conference on the Present and Future of Indoor Air Quality*, Brussels.

Proctor, C. J. (1989) A comparison of the volatile organic compounds present in the air of real-world environments with and without environmental tobacco smoke. *82nd Annual Meeting of the Air and Waste Management Association*, Anaheim, CA.

Quackenboss, J. J., Lebowitz, M. D., & Crutchfield, C. D. (1989) Indoor-outdoor relationships for particulate matter: exposure classifications and health effects. *Environ. Int.*, 15, 353–360.

Quant, F. R., Nelson, P. A., & Sem, G. J. (1982) Experimental measurements of aerosol concentrations in offices. *Environ. Int.*, 8, 223–227.

Raabe, O. G. (1980) Physical properties of aerosols affecting inhalation toxicology, *Proceedings of the 19th Annual Hanford Life Sciences Symposium—Pulmonary Toxicology of Respirable Particles, Series 53*, Richland, WA, pp. 1–28.

Ramakrishna, J., Durgaprasad, M. B., & Smith, K. R. (1989) Cooking in India: the impact of improved stoves on indoor air quality. *Environ. Int., 15,* 341–352.

Repace, J. L. & Lowrey, A. H. (1980) Indoor air pollution, tobacco smoke, and public health. *Science,* 208, 464–472.

Repace, J. L. & Lowrey, A. H. (1980) Indoor air pollution, tobacco smoke, and public health. *Science,* 208, 464–472.

Repace, J. L. (1987) Indoor concentrations of environmental tobacco smoke: models dealing with effects of ventilation and room size. In: O'Neill, 1. K., Brunnemann, K. D., Dodet, B., & Hoffmann, D., eds., *Environmental Carcinogens Methods of Analysis, Vol. 9 (IARC Scientific Publications No. 81),* International Agency for Research on Cancer, Lyon, pp. 25–41.

Repace, J. L. & Lowrey, A. H. (1982) Tobacco smoke, ventilation and indoor air quality. *ASHRAE Trans.,* 88, 894–914.

Ross, J. A., Sterling, E., Collett, C., & Kjono, N. E. (1996) Controlling environmental tobacco smoke in offices. *Heating/Piping/Air Conditioning,* May, pp. 76–83.

Spengler, J. D., Dockery, D. W., Turner, W. A., Wolfson, J. M., & Ferris, B. J., Jr. (1981) Long-term measurements of respirable sulfates and particles inside and outside homes. *Atmos. Environ., 15,* 23–30.

Spengler, J. D., Treitman, R. D., Tosteson, T. D., Mage, D. T., & Soczek, M. L. (1985) Personal exposure to respirable particulates and implications for air pollution epidemiology. *Environ. Sci. Technol., 19,* 700 706.

Sterling, T. D. & Mueller, B. (1988) Concentrations of nicotine, RSP, CO and CO_2 in nonsmoking areas of offices ventilated by air recirculated from smoking designated areas. *Am. Ind. Hyg. Assoc. J.,* 49(9), 423–426.

Sterling, T. D. & Sterling, E. M. (1984) Environmental tobacco smoke. 1.2. Investigations on the effect of regulating smoking on levels of indoor pollution and on the perception of health and comfort on office workers. *Eur. J. of Resp. Dis., 65* (Supplement 133), 17–32.

Stober, W. (1982) Generation, size distribution and composition of tobacco aerosols. In: Litzinger, E. F., Mattina, C. F., & Bush, L. P., eds., *Recent Advances in Tobacco Science, Vol. 8,* pp. 3–41.

Swift, D. L. (1983) Direct reading instruments for analyzing airborne particles. *Air Sampling Instruments for Evaluation of Atmospheric Contaminants, 6th Edition.* Lioy, P. J. & Lioy, M. J. Y., eds., Cincinnati, OH, pp. U2–U27.

Tang, H., Richards, G., Benner, C. L., Tuominen, J. P., Lee, M. L., & Lewis, E. A. (1990) Solanesol: A tracer for environmental tobacco smoke particles. *Environ. Sci. Technol.,* 24, 848–852.

Traynor, G. W., Apte, M. G., Carruthers, A. R., Dillworth, J. F., Grimsrud, D. T., & Gundel, L. A. (1987) Indoor air pollution due to emissions from wood-burning stoves. *Environ. Sci. Technol.,* 21, 691–697.

Turk, B. H., Brown, J. T., Geisling-Sobotka, K., Froehlich, D. A., Grimsrud, D. T., Harrison, J., Koonce, J. F., Prill, R. J., & Revzan, K. L. (1987) *Indoor Air Quality and Ventilation Measurements in 38 Pacific Northwest Commercial Buildings, Vol. I—Measurement Results and Interpretation* (LBL 22315 ½), *Vol. 11—Appendices* (LBL 22315 2/2), Lawrence Berkeley Laboratory, Berkeley, CA 94720.

Turner, S. & Binnie, P. W. H. (1990) An indoor air quality survey of twenty-six Swiss office buildings. *Proceedings of the 5th International Conference on Indoor Air Quality and Climate, Vol. 4,* Toronto, Canada, pp. 27–32.

Van der Wal, J. F., Moons, A. M. M. & Cornelissen, H. J. M. (1990) The indoor air quality in renovated Dutch homes. *Proceedings of the 5th International Conference on Indoor Air Quality and Climate, Vol. 2,* Toronto, Canada, pp. 441–446.

Weber, A. & Fischer, T. (1980) Passive smoking at work. *Int. Arch. Occup. Environ. Health,* 47, 209–221.

Zhao, B. C. & YuFeng, L. (1990) Characterization and situation of indoor coal-smoke pollution in endemic fluorosis areas. *Proceedings of the 5th International Conference on Indoor Air Quality and Climate, Vol. 1,* Toronto, Canada, pp. 98–102.

7

Field Studies—
Nicotine

INTRODUCTION

Nicotine [3-(1-methyl-2-pyrrolidinyl)-pyridine] is emitted to the ambient air in the sidestream smoke and the exhaled mainstream smoke of cigarettes. Nicotine has received and continues to receive major attention largely because of its potential utility (National Research Council 1986) as a marker for ETS exposure. Airborne nicotine is nearly unique to tobacco smoke in environments encountered by most people, is now readily quantified at ambient levels by means of well-defined procedures, and allows comparisons of exposure and dose by measurement of nicotine and its metabolites in physiological fluids. Nicotine itself is toxic at high doses, being readily adsorbed in the lungs or through the skin either as a vapor or a liquid (Budavari et al. 1989) but the interest in ETS nicotine is driven by its utility as a marker rather than its inherent toxicity. Concentrations of nicotine in ETS are much below those considered acutely toxic.

More than 120 studies have been reported in which nicotine has been used to detect or quantify ETS. More than 50 reports provide nicotine levels under actual field conditions, i.e., under prevailing ambient conditions without manipulating either smoking or environmental conditions. At least three dozen studies have been reported since the mid-1980s, several of them with more than 100 data points. Recent major studies are characterized by exceptional attention to detail as regards experimental design and analytical methodology used. Attention has also been given (Crouse and Oldaker 1990, Jenkins et al. 1991, Sterling et al. 1996, Jenkins and Counts 1999) to the comparability of results from area sampling versus personal sampling strategies. Hammond (1999) has recently summarized individual results from worker exposure studies.

The extraordinary attention given to nicotine is likely the combined result of several factors. First, nicotine is clearly a component of ETS and investigators often compute—unadvisedly or otherwise—the quantities of other constituents of cigarette smoke expected to be present based upon the observed nicotine concentration and the ratio of nicotine to these constituents in mainstream or sidestream smoke. Increasingly, investigators now measure both nicotine and other constituents to test the validity of this approach. Second, several well-documented sampling and analysis procedures are now available for routine application to large numbers of samples. Third, the increased concern about ETS exposure requires that exposure levels be measured under a wide variety of natural conditions.

OSHA lists an 8-h time-weighted average (TWA) exposure level for the workplace of 500 μg/m^3 for nicotine. The short-term exposure limit (STEL) is 1500 μg/m^3 and the level considered by NIOSH to be immediately dangerous to life and health (IDLH) is 35,000 μg/m^3 (NIOSH 1980). As discussed below, ETS nicotine concentrations are most frequently far less than 10 μg/m^3 and only seldom reach 100 μg/m^3.

In the mid- to late 1980s, it was learned through several studies (Eudy et al. 1985, Eatough et al. 1989, Benner et al. 1989, Eatough et al. 1988) that ETS nicotine resides primarily in the vapor phase. Five percent or less of ETS nicotine is associated with particulate matter. This understanding was crucial to the development of standardized and widely accepted sampling and analysis methodology for the determination of nicotine and another, now commonly used, marker of the ETS vapor phase, 3-ethenyl pyridine (3-EP), with detection limits typically less than 0.1 μg/m^3. The finding also had important implications for the utility of nicotine as a marker of ETS particulate phase constituents.

The utility and cautions of using nicotine as a marker for ETS are discussed in Chap. 4. Described below are the studies in which nicotine levels in a number of field situations have been reported. Field studies are distinguished from those studies that have been conducted in controlled settings by the absence of manipulation of the surrounding environment.

An important trend over the past 5–10 years has been the shift from measurement of *area* levels of ETS nicotine. Before 1990, only a handful of studies (Hammond et al. 1987, Thompson et al. 1989, Mattson et al. 1989) had been conducted in which actual personal exposure had been determined through breathing zone sampling of subjects in nonmanipulated environments. Since 1990, most, but not all, field studies have been directed toward personal exposure. There appears to be an increasing realization among investigators that breathing zone sampling better reflects actual human exposure to ETS.

SAMPLING AND ANALYSIS METHODS

Prior to the mid-1980s, ETS nicotine values had only been reported in four field studies. Little effort had been placed into the development of sampling and analysis procedures for trace quantities of ambient nicotine. Because OSHA had established TWA permissible exposure limits (PELs) for nicotine at 500 μg/m^3, analytical methods were geared to such levels. While 500 μg/m^3 may be relevant for workers in tobacco handling operations, it is clearly 1–3 orders of magnitude greater than those associated with ETS exposure. Many of the earlier studies were designed under the assumption that ETS nicotine was primarily associated with the particle phase, as is the nicotine in mainstream cigarette smoke (Guerin 1987). Thus, most of the earlier studies collected nicotine on filters designed to retain particulate matter (Weber and Fischer 1980, Harmsden and Effenberger 1957, Hinds and First 1975). Badre et al. (1978) relied on acidified ethanol bubblers to retain nicotine. However, a number of studies in the late 1980s (Eudy et al. 1985; Eatough et al. 1988, 1989; Benner et al. 1989) indicated that most of the nicotine is emitted into the sidestream smoke, and by the time it is diluted to ETS concentrations, is present in the vapor phase. Probably the most important factors contributing to the differences in physical state of nicotine include the pH (Guerin 1987) at which the two smokes—sidestream and mainstream—are formed, and the greater dilution that sidestream undergoes. Interestingly, the ambient nicotine levels determined in the earlier studies are not substantially different from those published in more recent studies. In retrospect, the collection of significant quantities of ETS nicotine on filters seems to have been fortuitously due to the high adsorbability of vapor phase nicotine on glass fibers of the filter. Nevertheless, it is obvious from the discussions in earlier studies that investigators were unaware of the now-well-confirmed difficulties associated with the *quantitative* collection and analysis of trace quantities of nicotine. In the decade beginning in 1984, considerable effort was directed toward developing precise, accurate analytical methods for ambient nicotine. Modern sampling strategies for ETS nicotine have considered the relatively low levels likely to be encountered under most situations by focusing efforts on concentrating the nicotine on or in some adsorbing medium. The medium is extracted, and the extract analyzed by gas chromatography. Recent analytical methods address the very high adsorbability of nicotine on surfaces within analytical systems, and take measures to prevent such. The current sampling and analysis methods are described in greater detail below. Note that there has been an important shift in the last ten years toward an increasing preference for direct measurement of personal breathing zone sampling over area monitoring. While studies utilizing the former approach are more complicated to conduct, resulting measurements provide a better accounting of actual exposure as an individual moves through a variety of micro-environments during a typical day.

Sampling—Active

Active sampling exists where mass transport to the adsorbing medium occurs primarily by convection, rather than diffusion (see Chap. 5). For nicotine in field situations, this is usually accomplished through the use of small, battery powered sampling pumps drawing the atmosphere through or across the collection medium. Target compounds are later released for analysis by heating the adsorptive resin, or extracting it with solvent. Most of the studies reported in recent years have relied on adsorption on one of three materials: either XAD-4 (a higher specific surface area version of the styrene-divinylbenzene copolymer XAD-2 resin used in the NIOSH method) (Ogden 1989a, Oldaker and Conrad 1987, Crouse et al. 1989, Jenkins et al. 1996, Phillips et al. 1997), Tenax [poly(p-2,6, diphenyl-phenylene) oxide] (Thompson et al. 1989, Kirk et al. 1988, Eatough et al. 1989a), or a Teflon-coated glass fiber filter treated with sodium bisulfate (Hammond and Leaderer 1987; Coultas et al. 1990). In the latter case, the vapor phase nicotine is chemically altered from its more volatile basic form (unprotonated) to its acidic form as it passes through and collects on the filter. The adsorptive resins are usually packed in small tubes, having dimensions of a few millimeters for outer diameter and up to 16 cm long. A method based on the retention of nicotine by a siloxane liquid coated diatomaceous earth (OV-1 on Uniport S) has been reported by Muramatsu and co-workers (Muramatsu et al. 1984). Bayer and Black (1987) have reported the use of Orbo-42 solid sorbent as a collection medium. Other techniques have been reported, including the use of impingers (Crouse et al. 1980), glass fiber filters (Weber and Fischer 1980), and Millipore filters (Hinds and First 1975). The development of a personal annular denuder for collection of vapor phase nicotine has been reported (Koutrakis et al. 1989), but its use for field measurements has not been reported. Currently, the method developed by Ogden and co-workers (i.e., ASTM 1997) appears to be the most widely used.

Relatively low air sampling flows, between 0.1 and 4 L/min., have been used for most of the measurements in which active sampling has been employed. There are important reasons for this approach. First, except for those cases in which chemically modified filters are employed, lower sampling flows are likely to result in less nicotine being evaporated from the collection media (Jenkins et al. 1982). Also, the noise associated with higher volume sampling can be quite obtrusive. In a field situation it is desirous to be unobtrusive to minimize any impact on the behavior of those individuals occupying the area being sampled.

One group of investigators developed a very sophisticated unobtrusive sampling system that has been used in a number of studies; the Portable Air Sampling System (PASS) (McConnell et al. 1987). Briefly, it consists of a briefcase that has been outfitted with an electrochemical carbon monoxide monitor (see Chap. 8), a temperature-measuring thermocouple, a sampling system for ambient nicotine, and one for particulate matter. In addition, the CO analyzer is attached to a data logger for recording of the real-time data. The unit is sufficiently sound-insulated so that it can be employed in even very quiet restaurants without notice by the general public. The advantage of such a unit is that it can obtain information about a number of ETS-related constituents simultaneously. Most other investigators have reported using small personal sampling pumps mounted in briefcases, or on individuals, to collect ambient nicotine (Jenkins et al. 1991). Another group has developed a somewhat different system (Eatough et al. 1989a) involving a combination of sorbent tubes and on-line monitors for both integrated and real-time monitoring. In addition to a series of 12 Tenax-filled micro-tubes and annular denuders for nicotine collection, temperature, relative humidity, oxides of nitrogen and carbon can be measured. Ogden et al. (1996) have developed a patented combination of sampling pumps and digital readouts for use in ETS sampling. The system, now commercialized, can be employed for both personal and area monitoring, and has been used in more than a dozen large studies.

Sampling—Passive

Passive sampling occurs when the primary mode of mass transport to the adsorbate or concentrating medium is diffusion (see Chap. 5). The primary advantage of the use of passive sampling is the ability to conduct longer term measurements without the requirement of a large active sampling pump. Typically, the adsorbent material is placed in an open-faced badge, which is usually worn by the individual whose exposure is being monitored, and the target compound diffuses toward the face of the badge. Because diffusion is inherently a slower process, effective sampling rates of the systems are much lower. For example, Hammond and Leaderer (1987) reported an effective sampling rate of 25 mL/min. (This acts to limit the volume of atmosphere sampled, relative to an active sampler; compare 1.5 L sampled in 1 h with 60 L for a typical active system.) There have been several reports of field measurements with passive systems, (e.g.,. Williams et al. 1984, Ogden et al. 1989b, Hammond and Leaderer 1991, Hammond et al. 1995, Heavner et al. 1995, Bergman et al. 1996, Scherer et al. 1996). Of the methodology used to date for passive sampling, collection on a sodium bisulfate-treated Teflon-coated glass fiber filter, developed by Hammond and Leaderer (1987), appears to be the most widely used approach.

Analysis

Most of the analytical methods reported for ambient nicotine involve a final determination by gas chromatography. Typically, chromatography has been performed on either a packed (Hammond et al. 1987; Muramatsu et al. 1984; Thompson et al. 1989) or capillary (Ogden 1989) column. Most of the packed columns used contain some potassium hydroxide in the stationary phase, to prevent the irreversible adsorption of nicotine on active sites in the column packing. The eluted nicotine is usually detected with a nitrogen-specific detector. Adsorption of trace quantities of nicotine throughout the analytical procedure must be prevented if quantitative results are to be achieved. Grubner et al. (1980) have reported in detail on the propensity of nicotine to adsorb on glass surfaces and employed ammonia to minimize the effect. Ogden (1989; Ogden et al. 1989a) used triethylamine (TEA) in ethyl acetate extraction solvent to prevent the adsorption of nicotine in glass sample containers following elution from XAD-4 resin beads. Others have reported the use of TEA to enhance the thermal desorption of nicotine from Tenax (Thompson et al. 1989). Muramatsu et al. (1984) employed ammonium hydroxide to enhance desorption of nicotine from OV-1 coated Uniport S. Table 7.1 summarizes the specifics of three of the most popular sampling and analysis methods for ambient nicotine. The reported limits of detection range from 0.07-2.0 μg/m^3 nicotine for a 1-h sample. Many other investigators have reported methods that vary slightly from those summarized in Table 7. 1. More recently, most investigators (Williams et al. 1993, Jenkins et al. 1996, Phillips et al. 1997) employ a megabore capillary column for the simultaneous analysis of nicotine and 3-EP.

Intercomparison Studies

Each method has inherent advantages and disadvantages. Direct thermal desorption has an inherent advantage of greater sensitivity than solvent extraction of a resin bed because in the former case, the entire sample collected on the resin bed is analyzed. In contrast, for solvent elution, only a small portion (5 μL of a 1 mL solution) is analyzed in the gas/liquid chromatograph. However, the reported detection limit of one of the solvent extraction procedures is quite similar to that reported for thermal desorption (see Table 7. 1), probably due to the use of a capillary GC column in the former case. Thermal desorption's disadvantage is an inability to repeat an analysis of a particular sample.

The benefit of diffusion badges filled with a sorbent bed or a treated filter is that no personal sampling pump is required. This reduces the burden to the individual being monitored. However, problems have been reported with nicotine adsorbing on the frames of such badges, and not reaching the targeted adsorbing material (Ogden et al. 1989b, Caka et al. 1989).

An extensive comparative study of various approaches to the sampling and analysis of ETS nicotine generated under controlled conditions was reported by Caka et al. (1989). In this study, a number of investigators employed different sampling technologies and analytical measurements to determine nicotine under controlled chamber conditions. The agreement among Tenax, treated filter, annular denuder, and XAD-4-based active samplers was good for vapor phase nicotine. Slopes of observed vs. predicted concentrations (ideally equal to 1.00) ranged from 0.89–1.16 for nicotine concentrations between 30 and 180 μg/m^3. Agreement was poorer for particle phase ETS nicotine, with slopes ranging from 0.65–1.24. The latter is not unexpected given the very small fraction of total ETS nicotine (a few percent), which the nicotine associated with the particle phase represents (Benner et al. 1989). Tenax- and XAD-4-based methods have been reported to exhibit identical results within statistical uncertainty at relatively low ETS levels $\leq$ 5 μg/m^3 nicotine) in chamber studies (Jenkins et al. 1988). Based on the number of instances that various technologies have been employed, collecting nicotine on XAD-4 and Tenax appear to be the most popular approaches. A greater

Table 7.1   Summary of Sampling and Analysis Parameters for Popular Methods for Ambient Nicotine

	Hammond and Leaderer, 1987	Ogden,1989	Thompson at al., 1989
Collection media	Bisulfate-treated filter	XAD-4 resin	Tenax resin
Sampling flow	1.7 L/min	1.0 L/min	170 mL/min
Pump type	Dupont P-4000	Portable, battery driven	Dupont Alpha 2
Desorption	Ammoniated heptane	Ethyl acetate	Thermal @ 250°C
GC column	Packed, 10% Apiezon L 3% KOH, Chromosorb W 6' x 0.125" O.D.	Capillary, DB-5 Coated 30m x 0.53 mm i.d.	Packed, 10% Carbowax 20M 2% KOH on Chromosorb W-AW 2m x 2mm i.d.
Detector	Nitrogen-specific	Nitrogen-specific	Nitrogen-specific
Analysis time	Not Specified	3 min	14 min
Limits of detection	2.0 μg/m^3 for h 0.2 μg/m^3 for 8 h	0.17 μg/m^3 for 1 h 0.02 μg/m^3 for 8 h	0.07 μg/m^3 for 1 h

degree of documentation exists on the XAD-4-based method, as an interlaboratory comparative study has been published (Ogden 1989) and an ASTM method has been reported (American Society of Testing and Materials 1997). Six different laboratories analyzed in duplicate XAD-4 cartridges that had been used to sample three concentration levels of ETS generated in a chamber. The results for the six laboratories averaged 2.28 ± 0.10, 4.14 ± 0.40, and 5.38 ± 0.42 $\mu g/m^3$ (mean $\pm$ one standard deviation). The data indicate that the variation among laboratories was very small, suggesting that laboratories using this method according to the specified protocol should achieve comparable results for identical levels of ambient nicotine. Ogden and Maiolo (1992) performed a detailed evaluation of the two most popular methods for nicotine and 3-ethenyl pyridine (3-EP) sampling: active sampling through XAD-4 and both active and passive sampling on sodium-bisulfate-treated, glass- fiber filters. Both the XAD-4 and bisulfate-treated filters were found to be equivalent for nicotine collection. However, the treated filter was not equivalent for active sampling of 3-EP, likely due to breakthrough.

RESULTS OF FIELD STUDIES

More than 50 separate studies have been reported in which the nicotine levels in well over 125 different environments were determined. Results from these studies are summarized in Appendix 2. Results are reported according to major categories, including private residences, offices, other work situations, public buildings, transportation situations, restaurants, bars, and other functions. Appendix 2 segregates area sampling studies and those devoted to personal monitoring studies, for ease of comparison. Before the early 1990s, most studies were directed toward area measurements in investigator-selected public sites. There are clearly good reasons for this. First, in contrast to personal exposure studies, area measurements are easier to conduct. They do not require the training or cooperation of individual subjects. Additional, short-duration area measurements using sampling pumps in unobtrusive packaging, such as a briefcase, can be conducted without notifying and receiving the cooperation of the facility being sampled. Also, depending on the number of constituents being sampled, the sampling system for an area measurement can be (but is not always) simpler. For example, stationary passive sampling badges have been employed for area monitoring (e.g., Hammond et al. 1995). Alternatively, small sampling pumps can be used, with little consideration being given to sound proofing or vibration damping, because the pumps do not have to be worn by humans. The focus of many of the earlier area sampling studies was nicotine levels in public or semipublic places, in contrast to the residential situation, even though the latter is a venue where individuals spend more or most of their time. Until recently, the number of studies of ETS nicotine (and 3-EP) levels in private residences had been relatively small.

Since the early 1990s, there has been a decreased emphasis on area sampling and an increased emphasis on personal monitoring studies. The realization among investigators that nonsmoking subjects spend only a fraction of their time in smoking environments has certainly been a factor in the increase in longer duration (e.g., 8-, 16-, 24-h or longer) time-integrated personal exposure sampling, despite the complexities of these studies. For studies involving individual subjects, the subjects must be trained, must be cooperative throughout the study, and usually must be compensated. An additional component of these newer studies has been both increased numbers of subjects (from a few hundred to more than 1500) and attempts to achieve a more demographically diverse sample population. The diversity is often achieved through randomized initial contacts (telephone or mail solicitation) or subject recruiting using databases where the demographic characteristics of a subject are already known. An additional emerging trend has been the simultaneous determination and reporting of 3-ethenyl pyridine levels along with those of nicotine. In many ways, 3-EP may be a better marker for the vapor phase of ETS. (See Chap. 4.) It is a pyrolysis product of nicotine, thus "tobacco specific," it is emitted preferentially into sidestream smoke, the predominant precursor of ETS, and it is much less prone to adsorption on surfaces than nicotine. From an analytical standpoint, the chromatographic characteristics are somewhat better for 3-EP than nicotine, which may lead to lower effective detection limits. The authors' recent experiences with low ETS concentration environments suggest that when nicotine levels may be below detectable levels, 3-EP can still be quantifiable.

In Appendix 2, data are provided both as to the reported ranges of the nicotine levels as well as the mean values, if such were published by the authors. In some cases, means have been calculated from individual levels reported in the original manuscripts. Also, some investigators choose not to report the absolute range, but an upper percentile of the levels. Because of the many differences among the studies, a cursory examination of Appendix 2 may lead to misinterpretation. For example, some of the listed data are for a reported sampling of one or just a few locations, whereas others represent anywhere from tens to hundreds of locations. In addition, reported detection limits range from < 0.1 μg/m^3 to > 13 μg/m^3. Thus, in some cases, detailed examinations of the original manuscript may be warranted. As to the data, most of the reported means are 10 μg/m^3 or less. The Kirk data (Kirk et al. 1988), despite its high limit of detection (LOD) of 13.6 μg/m^3, and its convention of reporting results below the LOD as 7 μg/m^3 have minimal impact on the distribution of the data. The larger studies, those reporting 15 or more data points or venues, tend to have lower mean values (see below).

The highest level which has been reported was 1010 μg/m^3, measured in a passenger car with the ventilation system shut off (Badre et al. 1978). Such a level is atypical of the maximal levels observed in more realistic environments. Rather, it seems to indicate that nearly any level of nicotine (and

ETS) can be produced in a small enclosed space. The lowest levels reported are values below the detection limit of the sampling and analytical method employed. Because the LOD is so dependent on both the duration and size of the sample taken, and the minimum mass of nicotine that the analytical method can detect, there is considerable variation in the range of detection limits reported, even for studies conducted in recent years. For example, for the collection of ETS nicotine on Tenax, followed by thermal desorption GC analysis, Thompson et al. (1989) reported limits of detection and quantitation of 0.07 and 0.17 $\mu g/m^3$, respectively, sampling at 170 mL/min for 1 h, while Kirk et al. (1988) report an LOD of 13.6 $\mu g/m^3$ sampling through Tenax at 100 mL/min for 1 h. Typical of modern 24-h sampling detection and quantification limits are those reported by Jenkins et al. (1996) and Phillips et al. (1998: Bremen). Jenkins et al. reported mean limits of detection for nicotine and 3-EP of 0.027 $\mu g/m^3$ and 0.011 $\mu g/m^3$, respectively, for 24-h sampling. Phillips et al. report limits of quantification for both constituents as 0.090 $\mu g/m^3$.

The extent to which "background" levels of nicotine exist in nonsmoking environments appears to depend on both the ability of the investigatory team or the subjects to understand the study criteria for nonsmoking conditions, and to accurately observe and report observations of smoking, as well as the effective detection limits for the constituents of interest. Occasionally, investigators will not report exact criteria for smoking and nonsmoking conditions. One approach to such criteria would be the apparent presence or absence (sight or smell) of burning tobacco during the sampling period. However, especially when subjects (as opposed to a professional sampling team) are required to record such observations, differences in sensitivity may cloud the accuracy of reporting. For example, subjects who live in residences where smoking is common may be less likely to report casual exposure to cigarette smoke in the workplace than a subject who has little or no ETS exposure outside of work. A particularly sensitive subject may report exposure to cigarette smoke from passing an active smoker in a large open area, when in reality, the short duration and low ETS concentration may have a trivial impact on overall exposure. Subjects may not understand criteria, or may simply misjudge the status of their surroundings. For example, in a large study of ETS exposure both inside and away from the workplace (Jenkins et al. 1996), 442 subjects reported themselves at the beginning of the study as working in an environment where smoking was permitted in some fashion. However, of the 442, only 331 actually reported some form of tobacco smoking in their presence in the workplace.

Clearly, although nicotine concentrations can decay quickly in indoor environments due to surface adsorption and removal through ventilation (see below), ETS can be present after all smoking has ceased. In a survey of 34 restaurants, Thompson et al., (1989) reported 3 cases in which no smokers were present, and yet nicotine levels ranged between 0.5 and 2.3 $\mu g/m^3$. Phillips et al. (1997: Turin) reported that for 33 subjects claiming to work in nonsmoking environments, median and 90th percentile at about 8-h shift length nicotine levels were 0.41 and 0.92, respectively. There can be many contributors to "background" levels of nicotine. Ventilation systems or normal convection currents can move air from smoking to nonsmoking areas. Also, because of nicotine's strong absorption on surfaces, followed by subsequent desorption, nicotine can be found in the air long after smoking has ceased. For example, nicotine levels measured several hours after a Boeing 767 aircraft had landed and was undergoing cabin cleaning (with no smoking permitted) have been found to range from 1.1–8.7 $\mu g/m^3$ (Nelson et al. 1990). Presumably, this was due to the volatilization of nicotine from fabric and cigarette butts in ashtrays. Nicotine was determined at a level of 0.09 $\mu g/m^3$ in the den of a nonsmoker's home two days after smoking had occurred in the room (Nelson et al. 1990). The same investigator has demonstrated that nicotine can desorb from a smoker's clothing and can contribute measurable levels of nicotine to an otherwise nicotine-free environment. This finding has been confirmed by another investigatory team (Piadé et al. 1999).

For several of the field studies reported in Appendix 2, the number of observations reported is fairly small, often fewer than 10. As a result, the data, while important contributions to the assessment of overall nicotine exposure of nonsmokers, may not accurately reflect values typical of the particular class of environment sampled. In Table 7.2 are summarized data from those studies reporting 15 or more observations for a particular type of environment. Sampling durations vary from 1 h –5 days. (Note that the Kirk data has not been reported in Table 7.2, due to problems with the nicotine analysis, as recently reported to the authors of this book by one of the study's investigators (Baek 1998).)

One important feature of the data presented in Table 7.2 is that, almost without exception, the median values are lower than the corresponding means, indicating that the data are skewed toward the lower values. While not strictly log-normal, the data approximate such a distribution. The distribution of 16-h away-from-work personal exposure nicotine levels for subjects living in smoking homes, presented in Fig. 7.1, illustrates such a distribution, in which virtually all of the values were above the LOD. For predominantly residential settings, median values in smoking environments ranged from 0.09 $\mu g/m^3$ for 16-h away-from-work personal monitoring samples obtained from employed subjects in Malaysia to 3.45 $\mu g/m^3$ for 3-h duration area samples obtained in Germany. Because of the variety of societies, and thus smoking practices, the differences between area and personal exposure samples, and the duration of sampling, it is difficult to strictly compare the values and assign a "typical" value for a median level, but such would appear to be in the range of 0.5 –2.0 $\mu g/m^3$. For mean values in residential settings, nicotine concentrations ranged from 0.26 $\mu g/m^3$ for 24-h TWA concentrations for house persons living in Hong Kong (Phillips et al. 1998) to 8.03 $\mu g/m^3$ for a 48-h duration area samples in smoking residences in the U.S. (Williams et al. 1993).

Table 7.2 Field Nicotine Concentrations Reported in Major Studies—Concentrations in $\mu g/m^3$

Data Type	Location, Use, Duration Info	Country	No. of Observations	Median	Mean	Std. Dev.	Range	Reference
Residences—smoking								
Area	Homes	U.S.	47	~ 1.2	2.17	2.43	0.1–9.4	Leaderer and Hammond 1991
Area	Homes—heavy use area120-h duration	U.S.	48	2.45				Ogden et al. 1993
Area	Homes—light use area 120-h duration	U.S.	48	0.56				Ogden et al. 1993
Area	Homes—5-day duration: adult smokers	Can.	33	-	4.5			Rickert 1995
Area	Homes—5-day duration: 1 adult smoker	Can.	137		1.7			Rickert 1995
Area	Homes—3-h duration	Ger.	20	3.45	7.33	-	0.35–35.2	Scherer et al. 1995
Area	Homes—168-h duration	Ger.	20	1.04	5.09	-	0.20–34.6	Scherer et al. 1995
Area	Homes—48-h duration	U.S.	42	3.08	8.03		< 0.05–94.2	Williams et al. 1993
Personal	Away from work—smoking homes: 16-h duration—employed subjects	U.S.	306	1.16	2.71	-	7.93[a]	Jenkins et al. 1996

Personal	Housepersons—24-h duration	France	48	0.52	0.93	-	2.4[b]	Phillips et al. 1998
Personal	Employed subjects while in the home only—24-h duration	Australia	30	0.30	0.67	-	1.6[b]	Phillips et al. 1998
Personal	Away from work—smoking homes 16-h duration	Ger.	23	0.40	0.65	-	1.7[b]	Phillips et al. 1998
Personal	Housepersons 24-h duration	Ger.	21	0.49	0.63	-	1.5[b]	Phillips et al. 1998
Personal	Away from work—smoking homes 16-h duration	France	56	0.68	1.2	-	2.9[b]	Phillips et al. 1998
Personal	Housepersons—24-h duration	Spain	43	0.74	1.4	-	2.8[b]	Phillips et al. 1997
Personal	Away from work—smoking homes 16-h duration	Spain	28	0.86	1.8	-	4.4[b]	Phillips et al. 1997
Personal	Housepersons—24-h duration	Switz.	26	0.60	0.70	-	1.5[b]	Phillips et al. 1999
Personal	Away from work—smoking homes 16-h duration	Switz.	37	0.39	1.1	-	3.0[b]	Phillips et al. 1999
Personal	Housepersons—24-h duration	Italy	36	1.1	1.9	-	4.9[b]	Phillips et al. 1997
Personal	Away from work—smoking homes 16-h duration	Italy	29	0.97	1.6	-	2.8[b]	Phillips 1997
Personal	Housepersons—24-h duration	Czech Rep.	51	0.72	1.3	-	3.1[b]	Phillips et al. 1998

Table 7.2 continued

Data Type	Location, Use, Duration Info	Country	No. of Observations	Median	Mean	Std. Dev.	Range	Reference
Personal	Away from work—smoking homes 16-h duration	Czech Rep.	72	0.78	1.3	-	3.3[b]	Phillips et al. 1998
Personal	Housepersons—24-h duration	Hong Kong	31	0.07	0.26	-	0.51	Phillips et al. 1998
Personal	Away from work—smoking homes 16-h duration	Hong Kong	48	0.31	0.95	-	2.1[b]	Phillips et al. 1998
Personal	Housepersons—24-h duration	Malaysia	40	0.18	0.65	-	1.3[b]	Phillips et al. 1998
Personal	Away from work—smoking homes 16-h duration	Malaysia	59	0.09	0.28	-	0.61[b]	Phillips et al. 1998
Personal	Housepersons—24-h duration	China	54	1.3	1.7	-	3.6[b]	Phillips et al. 1998
Personal	Housepersons—24-h duration	Portugal	24	0.19	0.59	-	1.2[b]	Phillips et al. 1998
Residences—Nonsmoking								
Personal	Away from work—nonsmokng homes 16-h duration employed subjects	U.S.	899	0.023	0.072	-	0.194[a]	Jenkins et al. 1996
Personal	Housepersons 24-h duration	China	44	0.15	0.38	-	0.72[b]	Phillips et al. 1998

Personal	Away from work—nonsmoking homes—16-h duration employed subjects	Czech Rep.	66	0.13	0.43	-	0.42[b]	Philips et al. 1997
Personal	Employed subjects while in the home only—24-h duration	Australia	46	< 0.05	0.06	-	< 0.10[b]	Phillips et al. 1998
Personal	Housepersons 24-h duration	France	42	0.13	0.18	-	0.29[b]	Phillips et al. 1998
Personal	Away from work—nonsmoking homes—16-h duration employed subjects	France	65	0.19	0.23	-	0.34[b]	Phillips et al. 1998
Personal	Housepersons 24-h duration	Spain	40	0.11	0.14	-	0.33[b]	Phillips et al. 1997
Personal	Away from work—nonsmoking homes—16-h duration employed subjects	Ger.	82	0.08	0.18	-	0.33[b]	Phillips et al. 1998
Personal	Housepersons—24-h duration	Portugal	53	0.05	0.14	-	0.31[b]	Phillips et al. 1998
Personal	Housepersons 24-h duration	Switz.	56	0.04	0.23	-	0.31[b]	Phillips et al. 1999
Personal	Away from work—nonsmoking homes—16-h duration employed subjects	Switz.	71	0.07	0.32	-	0.69[b]	Phillips et al. 1999
Personal	Housepersons 24-h duration	Ger.	59	0.05	0.10	-	0.22[b]	Phillips et al. 1998

Table 7.2 continued

Data Type	Location, Use, Duration Info	Country	No. of Observations	Median	Mean	Std. Dev.	Range	Reference
Personal	Away from work—nonsmoking homes—16-h duration employed subjects	Italy	75	0.19	0.27	-	0.40[b]	Phillips et al. 1997
Personal	Housepersons 24-h duration	Czech Rep.	38	0.15	0.31	-	0.51[b]	Phillips et al. 1997
Personal	Housepersons 24-h duration	Italy	47	0.14	0.32	-	0.60[b]	Phillips et al. 1997
Personal	Housepersons 24-h duration	Sweden	33	0.05	0.34	-	0.04–3.2	Phillips et al. 1996
Personal	Away from work—nonsmoking homes—16-h duration employed subjects	Sweden	119	0.07	0.19	-	0.05–9.3	Phillips et al. 1996
Personal	Housepersons 24-h duration	Hong Kong	33	0.05	0.10	-	0.27[b]	Phillips et al. 1998
Personal	Away from work—nonsmoking homes—16-h duration employed subjects	Hong Kong	68	0.08	0.11	-	0.17[b]	Phillips et al. 1998
Personal	Housepersons 24-h duration	Malaysia	51	0.05	0.24	-	0.24[b]	Phillips et al. 1998

		Area	n					Reference
Personal	Away from work—nonsmoking homes—16-h duration employed subjects	Malaysia	87	0.07	0.22	-	0.24[b]	Phillips et al. 1998
Personal	Away from work—nonsmoking homes—16-h duration employed subjects	Spain	41	0.17	0.48	-	0.61[b]	Phillips et al. 1997
Area	Homes 5-day duration	Can.	21		<0.05			Rickert 1995
Workplaces—smoking								
Area	Shop production areas and fire stations; restricted smoking	U.S.	54	0.7[d]	2.2[d]		4[b,c]	Hammond et al. 1995
Area	Open offices; unrestricted smoking	U.S.	61	8.6[d]	14.4[d]		34[b,c]	Hammond et al. 1995
Area	Open offices; restricted smoking	U.S.	35	1.3[d]	3.4[d]		9[b,c]	Hammond et al. 1995
Area	Shop production areas and fire stations: unrestricted smoking	U.S.	114	2.3[d]	4.4[d]		7.2[b,c]	Hammond et al. 1995
Area	Offices	U.S.	194	< 1.6	3.5	8.3	<1.6–71.5	HBI 1990
Area	6 office buildings; smoking restricted to rooms w/local air filtration	U.S.	22		44.2	3.6	-	Hedge et al. 1994
Area	Smoking offices	U.S., Can.	156	4.8[d]	-		0–69.7	Oldaker et al. 1990
Area	2 office buildings; unrestricted smoking	U.S.	16	2.1	2.2		3.9[a]	Sterling et al. 1996

Table 7.2 continued

Data Type	Location, Use, Duration Info	Country	No. of Observations	Median	Mean	Std. Dev.	Range	Reference
Area	Smoking offices—naturally ventilated	U.S.	17	-	10		0–41.9	Turner and Binnie 1990
Area	Offices	Switz.	44	-	1.1	2.1	0–16.0	Weber and Fischer 1980
Area	Miscellaneous work situations	U.S.	282	<1.6	4.3	11.8	<1.6–126	HBI 1990
Personal	Workers	U.S.	15	8.7	20.4	20.6	0–53.2	Coultas et al. 1990
Personal	Workers where smoking restricted to designated areas	U.S.	52	0.14	0.30		2.21[a]	Jenkins and Counts 1999
Personal	Workers in unrestricted smoking workplaces	U.S.	134	1.03	3.4		15.0[a]	Jenkins and Counts 1999
Personal	Workers	Malaysia	74	0.28	1.9		2.2[b]	Phillips et al. 1998
Personal	Office workers	Ger.	23	0.40	0.65		1.7[b]	Phillips et al. 1998
Personal	Workers: 24-h over approx. 3 work shifts	Australia	20	0.22	1.1		2.9[b]	Phillips et al. 1998
Personal	Workers	France	102	1.0	1.9		4.6[b]	Phillips et al. 1998
Personal	Workers	Italy	72	0.99	1.9		5.4[b]	Phillips et al. 1997

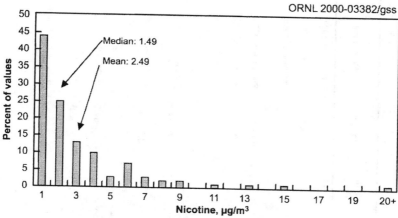

Fig. 7.1. Distribution of values for 16-h, away-from work, personal exposure of subjects living in homes where smoking is unrestricted. (Source: Jenkins et al. 1996.)

Not all investigators report the lower or upper end of the distribution of values they obtain in the same manner. Some report the upper 90th or 95th percentiles, others report the highest values. However, generally the 90th percentile levels appear to be in the range of 1–5 $\mu g/m^3$. Extreme values are around 100 $\mu g/m^3$.

Most of the larger studies that have focused on ETS levels in homes defined as nonsmoking have been personal exposure studies where ETS levels were measured away from work. While the time spent in a residence usually accounts for the bulk of the overall exposure duration, an away-from-work measurement still affords the opportunity for ETS exposure outside a residential setting (shopping, dining and entertainment, social situations). As a result, while most of the median and mean nicotine levels for subjects living in nonsmoking homes, or the one area sampling study reported in homes classified as nonsmoking (Rickert 1995), are at or below the limits of detection or quantification, some of the subjects do receive measurable ETS exposures. For example, the 90th percentile levels for house persons living in nonsmoking residences in China and the Czech Republic were 0.72 and 0.51 $\mu g/m^3$, respectively. Overall, the difference between 24-h TWA nicotine levels for house persons living in smoking vs nonsmoking homes appears to be a factor of 2–20 (see Fig. 7.2), with a median ratio of 6.

Table 7.2 continued

Data Type	Location, Use, Duration Info	Country	No. of Observations	Median	Mean	Std. Dev.	Range	Reference
Personal	Workers in nonsmoking environments	Switz.	41	0.14	0.20		0.26[b]	Phillips et al. 1999
Personal	Workers in nonsmoking environments	Italy	33	0.41	0.52		0.92[b]	Phillips et al. 1997
Personal	Workers in nonsmoking environments	Czech Rep.	32	0.35	0.90		1.3[b]	Phillips et al. 1998
Personal	Workers in nonsmoking environments	Sweden	80	0.15	0.23		0.10–1.8	Phillips et al. 1996
Personal	Workers in nonsmoking environments	Hong Kong	69	0.13	0.28		0.32[b]	Phillips et al. 1998
Personal	Workers in nonsmoking environments	Malaysia	71	0.15	0.37		0.54[b]	Phillips et al. 1998
Public Buildings								
Area	Restaurants	U.S.	21	4.2	6.3		0.3–24.8	Crouse and Oldaker 1990
Area	Supermarkets w/unrestricted smoking	U.S.	17	0.9	0.7		0.3–3.3	Crouse et al. 1989
Restaurants								
Area	Cafeterias	U.S.	62	3.4	7.5	12.9	< 1.6–84.5	HBI 1990
Area	Bank lobbies, courthouse areas, etc.	U.S.	45	<1.6	5.9	16.6	<1.6–90.8	HBI 1990
Area	Restaurants	U.S.	32	0.76	5.8		36[a]	Jenkins and Counts 1999

Area	Restaurants	U.S.	46		8.4			Oldaker et al. 1991
Area	Restaurants	U.S. & Hong Kong	170		5.1[d]		0.0–23.8	Oldaker, et al. 1990
Personal	Patrons—1-h duration	U.S.	34	4.1	5.4	6.4	0.5–37.2	Thompson et al. 1989
Personal	Patrons	U.S.	21	2.9	4.3		0.3–24.0	Crouse and Oldaker 1990
Personal	Restaurant Servers 4- to 8-h duration	U.S.	82	1.22	5.9		28.9[a]	Jenkins and Counts 1999
Bars and Taverns								
Area	Taverns and Nightclubs	Can.	31		38.6–58.0			Collett et al. 1992
Area	Bars and Bar areas in restaurants	U.S.	53	5.8	14.4		49.6[a]	Jenkins and Counts 1999
Personal	Bartenders 5-to 9-h work shift	U.S.	80	4.45	14.1		43.6[a]	Jenkins and Counts 1999
Transportation								
Area	Aircraft seats—tourist class nonsmoking	Europe	48		21	2.5	4.8–62.1	Malmfors et al. 1989
Area	Aircraft seats—tourist class smoking	Europe	48		32	3.5	15–98	Malmfors et al. 1989
Area	Aircraft seats—business class smoking	Europe	48		41	3.6	13–98	Malmfors et al. 1989
Area	Aircraft seats—business class nonsmoking	Europe	48		5.0	0.9	0.8–17.0	Malmfors et al. 1989

Table 7.2 continued

Data Type	Location, Use, Duration Info	Country	No. of Observations	Median	Mean	Std. Dev.	Range	Reference
Area	Aircraft seats—nonsmoking	Japan	19	-	5.3	-	-	Muramatsu et al. 1987
Area	Train seats—smoking	Japan	20		1.3		-	Muramatsu et al. 1987
Area	Train seats—nonsmoking	Japan	48		16.7		?–48.6	Muramatsu et al. 1987
Area	Aircraft seats—smoking	Japan	24	-	13.5	-	?–28.8	Muramatsu et al. 1987
Area	Aircraft seats—smoking	U.S.	69		13.4			Nagda et al. 1992
Area	Aircraft seats—nonsmoking	U.S.	69		0.04			Nagda et al. 1992
Area	Aircraft seats—boundary between smoking and nonsmoking sections	U.S.	69		0.26			Nagda et al. 1992
Area	Aircraft seats—smoking		17	-	25.0	15.3	5.3–55.4	Ogden et al. 1989b
Area	Aircraft seats—smoking		49	-	9.2		<0.03–112.4	Oldaker and Conrad 1987
Area	Aircraft seats—nonsmoking		26	-	5.5	-	<0.08–40.2	Oldaker and Conrad 1987

[a]95th percentile.
[b]90th percentile.
[c]Note: Due to unusual calculation procedure, value may be an overestimate by as much as a factor of 3.
[d]Geometric mean.
HBI=Healthy Buildings International.

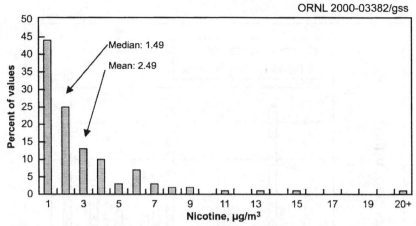

ORNL 2000-03382/gss

Fig. 7.1. Distribution of values for 16-h, away-from-work, personal exposure of subjects living in homes where smoking is unrestricted. (Source: Jenkins et al. 1996.)

Not all investigators report the lower or upper end of the distribution of values they obtain in the same manner. Some report the upper 90th or 95th percentiles, others report the highest values. However, generally the 90th percentile levels appear to be in the range of 1–5 μg/m^3. Extreme values are around 100 μg/m^3.

Most of the larger studies that have focused on ETS levels in homes defined as nonsmoking have been personal exposure studies where ETS levels were measured away from work. While the time spent in a residence usually accounts for the bulk of the overall exposure duration, an away-from-work measurement still affords the opportunity for ETS exposure outside a residential setting (shopping, dining and entertainment, social situations). As a result, while most of the median and mean nicotine levels for subjects living in nonsmoking homes, or the one area sampling study reported in homes classified as nonsmoking (Rickert 1995), are at or below the limits of detection or quantification, some of the subjects do receive measurable ETS exposures. For example, the 90th percentile levels for house persons living in nonsmoking residences in China and the Czech Republic were 0.72 and 0.51 μg/m^3, respectively. Overall, the difference between 24-h TWA nicotine levels for house persons living in smoking vs nonsmoking homes appears to be a factor of 2–20 (see Fig. 7.2), with a median ratio of 6.

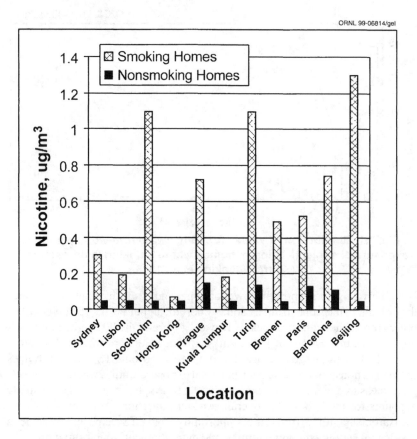

ORNL 99-06814/gel

Fig. 7.2. Comparison of 24-h, TWA median nicotine concentrations. Housepersons in smoking vs. nonsmoking homes. (Source: Phillips et al. 1997, 1998.)

Nearly every nonresidential enclosed space is someone's workplace. In part because of this, workplace and public facilities, until the last 6 years or so, have received the greatest attention from investigators of ETS levels. Table 7.2 lists the major workplace studies reported in the literature. In smoking workplaces, for area samples, medians range from 0.7–8.6 $\mu g/m^3$. (Note that in the Hammond et al. (1995) study, sampling was continuous over the course of a 7-day week, but effective concentrations were calculated by the authors on the basis of a 45-h work week. Given that nicotine is known to persist for hours in environments after smoking has ceased (Nelson, et al. 1990), this unconventional method of air concentration calculation may overestimate actual levels by as much as a factor of 2–4 (Ogden 1996; Daisey et al. 1998), bringing the actual level within the range of 2–4 $\mu g/m^3$.) Extreme levels measured are in the neighborhood of 70 $\mu g/m^3$. For personal exposure samples, median nicotine concentrations range from 0.2 $\mu g/m^3$ to 2.4 $\mu g/m^3$. A concentration of 15 $\mu g/m^3$ is the highest reported level for office-type workplaces, representing a 95th percentile value. An important exception to this is the study reported by Coultas et al. (1990), which reported at least one office worker exposed to a 5-h TWA nicotine level of 50 $\mu g/m^3$. However, many of the results of that study suggest problems with the nicotine collection or analysis. For example, in more than 25 % of the samples, the RSP:nicotine ratio was 2. Typical ratios range of 10–40. To say the least, many of the 15 subjects in the study would be considered to be working in very "nontypical" atmospheres.

In many countries, restrictions on smoking in workplaces that are not generally accessed by the public (e.g.,. offices, production facilities) have increased over the past 10 years. In the United States, it is increasingly difficult to find workplaces—especially offices—where smoking is unrestricted. Clearly, the restrictions have the effect of reducing overall exposure of nonsmokers to ETS. For example, data from Hammond et al. (1995), indicate that restrictions on smoking reduced reported median nicotine concentrations in offices and shop production areas by a factor of 3–6. Sampling by Jenkins and Counts (1999) reported a sevenfold reduction in personal exposure as a result of placing restrictions on smoking in a variety of workplaces. Results in an earlier, smaller study by Vaughan and Hammond (1990) suggested that such reductions could be achieved by segregation of smoking activity within the workplace.

Restaurants are environments where both nonsmoking servers and patrons are exposed to ETS, albeit for different durations, and several larger studies, both of area and personal monitoring levels, have been conducted. The area data suggest that earlier studies observed higher ETS levels than a more recent study, which may reflect diminished smoking incidence, longer sampling duration, and improvements in ventilation (or newer facilities). Similar differences exist between earlier and more recent personal exposure studies. The highest ambient nicotine levels have been observed in bars and taverns (sometimes above 50 $\mu g/m^3$), or small enclosed spaces where ventilation is minimal. The latter category can include a number of conditions, including small offices, enclosed vehicles or transportation systems. There has been a good deal of investigation regarding nicotine levels in commercial aircraft, probably due to then-proposed (and now enacted) legislation designed to ban smoking on these conveyances. Typically, nicotine levels are greater in smoking than in nonsmoking sections of the aircraft. However, in one instance, substantial levels of nicotine have been reported in nonsmoking sections. In a study of 48 flights in Europe, nicotine levels averaged 21 $\mu g/m^3$ in the tourist nonsmoking section (Malmfors et al. 1989). This section was immediately behind the smoking seats of the business class cabin in a jet aircraft. In a U.S. Department of Transportation conducted, detailed study of ETS nicotine levels, which was designed to be statistically representative of departure

Table 7.3 24-h TWA Nicotine and 3-Ethenyl Pyridine Concentrations as Determined from Personal Exposure Sampling Workers Living and Working in Smoking Environments (Home, Work, and Elsewhere) in Selected Recent Studies (Concentrations in $\mu g/m^3$)

Study	Number of Subjects	Pumps or Samplers	Nicotine				3-Ethenyl Pyridine			
			Median	Arithmetic Mean	90th Percentile	95th Percentile	Median	Arithmetic Mean	90th Percentile	95th Percentile
Phillips et al. 1998 Lisbon	27	2	0.58	1.2	2.8		0.36	0.58	1.1	
Jenkins et al. 1996 USA	157	2	1.47	2.98		9	0.8	1.19		3.45
Phillips et al. 1999 Switzerland	23	2	0.9	2.4	4.5		0.42	0.70	1.4	
Phillips et al. 1998 Hong Kong	18	2	0.44	3.2	7		0.45	0.92	2.0	
Phillips et al. 1998 Prague	57	2	1.6	2.3	4.9		0.77	0.91	1.7	
Phillips et al. 1998 Kuala Lumpur	29	2	0.3	0.66	1.2		0.18	0.28	0.62	
Phillips et al. 1997 Turin	21	2	1.3	1.6	3.6		0.79	0.83	1.8	
Phillips et al. 1998 Bremen	17	2	0.69	1.2	2.1		0.61	0.69	1.3	
Phillips et al. 1998 Paris	41	2	1.4	1.8	4.1		0.66	0.84	1.7	
Phillips et al. 1998 Beijing	21	2	1.8	2.5	4.6		0.53	0.82	1.3	

NR=Not reported for this data compilation. However, data may have been reported for other summaries. See original reference.

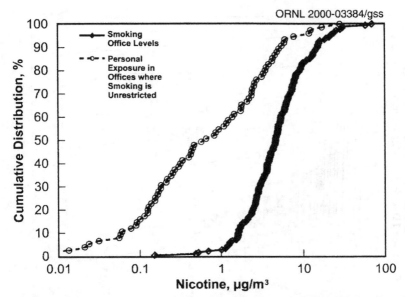

ORNL 2000-03384/gss

Fig. 7.3. Comparison of area monitoring vs. personal exposure
sampling; smoking offices vs. subjects working in office
environments where smoking was not restricted.
(Source: Oldaker et al. and Jenkins et al. 1996.)

95th percentiles, values ranged up to 60 μg/m^3. However, nearly 25% of the
reported nicotine levels in this category were less that 2.5 μg/m^3, and nearly
60% of the values fell between 2.5–15 μg/m^3. The data suggest that factors
other than smoking intensity must play a major role in controlling ETS and
concomitant nicotine levels. Such factors are likely to include ventilation rates,
the degree of adsorbability of the surfaces present, and temporal and spatial
distribution of smoking.

An alternative approach for describing the distribution of observed
nicotine levels involves determining the fraction of reported values that fall at
or below a specified nicotine level. In Fig. 7.5 are summarized the data from
several relatively large studies (N: 24–577) for which the individual data
points were available (Oldaker et al. 1990, 1990a, 1991; Crouse and Oldaker
1990; Crouse and Carson 1989; Oldaker and McBride 1988; Oldaker and
Conrad 1987; Healthy Buildings International 1990; Thompson et al. 1989).
Only the two aircraft situations, which represent relatively enclosed
environments, were more than 30% of the reported concentrations greater than
10 μg/m^3. Less than 10% of the values were greater than 20 μg/m^3.

Based on the data in Appendix 2 and discussed above, it is possible to
encounter concentrations of ETS nicotine of as much as 100 μg/m^3 in selected
situations. Such venues would likely be bars or taverns, or airline smoking
seats, where relatively large numbers of smokers are likely to be present in a

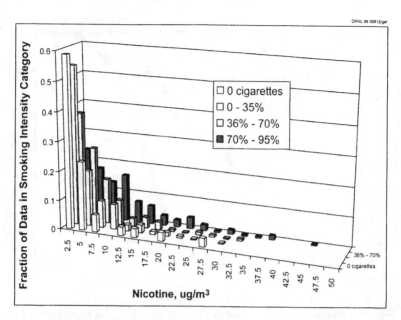

Fig. 7.4. Distribution of ambient nicotine concentrations as a function of smoking intensity level (cigarettes smoked per cubic meter per hour). Bars refer to the percentile range of smoking intensity level. For example, nicotine levels ≤2.5 mg/m³ were measured in slightly less than 60% of the cases where no cigarettes were observed to have been smoked, while the same nicotine levels were measured in only 25% of the highest range of smoking intensities (70th to 90th percentiles).

small volume. Perhaps more important than the concentrations encountered is the exposures that individuals receive. This is clearly one of the drivers behind the increase in personal exposure studies over the last 10 years. Exposure is the product of concentration and time. The concept of exposure, in contrast to concentration, recognizes the fact that nonsmokers may encounter dozens of micro-environments throughout a given day, each with its particular level of ETS. Exposure also reflects that the relative importance of a long duration/low concentration experience and a short duration/high concentration experience can be comparable. Strictly speaking, exposure values should be reported in units of $\mu g/h/m^3$. However, many investigators choose to report exposures as 24-h TWA concentrations, perhaps because most readers are more comfortable with the concept of a time averaged concentration than comparisons of the less frequently used $\mu g/h/m^3$. It is clear that for situations where individuals live and work in smoking environments, if the individuals have some degree of control over their proximity to smokers, 24-h TWA levels (or exposures) are not particularly high. For example, Table 7.3 shows 24-h TWA levels of

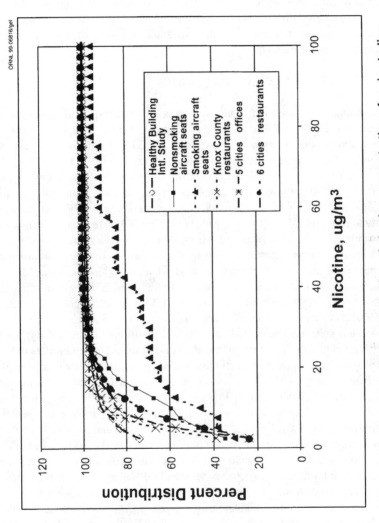

Fig. 7.5. Observed cumulative distribution of nicotine concentrations for six studies. See discussion of Fig. 7.5 in text.

nicotine and 3-ethenyl pyridine for subjects who both lived and worked in smoking environments from some selected studies with comparable experimental designs (workplace and away-from-work sampling pumps collecting personal breathing zone samples over a total of 24 h). Median levels of nicotine range from 0.3 $\mu g/m^3$ in Kuala Lumpur to 1.8 $\mu g/m^3$ in Beijing. 3-EP levels were comparable to one-third those of nicotine. All of the 24-h TWA 90th or 95th percentile levels were < 10 $\mu g/m^3$. This suggests the vast majority of nonsmokers must spend a relatively small fraction of time in high concentration environments.

NICOTINE UPTAKE BY NONSMOKERS

A number of studies in the past two decades, involving the determination of nicotine and/or its metabolites in physiological media (blood, urine, saliva) have indicated that nonsmokers exposed to ETS do in fact receive a measurable dose (Russell et al. 1986, Henderson et al. 1989, Feyerabend et al. 1982, Proctor et al. 1991, Scherer et al. 1989, Jarvis et al. 1984, Jarczyk et al. 1989). Most of the physiological levels of salivary or serum cotinine, one of the major metabolites of nicotine and the one most frequently measured in nonsmokers, range up to 10–15 ng/mL (Etzel 1990, Pirkle et al. 1996), with urinary cotinine levels a factor of 10 or greater. These studies have been conducted under experimentally controlled (test chamber or other manipulated environments) and field conditions. Iwase et al. (1991), employing direct measurement of nicotine levels in inspired and expired ETS, have confirmed that nonsmokers retain 60–80% of inspired ETS nicotine. Benowitz (1996) recommends using a value of 71% for estimation of ETS dose.

Under selected controlled and field conditions, investigators have demonstrated that the quantities of nicotine and/or its major metabolites in the physiological fluid appear proportional to the degree of exposure (Jarvis et al. 1984, Roussel et al. 1991, Henderson et al. 1989). And there have been at least two larger studies (Pirkle et al. 1996, Jenkins and Counts 1999) indicating that the overall level of serum or saliva cotinine is proportional to the time spent in smoking environments. However, there is emerging data that suggest that while group average salivary cotinine levels may be useful predictors of group ETS nicotine exposure, on an individual basis, salivary cotinine may be a poor predictor of ETS nicotine exposure (Jenkins and Counts 1999b). Data from several large studies indicate that in general, functional correlation between salivary cotinine and direct ETS nicotine exposure, while perhaps statistically significant, is very low (Jenkins and Counts 1999; Phillips et al. 1997 (Turin); Phillips et al. 1998 (Sydney, Kuala Lumpur, Bremen, Prague, Hong Kong, and Lisbon) and is not useful for quantifying actual exposure. There may be several reasons for this. First, the metabolism of nicotine is very complex, with nicotine, cotinine, trans-3-hydroxycotinine, as well as their respective glucuronide conjugates, being present in physiological fluids (Curvall et al.

1989; Byrd et al. 1992). Also, there are significant differences among nonsmokers in their metabolic pathways and excretion rates of these various metabolites (Winiwarter et al. 1990). Perhaps it is not surprising that Jenkins and Counts (1999b) have reported that for a group of more highly ETS exposed nonsmokers, employing salivary cotinine levels in at least one model results in overestimation of ETS nicotine exposures by as much as a factor of 10 or more.

To attempt to use physiological levels of nicotine in order to quantitatively relate ETS exposure in nonsmokers with mainstream smoke exposure in smokers may be even more difficult. First, ETS nicotine is primarily in the vapor phase, whereas mainstream nicotine is associated with the particle phase (see Chap. 3). Thus, uptake of the two forms of nicotine may be different. Secondly, nonsmokers metabolize nicotine differently than smokers (Haley et al. 1988). There are also racial differences for cotinine metabolism (Perez-Stable et al. 1998). These concerns notwithstanding, Jarvis (1989) has estimated that, based on physiological levels of nicotine and cotinine, nonsmokers exposed to ETS receive a dose of nicotine ranging from 0.5% to 2% that of a heavy smoker. More recent studies suggest that this estimate of relative exposure may be high. Ogden et al. (1997) recently reported in a U.S. study that the mean smoker level of salivary cotinine is ca. 350 ng/mL. In contrast, unless subjects are routinely exposed to ETS in both the home and workplace, median serum and saliva cotinine levels in two U.S. studies were less than 1.0 ng/mL (Pirkle et al. 1996, Jenkins and Counts 1999b). Median serum and saliva levels for those subjects who both lived and worked in smoking environments ranged from 1–2 ng/mL, suggesting a nonsmoker/smoker nicotine exposure fraction of 1% or less. Haley et al. (1988), Reasor (1990), the National Research Council (1986), and Benowitz (1996, 1999) have reviewed the uptake of nicotine in passively exposed nonsmokers.

SUMMARY

Sampling and analysis methods are now available that allow the confident quantitative determination of nicotine at commonly observed concentrations with a precision (relative standard deviation) of ± 10% or better. Particle phase nicotine is less readily determined but generally constitutes less than 5 percent of the total in ETS. The data from a large number of studies indicate that most individuals in the most common public exposure settings (e.g., offices, restaurants, public buildings) encounter nicotine concentrations of 10 $\mu g/m^3$ or less. Higher concentration settings (e.g., bars, smoking sections of conveyances) yield nicotine levels 2–3 times greater. Conditions exist (e.g., smoking in a closed unventilated automobile, poorly or unventilated smoking lounges) which can yield nicotine concentrations 5–50 times greater than are commonly encountered. True background concentrations of nicotine in environments where smoking has never occurred are expected to be 0, and many studies confirm this within the LOD of the analytical method employed. Concentrations of nicotine in environments thought, however, to be background environments (e.g., nonsmoking areas and locations where smoking is not observed during sampling) frequently contain measurable and, in some cases, inexplicably high nicotine concentrations. This is likely due to infiltration from smoking areas, reintroduction from heat recovery ventilation systems, or out-gassing of nicotine deposited on surfaces from earlier smoking in the area.

Extensive data are available for offices, restaurants, public buildings, and transportation facilities. Comparative data are also available for expected high exposure locations such as bars and areas zoned for smoking. Nicotine concentrations typically encountered are generally at least 1 and up to 3 orders of magnitude lower than the 8-h, time-weighted PEL of 500 $\mu g/m^3$ specified by OSHA for workplace exposure. In the last decade, there have been more than a dozen major personal exposure ETS studies. These studies have determined both residential and occupational exposure. Combined with earlier studies, findings from the recent studies indicate that although nonsmokers may encounter higher concentrations of ETS, overall 24-h exposures are, in general, lower than would be predicted from direct extrapolation from short duration area sampling measurements. Data which is as yet unavailable in the scientific literature relates to the day-to-day repeatability of measured exposures. Such data would provide greater confidence in risk assessments regarding ETS exposures.

REFERENCES

American Society for Testing and Materials, 1997 D5075-Standard Test Method for Nicotine and 3-Ethenylpyridine in Indoor Air. In: *Annual Book of ASTM Standards*. West Conshohocken, PA: American Society for Testing and Materials: 400-407.

Badre, R., Guillerm, R., Abran, N., Bourdin, M., & Dumas, C. (1978) Atmospheric pollution by smoking. *Annales Pharmaceutiques Francaises,* 36 (9–10), 443–452.

Baek, S.O. (1998) Personal communication to R.A. Jenkins, Taegu, Korea, August 1998.

Baek, S.O., Kim U.S., and Perry R. (1997) Indoor air quality in homes, offices, and restaurants in Korean urban areas—indoor/outdoor relationships. *Atmos. Environ.*, 31, 529-544.

Bayer, C. W. & Black, M. S. (1987) Thermal desorption/gas chromatographic/mass spectrometric analysis of volatile organic compounds in the offices of smokers and nonsmokers. *Biomed. Environ. Mass Spectrom.,* 14, 363–367.

Benner, C. L., Bayona, J. M., Caka, F. M., Tang, H., Lewis, L., Crawford, J., Lamb, J. D., Lee, M. L., Lewis, E. A., Hansen, L. D., & Eatough, D. J. (1989) Chemical composition of environmental tobacco smoke. 2. Particulate-phase compounds. *Environ. Sci. Technol., 23,* 688–699.

Benowitz N. L. (1999) Biomarkers of environmental tobacco smoke exposure. *Environ. Health Perspect.,* 107 (Suppl 2), 349–55

Benowitz, N. L. (1996) Cotinine as a biomarker of environmental tobacco smoke exposure, *Epidem. Rev.,* 18, 188–204.

Bergman, T. A., Johnson, D. L., Boatright, D. T., Smallwood, K. G., and Rando, R. J. (1996) "Occupational Exposure of Nonsmoking Nightclub Musicians to Environmental Tobacco Smoke." American Industrial Hygiene Association Journal, 57:746–752.

Budavari, S., O'Neil M. J., Smith, A., & Heckelman, P. E., eds. (1989) *The Merck Index,* Eleventh Edition, Merck & Company, Inc., Publishers. p. 1030.

Byrd, G. D., Chang K. D., Greene J. M., Debethizy J. D. (1992) Evidence for urinary excretion of glucuronide conjugates of nicotine, cotinine, and trans-3'-hydroxycotinine in smokers. *Drug Metabolism and Disposition,* 20, 192–197

Caka, F. M., Eatough, D. J., Lewis, E. A., & Tang, H., (1989) A comparison study of sampling techniques for nicotine in indoor environments. *Proc., 1989 EPA/AWMA International Symposium on Measurement of Toxic and Related Air Pollutants,* pp. 525–541.

Chan C. C., Chen S. C., & Wang J. D. (1995) Relationship between indoor nicotine concentrations, time-activity data, and urine cotinine-creatinine ratios in evaluating children's exposure to environmental tobacco smoke. *Arch. Environ. Health,* 50, 230-234.

Collett, C. W., Ross J. A., & Levine K. B. (1992) Nicotine, RSP, and CO2 levels in bars and nightclubs. *Environ. Int.,* 18, 347–352.

Coultas, D. B., Samet, J. M., McCarthy, J. F., & Spengler J. D. (1990) A personal monitoring study to assess workplace exposure to environmental tobacco smoke. *Am. J. Public Health,* 80, 988-990.

Crouse, W. E., Ireland, M. S., Striegel, R. M., Jr., Williard, C. S., & DeLuca, T. C. (1989) Results from a survey of nicotine on supermarkets. *Proc., 1989 EPA/AWMA International Symposium on Measurement of Toxic and Related Air Pollutants,* Raleigh, NC, pp. 583–589.

Crouse, W. E., & Oldaker, G. B., III (1990) Comparison of area and personal sampling methods for determining nicotine in environmental tobacco smoke. *Proceedings of the 1990 EPA/AWMA International Symposium Toxic and Related Air Pollutants,* Raleigh, NC. pp. 562-566.

Crouse, W. E., Johnson, L. F., & Marmor, R. S. (1980) A convenient method for the determination of ambient nicotine. *Beitr. Tabakforsch. Int.* (2), 111–113.

Crouse, W. E. & Carson, J. R. (1989) Surveys of environmental tobacco smoke (ETS) in Washington, D.C. offices and restaurants. *43rd Tobacco Chemists' Research Conference,* Richmond, VA.

Curvall, M., Vala, E. K., Englund, G., & Enzell, C. R. (1989) Urinary excretion of nicotine and its major metabolites. *43rd Tobacco Chemists' Research Conference,* Richmond, VA.

Daisey, J. M., Mahanama, K. R. R., Hodgson, A. T. (1998) Toxic volatile organic compounds in simulated environmental tobacco smoke: emission factors for exposure assessment. *J. Expo. Anal. Environ. Epidemiol.,* 8, 313–334.

Eatough, D. J., Caka, F. M., Wall, K., Crawford, J., Hansen, L. D., & Lewis, E. A. (1989a) An automated sampling system for the collection of environmental tobacco smoke constituents in commercial aircraft. *Proceedings of the 1989 EPA/AWMA International Symposium on Measurement of Toxic and Related Air Pollutants,* Raleigh, NC, pp. 564–575.

Eatough, D. J., Benner, C. J., Bayona, J. M., Richards, G., Lamb, J. D., Lee, M. L., Lewis, E. A., & Hansen, L. D. (1989) Chemical composition of environmental tobacco smoke. 1. Gas-phase acids and bases. *Environ. Sci. Technol.,* 23, 679–687.

Eatough, D. J., Lewis, L., Lamb, J. D., Crawford, J., Lewis, E. A. and Hansen, L. D., (1988) Nitric and nitrous acids in environmental tobacco smoke. *Proc., 1988 EPA/AWMA International Symposium on Measurement of Toxic and Related Air Pollutants,* pp. 104–112.

Etzel, R.A. (1990) A review of the use of salivary cotinine as a marker of tobacco smoke exposure, *Prev. Med.,* 19, 190–197.

Eudy, L. W., Thorne, F. A., Heavner, D. L., Green, C. R., & Ingebrethesen, B. J. (1985) Studies on the vapor-particulate phase distribution of environmental nicotine by selected trapping and detection methods. *39th Tobacco Chemists' Research Conference,* Montreal, Quebec, Canada.

Feyerabend, C., Higenbottam, T., & Russell, M. A. H. (1982) Nicotine concentrations in urine and saliva of smokers and non-smokers. *Brit. Med. J.,* 284, 1002–1004.

Grubner, O., First, M. W., & Huber, G. L. (1980) Gas chromatographic determination of nicotine in gases and liquids with suppression of adsorption effects. *Anal. Chem.,* 52, 1755–1758.

Guerin, M. R. (1987) *Formation and physicochemical nature of sidestream smoke.* In: O'Neill, I. K., Brunnemann, K. D., Dodet, B. & Hoffmann, D., eds., *Environmental Carcinogens Methods of Analysis and Exposure Measurement, Vol. 9— Passive Smoking (IARC Publications No. 81),* Lyon, International Agency for Research on Cancer, pp. 11–23.

Haley, N. J., Sepkovic, D. W., Brunnemann, K., & Hoffmann, D. (1988) Biomarkers for assessing environmental tobacco smoke uptake. *Proceedings, Air Pollution Control Association,* Niagara Falls, New York.

Hammond S. K. (1999) Exposure of U.S. Workers to Environmental Tobacco *Smoke Environ. Health Perspec.,* 107 (Suppl. 2), 329-340.

Hammond, S. K. & Leaderer, B. P. (1987) A diffusion monitor to measure exposure to passive smoking. *Environ. Sci. Technol.,* 21, 494–497.

Hammond, S. K., Coghlin, J., & Leaderer, B. P. (1987) Field study of passive smoking exposure with passive sampler. *Proceedings, 4ᵗʰ International Conference on Indoor Air Quality and Climate, Vol. 2.* West Berlin, pp. 131–136.

Hammond, S. K., Sorenson, G., Youngstrom, R., Ockene, J. K. (1995) Occupational Exposure to Environmental Tobacco Smoke. *JAMA,* 274, 956–960.

Harmsden, H. & Effenberger, E. (1957) Tabakrauch in verkehrsmitteln, wohn-, und arbeitsraumen [Tobacco smoke in transportation vehicles, living, and working rooms]. *Archiv fur Hygiene und Bakteriologie,* 14, 383–400.

HBI (Healthy Buildings International) USA (1990) *Measurements of Environmental Tobacco Smoke in General Office Areas,* Fairfax, VA.

Heavner, D. L., Morgan, W. T., & Ogden, M. W. (1995). Determination of volatile organic compounds and ETS apportionment in 49 homes. *Environ. Int.,* 21, 3–21.

Hedge A., Erickson W.A., & Rubin G. (1994) The Effects of Alternative Smoking Policies on Indoor Air Quality in 27 Smoking Office Buildings. *Ann. Occup. Hyg.,* 38, 265-278.

Henderson, F. W., Reid, H. F., Morris, R., Wang, O.-L., Hu, P. C., Helms, R. W., Forehand, L., Mumford, J., Lewtas, J., Haley, N. J., & Hammond, S. K. (1989) Home air nicotine levels and urinary cotinine excretion in preschool children. *Am. Rev. Respir. Dis.,* 140, 197–201.

Hinds, W. C. & First, M. W. (1975) Concentrations of nicotine and tobacco smoke in public places. *New Eng. J. Med.,* 292(16), 844–845.

Iwase, A., Aiba, M., & Kira, S. (1991) Respiratory nicotine absorption in non-smoking females during passive smoking. *Int. Arch. Occup. Environ. Health,* 63, 139–143.

Jarczyk, L., Scherer, G., von Maltzan, C., Luu, H. t., & Adlkofer, F. (1989) Comparison of intake of nicotine from environmental tobacco smoke between nose and mouth breathers. *Environ. Int.,* 15, 35–40.

Jarvis, M., Tunstall-Pedoe, H., Feyerabend, C., Vesey, C., & Salloojee, Y. (1984) Biochemical markers of smoke absorption and self reported exposure to passive smoking. *J. Epidemiol. Community Health,* 38, 335–339.

Jarvis, M. J. (1989) Application of biochemical intake markers to passive smoking measurement and risk estimation. *Mut. Res.,* 222, 101–110.

Jenkins, R. A., Gayle, T. M., Wike, J. S., & Manning, D. L. (1982) Sampling and chemical characterization of concentrated smokes. *Toxic Materials in the Atmosphere, ASTM STP 786,* American Society for Testing and Materials, pp. 153–166.

Jenkins, R. A., Palausky , A., Counts, R. W., Bayne C. K., Dindal, A. B., Guerin, M. R. (1996) Exposure to environmental tobacco smoke in sixteen cities in the United States as determined by personal breathing zone air sampling. *J. Expo. Anal. Environ. Epidemiol.,* 6, 473–502.

Jenkins, R. A., Thompson, C. V., & Higgins, C. E. (1988) Development and application of a thermal desorption-based method for the determination of nicotine in indoor environments. *Proceedings, International Conference on Indoor & Ambient Air Quality,* London, pp. 557–566.

Jenkins, R. A. & Counts, R. W. (1999a) Occupational exposure to environmental tobacco smoke: results of two personal exposure studies. *Environ. Health Perspect.,* 107 (Suppl. 2), 341-348.

Jenkins R. A. & Counts R.W. (1999b) Personal exposure to environmental tobacco smoke: salivary cotinine, airborne nicotine, and non-smoker misclassification. *J. Expo. Anal. Environ. Epidem.,* 352-363.

Jenkins, R. A., Moody, R. L., Higgins, C. E., & Moneyhun, J. H. (1991) Nicotine in environmental tobacco smoke (ETS): comparison of mobile personal and stationary area sampling. *Proceedings of the 1991 EPA /AWMA International Symposium on Measurement of Toxic and Related Air Pollutants,* Durham, NC. pp. 437-442.

Kirk, P. W. W., Hunter, M., Back, S. 0., Lester, J. N., & Perry, R. (1988) Environmental tobacco smoke in indoor air. *Proceedings, Indoor Ambient Air Quality Conference,* London, pp. 99–112.

Koutrakis, P., Fasano, A. M., Slater, J. L., Spengler, J. D., McCarthy, J. F., & Leaderer, B. P. (1989) Design of a personal annular denuder sampler to measure atmospheric aerosols and gases. *Atmos. Environ.,* 12, 2767–2773.

LaKind, J. S., Graves, C. G., Ginevan, M.E., Jenkins, R. A., Naiman, D. Q., & Tardiff, R. G. (1999) Exposure to environmental tobacco smoke in the workplace and the impact of away-from-work exposure. *Risk Analysis: An International Journal,* 19, 349–358.

Lambert, W. E., Samet, J. M., Spengler, J. D. (1993) Environmental tobacco smoke concentrations in no-smoking and smoking sections of restaurants. *Am. J. Public Health,* 83, 1339–1341.

Leaderer, B. P. & Hammond, S. K. (1991) Evaluation of vapor-phase nicotine and respirable suspended particle mass as markers for environmental tobacco smoke. *Environ. Sci. Technol.,* 25, 770–777.

Malmfors, T., Thorburn, D. & Westlin, A. (1989) Air quality in passenger cabins of DC-9 and MD-80 aircraft. *Environ. Tech. Letts.,* 10, 613–628.

Maskarinec, M. P., Jenkins, R. A., Counts, R. W., & Dindal, A. B., (1999) Determination of exposure to environmental tobacco smoke in restaurant and tavern workers in one U.S. city. *J. Expo. Anal. Environ. Epidem.*, in press.

Mattson, M. E., Boyd, G., Byar, D., Brown, C., Callahan, J. F., Corte, D., Cullen, J. W., Greenblatt, J., Haley, N. J., Hammond, S. K., Lewtas, J., and Reeves, W. (1989) Passive smoking on commercial airline flights. *JAMA*, 261(6), 867–871.

McConnell, B. C., Perfetti, P. F., Walsh, R. F., Oldaker, G. B., III, Conrad, F. W., Jr., Heavner, D. L., Conner, J. M., Ingebrethsen, B. J., Eudy, L. W., Ogden, M. W., & Stancill,. M. W. (1987) Development and evaluation of portable air sampling system (PASS) for environmental tobacco smoke (ETS). *41st Tobacco Chemists' Research Conference*, Greensboro, NC.

Mumford, J. L., Lewtas, J., Burton, R. M., Henderson, F. W., Forehand, L., Allison, J. C., Hammond, S. K. (1989) Assessing environmental tobacco smoke exposure of pre-school children in homes by monitoring air particles, mutagenicity, and nicotine. *Proceedings of the 1989 EPA/A&WMA International Symposium Measurement of Toxic and Related Air Pollutants*, Raleigh, NC, pp. 606-610.

Muramatsu, M., Umemura, S., Okada, T., & Tomita, H. (1984) Estimation of personal exposure to tobacco smoke with a newly developed nicotine personal monitor. *Environ. Res.*, 35, 218–227.

Muramatsu, M., Umemura, S., Fukui, J., Arai, T., & Kira, S. (1987) Estimation of personal exposure to ambient nicotine in daily environment. *Int. Arch. Occup. Environ. Health*, 59, 545–550.

Nagda, N. L., Koontz, M. D., Konheim, A. G., & Hammond, S. K. (1992) Measurement of cabin air quality aboard commercial airlines. *Atmos. Environ.*, 26A, 2203-2310.

Nagda, N., Fortmann, R., Koontz, M. & Konheim, A. (1990) Investigation of cabin air quality aboard commercial airliners. *Proceedings of the 5th International Conference on Indoor Air Quality and Climate, Vol. 2*, Toronto, Canada, pp. 245–250.

National Research Council (1986) Assessing exposures to environmental tobacco smoke using biological markers. *Environmental Tobacco Smoke—Measuring Exposures and Assessing Health Effects*, pp. 133–159.

Nelson, P. R., Heavner, D. L., Oldaker, G. B., III (1990) Problems with the use of nicotine as a predictive environmental tobacco smoke marker. *1990 EPA/AWMA International Symposium on Measurement of Toxic and Related Air Pollutants*, Raleigh, NC. pp 550-555.

NIOSH National Institute for Occupational Safety & Health (1980) *Registry of Toxic Effects of Chemical Substances, 1979 Edition*, Vol. Two.

Ogden, M. W. & Maiolo, K. C. (1992) Comparative evaluation of diffusive and active sampling systems for determining airborne nicotine and 3-ethenylpyridine. *Environ. Sci. Technol.*, 26, 1226-1234.

Ogden, M. W. (1996) Occupational exposure to environmental tobacco smoke. *JAMA*, 275, 441.

Ogden, M. W., Davis, R. A., Maiolo, K. C., Stiles, M. F., Heavner, D. L., Hege, R. B., et al. (1993) Multiple measures of personal ETS exposure in a population-based survey of nonsmoking women in Columbus, Ohio. In: Jaakkola, J. J. K., Ilmarinen, R, Seppänen, O., eds. Indoor Air '93. *Proceedings of the 6th International Conference on Indoor Air Quality and Climate. Helsinki, Finland: Indoor Air '93*, 1993, 523–528.

Ogden, M. W., Morgan, W. T., Heavner, D. L., Davis, R. A., & Steichen, T. J. (1997) National incidence of smoking and misclassification among the U.S. married female population. *Clin. Epidem.*, 50, 253–263.

Ogden, M. W., Heavner, D. L., Foster, T. L., Maiolo, K. C. Cash, S. L., Richardson, J. D., Martin, P., Simmons, P. S., Conrad, F. W., & Nelson, P. R. (1996a) Personal monitoring system for measuring environmental tobacco smoke exposure. *Environ. Technol.*, 17, 239–250.

Ogden, M. W., Nystrom, C. W., Oldaker, G. B., III, & Conrad, F. W., Jr. (1989b) Evaluation of a personal passive sampling device for determining exposure to nicotine in environmental tobacco smoke. 1989 *EPA/AWMA International Symposium on Measurement of Toxic and Related Air Pollutants,* Raleigh, NC.

Ogden, M. W., Eduy, L. W., Heavner, D. L., Conrad, F. W., Jr., & Green, C. R. (1989a) Improved gas chromatographic determination of nicotine in environmental tobacco smoke. *Analyst,* 114, 1005–1008.

Ogden, M. W. (1989) Gas chromatographic determination of nicotine in environmental tobacco smoke: collaborative study. *J. Assoc. Off. Anal. Chem.,* 72(6), 1002–1006.

Oldaker, G. B., III, Ogden, M. W., Maiolo, K. C., Conner, J. M., Conrad, F. W., Jr., Stancill, M. W., & DeLuca, P. O. (1990a) Results from surveys of environmental tobacco smoke in restaurants in Winston-Salem, North Carolina. *Proceedings of the 5th International Conference on Indoor Air Quality and Climate, Vol. 2*, pp. 281–285.

Oldaker, G. B., III, Perfetti, P. F., Conrad, F. W., Jr., Conner, J. M., & McBride, R. L. (1990) Results from surveys of environmental tobacco smoke in offices and restaurants. *Int. Arch. Occup. Environ. Health,* 99–104.

Oldaker, G. B., III, Stancill, M. W., Conrad, F. W., Jr., Morgan, W. T., Collie, B. B., Fenner, R. A., Levhardt. J. O., Baker. P. G., Lyons-Hart, Jimmy. & Parrish, M. E. (1991) Results from a survey of environmental tobacco smoke in Hong Kong restaurants. *Unpublished work. Personal Communication.*

Oldaker, G. B., III & Conrad, F. W., Jr. (1987) Estimation of effect of environmental tobacco smoke on air quality within passenger cabins of commercial aircraft. *Environ. Sci. Technol.,* 21 (10), 994–999.

Oldaker, G. B., III, Crouse, W. E., & De Pinto, R. M. (1989) On the use of environmental tobacco smoke component ratios. In: *Exerpta Medica Int. Congress Series, Present and future indoor air quality.* C. J. Bieva, Y. Courtois, and M. Govaerts, eds., pp. 287-290.

Oldaker, G. B., III & McBride, R. L. (1988) Portable air sampling system for surveying levels of environmental tobacco smoke in public places. *Symposium on Environment and Heritage, World Environment Day Hong Kong, 1988,* Hong Kong University, Hong Kong, June 6, 1988.

Oldaker, G. B., Taylor, W. D., & Parrish, K. B. (1995) Investigations of ventilation rate, smoking activity and indoor air quality at four large office buildings. *Environ. Technol.,* 16, 173–180.

Perez-Stable, E. J., Herrera, B., Jacob, P., 3rd, & Benowitz, N. L. (1998) Nicotine metabolism and intake in black and white smokers. *J. Am. Med. Assoc.,* 280(2), 152–156.

Phillips, K., Howard, D. A., Bentley, M. C., & Alvan, G. (1998) Assessment of environmental tobacco smoke and respirable suspended particle exposures for nonsmokers in Hong Kong using personal monitoring. *Environ. Int.,* 24, 851–870.

Phillips, K., Bentley, M. C., Howard, D. A., & Alvan, G. (1996) Assessment of air quality in Stockholm by personal monitoring of nonsmokers for respirable suspended particles and environmental tobacco smoke. *Scan. J. Work. Environ. Health,* 22, 1–24.

Phillips, K., Howard, D. A., Bentley, M., & Alvan, G. (1998) Measured exposures by personal monitoring for respirable suspended particles and environmental tobacco smoke of housewives and office workers resident in Bremen, Ger. *Int. Arch. Occup. Environ. Health,* 71, 201–212.

Phillips, K., Howard, D. A., Bentley, M. C., & Alvan, G. (1998) Assessment by personal monitoring of respirable suspended particles and environmental tobacco smoke exposure for non-smokers in Sydney, Australia. *Indoor Built Environ.,* 7, 188–203.

Phillips, K., Howard, D. A., Browne, D. & Lewsley, J. M. (1994) Assessment of personal exposures to environmental tobacco smoke in British nonsmokers. *Environ. Int.,* 20, 693-712.

Phillips, K., Bentley, M., Howard, D., & Alvan, G. (1998) Assessment of environmental tobacco smoke and respirable suspended particle exposures for nonsmokers in Prague using personal monitoring. *Int. Arch. Occup. Environ. Health* , 71, 379–390.

Phillips, K., Howard, D.A., Bentley, M., & Alvan, G. (1999) Assessment of environmental tobacco smoke and respirable suspended particle exposures for nonsmokers in Basel by personal monitoring. *Atmos. Env.*, 33, 1889-1904.

Phillips, K., Howard, D. A., Bentley, M. C., and Alvan, G. (1998). Environmental tobacco smoke and respirable suspended particle exposures for non-smokers in Beijing. *Indoor Built Environ.*, 7, 254–269.

Phillips, K., Bentley, M. C., Howard, D. A., & Alvan, G. (1998) Assessment of air quality in Paris by personal monitoring of nonsmokers for respirable suspended particles and environmental tobacco smoke. *Environ. Int.*, 24, 405–425.

Phillips, K, Bentley, M. C., Howard, D., & Alvan, G. (1998) Assessment of environmental tobacco smoke and respirable suspended particle exposures for nonsmokers in Kuala Lumpur using personal monitoring. *J. Expo. Anal. Environ. Epidem.*, 8, 519–541.

Phillips, K., Bentley, M. C., Howard, D. A., Alvan, G, & Huici, A. (1997) Assessment of air quality in Barcelona by personal monitoring of nonsmokers for respirable suspended particles and environmental tobacco smoke. *Environ. Int.*, 23, 173–196.

Phillips, K., Howard, D. A., Bentley, M. C., & Alvan, G. (1998) Assessment of environmental tobacco smoke and respirable suspended particle exposure of nonsmokers in Lisbon by personal monitoring. *Environ. Int.*, 24, 301–324.

Phillips, K, Howard, D. A., Bentley, M. C. (1997) Assessment of air quality in Turin by personal monitoring of nonsmokers for respirable suspended particles and environmental tobacco smoke. *Environ. Int.*, 23, 851–871.

Piadé, J. J., D'Andres, S., Sanders, E. B. (1999) Sorption phenomena of nicotine and ethenylpyridine vapours on different materials in a test chamber. *Environ. Sci. Technol.*, 33, 2046-2052.

Pirkle, J. L., Flegal, K. M., Bernert, J. T., Brody, D. J., Etzel, R. A., & Maurer, K. R. (1996) Exposure of the U.S. population to environmental tobacco smoke: The third national health and nutrition examination survey, 1988–1991, *JAMA*, 275, 1233–1240.

Proctor, C. J., Warren, N. D., Bevan, M. A. J., & Baker-Rogers, J. (1991) A comparison of methods of assessing exposure to environmental tobacco smoke in nonsmoking British women. *Environ. Int.*, 17, 287–297.

Proctor, C. J. (1989) A comparison of the volatile organic compounds present in the air of real-world environments with and without environmental tobacco smoke. *82nd Annual Meeting of the Air and Waste Management Association,* Anaheim, CA.

Proctor, C. J., Warren, N. D., & Bevan, M. A. J. (1989) Measurements of environmental tobacco smoke in an air-conditioned office building. *Proceedings, Conference on the Present and Future of Indoor Air Quality,* Brussels, Poster P- 1.

Ramakrishna, J., Durgaprasad, M. B., & Smith, K. R. (1989) Cooking in India: the impact of improved stoves on indoor air quality. *Environ. Int.,* 15, 341–352.

Reasor, M. J. (1990) Biological markers in assessing exposure to environmental tobacco smoke. In: Ecobichon, D. J. & Wu, J. M., eds., *Environmental Tobacco Smoke,* Lexington Books, Lexington, MA, pp. 69–77.

Rickert, W. S. (1995) Levels of ETS Particulates, Nicotine, Solanesol, and Carbonyls in a 'Random' Selection of Homes in a Midsize Canadian City *Presented at the 49th Tobacco Chemists' Research Conference*, September 24–27, 1995 Lexington, KY.

Roussel, G., Quang, L. E., Migueres, M. L., Roche, D., Mongin-Charpin, D., Chretien, J., & Ekindjian, O. G. (1991) An interpretation of urinary cotinine values in smokers and non-smokers. *Rev. Mal. Resp.,* 8, 225–232.

Russell, M. A. H., Jarvis, M. J., & West, R. J. (1986) Use of urinary nicotine concentrations to estimate exposure and mortality from passive smoking in non-smokers. *British J. of Addiction,* 81, 275–281.

Scherer, G., Westphal, K., Adlkofer, F., & Sorsa, M. (1989) Biomonitoring of exposure to potentially genotoxic substances from environmental tobacco smoke. *Environ. Int.,* 15, 49–56.

Scherer, G. et al. (1995) Contribution of Tobacco Smoke to Environmental Benzene Exposure in Ger. *Environ. Int.* 21, 779–789.

Sterling E.M., Collett C. W., Ross, J. A. (1996) Assessment of non-smokers' exposure to environmental tobacco smoke using personal-exposure and fixed location monitoring. *Indoor Built Environ.,* 5, 112– 125.

Thompson, C. V., Jenkins, R. A., & Higgins, C. E. (1989) A thermal desorption method for the determination of nicotine in indoor environments. *Environ. Sci. Technol.,* 23, 429–435.

Trout, D., Decker, J., Mueller, C., Bernert, J. T., and Pirkle, J. (1998) Exposure of casino employees to environmental tobacco smoke, *J. Occup. Environ. Med.,* 40(3) 270–276.

Turner, S. & Binnie, P. W. H. (1990) An indoor air quality survey of twenty-six Swiss office buildings. *Proceedings of the 5th International Conference on Indoor Air Quality and Climate, Volume 4,* Toronto, Canada, pp. 27–32.

Turner S., Cyr, L. and Gross, A. J. (1992) The measurement of environmental tobacco smoking in 585 office environments. *Environ. Int.,* 18, 19-28.

Weber, A. & Fischer, T. (1980). Passive smoking at work. *Int. Arch. Occup. Environ. Health,* 47, 209–221.

Williams, R., Collier, A., & Lewtas, J. (1993) Environmental tobacco smoke exposure of young children as assessed using a passive diffusion device for nicotine. *Indoor Environ.*, 2, 98–104.

Williams, D. C., Whitaker, J. R., & Jennings, W. (1984) Air monitoring for nicotine contamination. *J. Chromatogr. Sci.,* 22, 259–261.

Winiwarter, W., Gunther, K., Belnap, D., Tang, H., Crawford, J., Hansen, L. D., Eatough, D. J., & Lewis, E. A. (1990) Elimination of nicotine and cotinine by nonsmokers exposed to sidestream tobacco smoke under controlled conditions. *Proceedings of the Conference on Total Exposure Assessment Methodology,* Las Vegas, NV, pp. 526–540.

8

Field Studies— Carbon Monoxide

INTRODUCTION

Carbon monoxide (CO) has been among the most extensively studied constituents of environmental tobacco smoke (ETS). Popular U.S. commercial filter cigarettes deliver approximately 15 mg of CO in mainstream smoke and an additional 50 mg in sidestream smoke. This suggests that ETS would be a major contributor to ambient CO concentrations.[1] CO is also a commonly recognized toxin capable of being lethal at very high concentrations, being a contributor to cardiovascular and other health effects with prolonged exposure to elevated concentrations, and possibly contributing to general ill health and discomfort associated with the Sick Building Syndrome. CO is also a potentially attractive constituent to monitor in studies that seek to establish relationships between exposure (ambient-air concentrations) and dose (the quantity taken up by the body) because both ambient CO levels and the blood's carboxyhemoglobin (COHb), a CO complex formed with the blood's hemoglobin and an indicator of CO dose, can be measured easily.

CO concentrations have been measured in numerous environments with a great variety of experimental designs. Studies range from simple surveys (e.g., Sebben et al. 1977), to short-term integrated measurements of 20- to 30-min exposure periods (e.g., Harke and Peters 1974, Badre et al. 1978), to longer-term (6- to 7-h exposure assessments of selected locations (DOT 1971), to detailed 24-h integrated personal exposures (e.g., Akland et al. 1985). The relative ease of measuring ambient CO has undoubtedly contributed to the very large CO database.

The universal assumption is that CO concentrations are greater in environments that contain ETS than they are in the same environments in the absence of ETS, but that the absolute concentration is similar to (and is frequently exceeded by) non-ETS environments containing other sources of CO. Because other sources are so common and so numerous (e.g., cooking appliances, heating appliances, vehicle exhaust), and because contact with

---

[1]The large number of cigarettes smoked worldwide coupled with the relatively large CO production by smoking raised the issue in the early 1970s of smoking's contribution to global CO. Calculations showed (Weinstock and Niki 1972; Maugh 1972) that 0.6 million tons of CO were produced annually by smoking. This was compared with 270 million tons by fossil fuel combustion and 5000 million tons by the atmospheric oxidation of methane.

high-concentration CO micro-environments (e.g., underground parking garages, automotive travel) are common, the contribution of ETS to total CO exposure is often difficult to confidently detect. This factor is likely the cause for the decline in interest over the past few years for codetermination of CO with other ETS-related constituents.

Health effects of commonly encountered indoor air levels of CO can be significant, but are seldom life threatening. Exposure to 15 ppm CO for 10 h is sufficient to reach a 2.5% COHb blood content, a level at which physiologically adverse effects have been cited to occur in humans (NRC 1977, cited in Turk et al. 1987). However, some acute adverse health effects of CO observed at concentrations lower than 15 ppm have been noted. These include interference with alertness, vision, and perception; dizziness; drowsiness; and headaches. Schievelbein and Richter (1984) summarized several studies describing the ranges of COHb in unexposed smokers before and after exposure to ETS. COHb percentages for nonsmokers range from 0.1 to 3.7%, as shown in nine studies that used between 78 and 16,000 subjects. The percentages for nonsmokers exposed to ETS range from 0.5 to 2.6%, based on nine studies involving between 7 and 47 subjects. Hence, studies that quantified the percentage COHb in nonsmokers show very little difference in nonsmokers who were or were not exposed to ETS.

SAMPLING AND ANALYSIS

Both active and passive sampling can be employed to determine CO concentrations (see Chap. 5). Passive sampling is convenient because it requires no power source, and is useful for obtaining an estimate of long-term (e.g., 24-h) integrated exposures. Active sampling is more popular, however, both because it allows both time-integrated and temporal concentrations to be measured and it is generally more accurate. Active sampling combined with instrumentation providing instantaneous and/or integrated measurements of CO concentration is the currently preferred method of determining CO exposures. Several measurement methods are described below.

Electrochemical Oxidation Detectors. A frequently used method of detecting CO in air employs electrochemical oxidation detectors (Jenkins and Guerin 1984). These instruments oxidize CO to CO_2 at a catalytically active electrode such as platinum in an aqueous electrolyte according to the equation $CO + H_2O = CO_2 + 2 H+ + 2 e-$. When a filtered air sample is pumped into the sensor, it is passed over the back (gas) side of the anode. The CO diffuses to the catalytic surface, where it is oxidized. The resulting current flowing between the sensing and counter electrodes is proportional to the CO concentration in the sampled air. The CO_2 formed by the electrochemical reaction is removed from the sensor cell continuously, at the same rate that it is formed. Many electrochemical-based monitors are commercially available, in both bench-top and hand-held configurations. They may be prone to positive responses to ethanol and petroleum hydrocarbon vapors, and individual vendors should be queried about potential interferences for their units.

Nondispersive Infrared (IR) Detectors. Nondispersive infrared (NDIR) measurement of ETS CO levels seems to have received much less attention than have electrochemical oxidation detectors, perhaps because the instrumentation is somewhat less portable and is 4 to 6 times more expensive. NDIR is superior to the electrochemical analyzers because the former exhibits improved specificity and sensitivity (Jenkins and Guerin 1984). The physical size of the NDIR instruments should be considered carefully for field operations because typical instruments are substantially larger that the hand-held CO monitors. However, they are less prone to chemical interferences than the electrochemical systems. The operational theory of NDIR analyzers is described in detail by Jenkins and Guerin (1984). Briefly, IR radiation from two separate energy sources is chopped at 10 Hz and passed through optical filters to reduce background interference from other IR-absorbing components. The infrared beams pass through a reference cell (which contains a nonabsorbing gas) and a sample cell (in which the test gas circulates freely), where a portion of the IR radiation is absorbed by the component of interest. The detector converts the difference in energy between sample and reference cells to a capacitance change that is proportional to component concentration. This capacitance change is then amplified and indicated on a meter and, if desired, is used to drive a recorder, controller, or computerized data logger.

Gas Filter Correlation Instruments. Gas filter correlation instruments work somewhat similarly to NDIR systems. Radiation from an IR source is chopped and passed through a gas filter that alternates between CO and nitrogen. IR radiation passes through a narrow bandpass filter and a multiport optical cell filled with the sample gas. If the sample gas contains CO, the chopped signal will be modulated as the filter cycles between N_2 and CO. These units have advantages and disadvantages similar to those of NDIR systems.

Gas Detector Tube. The success of the instrumental electrochemical oxidation and NDIR detector has led to their general acceptance and, to a degree, to a decreased interest in gas detector tubes for monitoring CO. The latter do exist, however, and are useful for personal-monitoring situations or field sampling. Some are marketed commercially by several manufacturers.

Gas detector tubes are usually made of glass and contain a packing material designed to give a selective color reaction for the analyte of interest. The tubes are precalibrated by the manufacturer so that the analyte concentration may be read directly from the tube barrel, assuming a specified volume of air has been sampled, often by means of a small hand-held pump. Usually, one of three reactions is employed to allow selective detection of CO. In the first, CO reacts with iodine pentoxide in the presence of selenium dioxide and fuming sulfuric acid to produce iodine. In the second, CO reduces potassium palladosulfite to liberate metallic palladium. In the third, CO reacts with both palladium sulfate and molybdate to yield a characteristic molybdenum blue color. Interferences that produce a positive response include carbon disulfide, hydrogen, acetylene, hydrogen sulfide, mercaptans, phosphine, phosgene, and sulfur dioxide at concentrations ranging from one-fiftieth to one-fifth that of CO. Nitrogen dioxide may produce a negative interference. Under most normal conditions for monitoring CO in indoor air, none of these species at their usual concentrations should present a significant problem.

FIELD STUDIES

CO is among the most widely monitored constituents of ETS. At least 50 investigators have examined the concentration of CO under a variety of indoor circumstances. More than a dozen of the studies examined fifteen or more locations. Examples include offices, workplaces, shops, bars, restaurants, and transportation. Studies range in complexity from those designed to assess excess general-population exposure due to various sources (e.g., Akland et al. 1985), to detailed surveys of selected classes of environments (e.g., the R. J. Reynolds/Lorillard Restaurant Study), to many studies of single environments at a given time addressing a given issue. Results from a representative number of these studies are tabulated in Appendix 3.

It seems logical that nonsmokers in ETS environments would be exposed to elevated levels of CO. CO is emitted into ambient air in measurable quantities from smoldering cigarettes, and individuals who inhale CO retain a substantial fraction of it. The CO is present in the body as COHb. However, being able to discern a statistically significant increase in COHb due to ETS exposure in a large population of individuals is problematic. First, there can be important individual differences in the rate of exhalation of CO due to differences in metabolism, exercise rates, etc. Secondly, there are bound to be differences in background levels of CO because of contributions from other sources.

Elevated COHb levels due to ETS exposure have been observed under controlled exposure situations, where ETS levels (and thus corresponding CO levels) are high. For example, ten nonsmokers exposed under controlled conditions to ETS with CO levels of 10 and 25 ppm showed average increases in COHb levels of 0.7 and 2.1%, respectively (Scherer et al. 1988). It has been possible to confirm human uptake of CO in natural settings that had high levels of ETS present and when careful determination was made of pre- and post-exposure levels of COHb (or CO concentrations in expired air). For example, Jarvis et al. (1983) observed increases in expired-air CO levels of nearly 6 ppm in seven nonsmokers who had sat for 2 h in a pub where CO levels rose from 2 to 13 ppm over the course of the exposure. However, elevated COHb or expired-air CO levels are difficult to discern for general-population studies in which confounding factors may play a greater role. For example, Jarvis et al. (1984) indicated no statistically significant differences among subgroups of a population of 100 smokers, some of whom reported no exposure to ETS, while others reported moderate or heavy exposure.

Schievelbein and Richter (1984) summarized several studies describing the ranges of COHb in unexposed smokers before and after exposure to ETS. COHb percentages for nonsmokers range from 0.1 to 3.7%, as shown in nine studies that used between 78 and 16,000 subjects. The COHb levels for nonsmokers exposed to ETS ranged from 0.5 to 2.6%, based on nine studies involving between 7 and 47 subjects. Hence, studies that quantified the percentage COHb in nonsmokers show very little difference in nonsmokers who were or were not exposed to ETS. Thus, it is not surprising that many investigators (e.g., Haley et al. 1988) have concluded that CO lacks the appropriate specificity as an ETS uptake marker for epidemiological studies.

Table 8.1 summarizes CO concentrations found in various environments containing ETS and in their corresponding "control" environments. Control environments are variously defined in individual studies as representing the ETS-containing environment in the absence of ETS. Definitions of a control environment range from nearby outdoor air, the same indoor environment but known to be free of ETS (e.g., a bar before it is opened for business), or the same indoor environment in the absence of visible smoking. Chapter 2 discusses the difficulty of identifying a true control environment.

Concentrations of CO in both ETS and control environments most commonly range between 1 and 5 ppm. A similar range is observed for truly ETS-free environments (see Chap. 2). The similarity of CO concentrations found in environments known to contain ETS and those known not to contain ETS reflect the variety of non-ETS sources that contribute to ambient CO levels. It is obviously very important that studies of CO exposure due to ETS be accompanied by careful measurements of background (ETS-free) CO levels. Table 8.2 summarizes background CO concentrations accompanying many studies of CO exposure due to ETS.

Results in Akland et al. (1985) provide background CO concentrations and source information (see Table 8.3) The study employed a stratified probability sample of subjects selected to address urban exposure to CO with particular attention to combustion sources and, especially, to mobile-source emissions (i.e., vehicle emissions). Subjects in Washington, D.C., and in Denver, Colorado, were monitored for 24-h periods during the 1982–83 winter. The subjects carried personal-exposure monitors that automatically monitored the ambient CO concentration every hour on the hour and that also allowed the subjects to activate the monitor at will. Individual control allowed the subjects to sample during periods of time when they suspected that they were near target sources of CO exposure. This is an informative study in which several very important points are made; however, it is excluded from Appendix 3 and from summary Table 8.2 because it presents insufficient detail to identify specific ETS and ETS-free data points confidently.   Results of the study (see Table 8.3) show that personal-monitoring concentrations of CO most commonly range from approximately 1–5 ppm, as is the case for commonly reported background concentrations by others (Table 8.1). The results also demonstrate the importance of vehicle exhaust to CO exposure and the importance of assessing time and concentration to determine exposure. Residential and office concentrations of CO were typically between 1–3 ppm, while concentrations in indoor parking garages ranged from 10–19 ppm. However, exposure periods are one hundred times greater for the residential plus office environments than those for the parking garage. A particularly relevant finding from the Denver residential cohort (unfortunately, mentioned but not discussed) is that exposure to smoking increased mean exposure by 84% (1.59 ppm). Similarly, the average mean exposure to CO was increased by 134% (2.59 ppm) by gas-stove operation and 22% (0.41 ppm) by the presence of an attached garage.

Table 8.1 Summary of Field Measurements[a] for CO

Site	Ranges of CO Concentrations, ppm	
	Means	All Data
Offices		
Smoking	1.0–2.8	<0.1–9.0
Control	1.2–2.5	< 0.1–5.8
Other Workplaces		
Smoking	1.4–4.2	0–32[b]
Control	1.7–3.5	0–22[b]
Functions and Public Gatherings		
Smoking	3.0–25[b]	3.5–11.5
Control	3–9.4	2.2–9.0
Transportation		
Smoking	1.6– 33[b]	0–40[b]
Control	0–5.9	1–15
Restaurants and Cafeterias		
Smoking	1.2–9.9	0.3–90[b]
Control	0.5–7.1	0.3–6.6
Bars and Taverns		
Smoking	3–17[b]	3–42[b]
Control	~1–9.2	1–35[b]

[a]From data in Appendix 3.
[b]Extreme conditions, as discussed in the text.

The common observation of most studies (e.g., Appendix 3, Table 8.1) is that CO concentrations are higher in the presence of ETS than in the absence of ETS in any given environment, all other conditions being equal. The absolute concentration of CO found, however, is equaled to or exceeded by concentrations encountered in many ETS-free environments. This fact is often obscured by tabulations of results that include the full range of concentrations observed in any study (including those used in this document). The full range is reported for completeness and to illustrate extremes that can be encountered, but it is misleading because the extremes appear to be of greater significance than the more frequently encountered mean. The "mean values" or "all data" in Table 8.1 that exceed approximately 15 ppm should be interpreted as both "real" and "extreme" CO concentrations. These values occur infrequently, under unusual circumstances, and should not be interpreted as "typical" CO concentrations under normal conditions.

As an example, Kirk et al. (1988) reported maximum CO concentrations from a 30-week study in the U.K. representing 221 smoking situations and 450 nonsmoking situations as 31.9 and 21.9 ppm, respectively. However, the values at the ninety-fifth percentile for smoking situations and nonsmoking situations were 7.2 and 5.9 ppm, respectively, while those at the ninety-ninth percentile were 12.6 and 11.5 ppm, respectively. Sterling et al. (1987) reported a similar situation, where the range of CO for 194 smoking situations ranged between "not detected" and 242 ppm, but the median value was only 3.1 ppm. "Extreme" values may reflect transportation sources where air circulation is limited, such as in an enclosed automobile. Elliot and Rowe (1975) observed 25 ppm CO in an arena containing 2000 spectators. Pennanen et al. (1997) reported CO levels up to 29 ppm in indoor ice arenas. Spengler and Sexton (1983) reported up to 100 ppm CO in ice-skating rinks, apartments, and offices

Table 8.2 Background Indoor and Outdoor Concentrations for Carbon Monoxide (CO)

Location/Situation	Reference	CO Concentrations, ppm		Comment
		Mean	Range	
Outdoor Background				
Mean total hourly CO, all days	Houck et al. 1988	-	1.0–3.3	6 western U.S. cities
Mean total hourly CO, cold days	Houck et al. 1988	-	1.2–4.3	6 western U.S. cities
Mean total hourly CO, weekdays	Houck et al. 1988	-	1.1–3.5	6 western U.S. cities
Mean total hourly CO, weekends	Houck et al. 1988	-	0.9–2.8	6 western U.S. cities
Outdoors	Lebret at al., 1987	<1–2.2	-	
18 Outdoors, heart patients	Colome et al. 1987	3.5 ± 2.4[a]	-	
35 Outdoors	Sterling et al. 1987	2	ND–32	Median, not mean, value
Outdoor background, woodstoves	Traynor et al. 1987	-	0.0–1.1	Airtight woodstoves
Outdoor background, woodstoves	Traynor et al. 1987	-	0.0–1.0	Non-airtight woodstoves
Outside air background	First 1983	-	2.0–2.5	
Background	Jackson et al. 1988	-	0.7–1.5	
Background	Jackson et al. 1988	-	0.8–1.8	
Outdoors, outside residences	Baek et al. 1997	2.1	0.2–7.0	Korean summer and winter
Outdoors, outside offices	Baek et al. 1997	2.2	0.5–6.1	Korean summer and winter
Outdoors, outside restaurants	Baek et al. 1997	2.4	0.4–6.6	Korean summer and winter
Outdoors, urban	Rowe et al. 1989	-	21–28	
Outdoors, rural	Rowe et al. 1989	-	3	
6 Background	Turk et al. 1987	-	2–5	Outside urban office bldgs.
12 Kitchens w/gas cookers	Lebret et al. 1987	1–4.6	3.4–64	Ranges are 1-h averages
12 Living rooms w/gas cookers	Lebret et al. 1987	1–4.6	2.2–30	Ranges are 1-h averages
12 Bedrooms w/gas cookers	Lebret et al. 1987	<1–4.6	1–30	Ranges are 1-h averages
12 Residences	Baek et al. 1997	2.1	0.1–6.2	Korean summer and winter
13 "At home," winter	Fugas et al. 1988	5.4	0.1–13	
12 "At home," summer	Fugas et al. 1988	1.7	0.1–3.6	
23 Bedrooms, heart patients	Colome et al. 1987	2.9 ± 2.4[a]	-	
19 Kitchens, heart patients	Colome at al., 1987	4.2 ± 2.0[a]	-	
22 Living rooms, heart patients	Colome et al. 1987	3.6 ± 1.8[a]	-	
Airtight woodstoves	Traynor et al. 1987	0.7–2.8	1.2–3.8	0.4-0.9 ACH

Table 8.2 continued

Location/Situation	Reference	CO Concentrations, ppm Mean	Range	Comment
Non-airtight woodstoves	Traynor et al. 1987	1.8–14	3.5–43	0.6-0.7 ACH
C/R gas heater present	Jackson et al. 1988	-	1.2–4.7	Convective/radiant heater
R/R gas heater present	Jackson et al. 1988	-	4.7–12.6	Radiant/radiant heater
From two parties	Bridge and Corn, 1972	-	1–2	
Indoors, urban	Rowe et al. 1989	6–16	-	8-h average
Workplaces				
"Worst" environments	Nguyen and Goyer, 1988	-	1–5	Selected by workers
13 "At work", winter	Fugas et al. 1988	3.7	2.6–4.5	
12 "At work", summer	Fugas et al. 1988	2.2	0.1–4.5	
3 Office towers	Nguyen and Martel, 1987	-	2.2–5.7	Survey of "tight" buildings
44 Light-duty workrooms	Weber and Fischer, 1980	2.8 ± 1.6[a]	Max 8.9	490 measurements
44 Light-duty workroom	Weber and Fischer, 1980	1.1 ± 1.3[a]	Max 6.5	353 measurements, corrected for unoccupied rooms
44 Light-duty workrooms	Weber and Fischer, 1980	0.8 ± 1.1[a]	Max 6.2	485 measurements, corrected for outdoors
12 Restaurants	Baek et al. 1997	11.3	0.7– 89.9	Korean summer and winter
23 Bars and shops with smokers	Valerio et al. 1997	7.0		Street-level businesses
15 Bars and shops without smokers	Valerio et al. 1997	6.2		Street-level businesses
Office buildings				
Office buildings	E. Sterling et al. 1988	2.65	Max 245	Median of 241 measurements
10 Nonsmoking offices	Proctor, 1989	1.2	0–7.4	Median value 1.2 ppm
6 Office bldgs.	Turk et al. 1987		2.1–7.0	Some with underground parking
12 Offices	Baek et al. 1997	2.4	0.4 – 6.8	Korean summer and winter
4 Office bldgs.—nonsmoking	Hedge et al. 1994	0.3	0.2 – 0.7	20 locations: restricted smoking to separately ventilated areas
4 Office bldgs.— smoking	Hedge et al. 1994	1.5	0.7 – 3.4	12 locations: Restricted smoking to separately ventilated areas
6 Office bldgs.—nonsmoking	Hedge et al. 1994	0.2	0.1–0.3	26 locations: Restricted smoking to rooms with local air filtration

6	Office bldgs.—smoking	Hedge et al. 1994	1.2	0.7–2.1	22 locations Restricted smoking to rooms with local air filtration
5	Office bldgs.—nonsmoking	Hedge et al. 1994	0.4	0.1–2.1	23 locations: Restricted smoking to rooms with no additional. air treatment
5	Office bldgs.—smoking	Hedge et al. 1994	1.0	0.4–2.2	13 locations: Restricted smoking to rooms with no additional air treatment
Transportation					
23	Nonsmoking commercial aircraft flights	Nagda et al. 1992	0.6		
69	Commercial aircraft flights	Nagda et al. 1992	1.4		Smoking section
69	Commercial aircraft flights	Nagda et al. 1992	0.8		Remote area of nonsmoking section
13	"Commuting," winter	Fugas et al. 1988	9.9	4.8–19	
12	"Commuting," summer	Fugas et al. 1988	5.8	2.8–10.4	
10	Nonsmoking train compartments	Proctor, 1989	1.3 ±0.7	0.5–2.9	Median value 1.3 ppm

[a]Mean ± standard deviation.

h = hour

Table 8.3 Personal Exposure Monitoring of CO[a]

Location	Denver, CO			Washington, D.C.		
	Sample Size person-days	Mean CO ppm	Median Time minutes	Sample Size person-days	Mean CO ppm	Median Time minutes
Indoors, parking garage	31	18.8	14	59	10.4	11
In transit, automobile	634	8.0	71	592	5.0	79
In transit, other	107	7.9	66	130	3.6	49
Outdoors, near roadway	188	3.8	33	164	2.6	20
In transit, walking	171	4.2	28	226	2.4	32
Indoors, restaurant	205	4.2	58	170	2.1	45
Indoors, office	283	3.0	478	349	1.9	428
Indoors, store/mall	243	3.6	50	225	2.5	36
Indoors, residence	776	1.7	975	705	1.2	1048

[a]From Akland et al. 1985.

with attached or underground garages. Measurements in the ice-skating rink are further confounded by routine use of the gasoline-powered ice-resurfacing machine. The elevated concentrations reported under "transportation" sources reflect a train car with 1 to 18 smoking occupants (Harmsden and Effenberger 1957) and an intercity bus with 23 cigarettes smoked continuously (DOT 1973). Bars, taverns, restaurants, and cafeterias are subject to CO sources discussed above as well as to two additional sources: cooking (particularly with a gas-fired stove) and proximity to vehicular-generated CO. The latter arises because these establishments are frequently located on major thoroughfares where traffic is heavy. Such considerations may explain the approximate agreement between extreme smoking and control values for bars and taverns (Sebben et al. 1977).

Numerous major studies illustrate the range of CO concentrations encountered in ETS-containing environments. The following paragraphs summarize observations from four of these studies.

Healthy Buildings International (HBI) Study. Table 8.4 presents summary statistics from a study (Healthy Buildings International 1990, Turner et al. 1992) that described indoor air measurements of particulate matter, CO, CO_2, and nicotine for nearly 600 situations. In general, the ranges and means of the data in each situation fall within those listed for Table 8.1; the two mean values for offices are slightly higher in the present study, but certainly reasonable. Because individual data values were available, it was possible to calculate the values presented in Table 8.4 and to perform a Student's t-test comparison of the means for smoking vs. control (no cigarette smoking observed and nicotine below detection limits; assumed to be a nonsmoking condition) in all situations except for bars (only two values available, both for smoking). In all situations, mean CO levels were higher for the smoking situations, but the differences were not statistically significant from the control situations at the 95% confidence level.

Pacific Northwest Commercial Buildings Study. Turk et al. (1987) performed extensive measurements of pollutants in 38 commercial buildings in the Pacific Northwest. Their work included CO measurements from building-wide averages, in specific areas of given buildings, and background. Indoor sampling for CO was performed at 32 buildings while outdoor CO samples were taken at 11 buildings. Additional ambient CO concentrations measured independently by local air pollution control agencies were available for 26 buildings. The authors performed their measurements using an LBL ConstantFlow Gas Collection Bag and a GE Model 15EC53COI Electrochemical Analyzer.

The 32-building indoor results yielded only 6 building-average values higher than 2 ppm, the minimum detectable level of the sampling system. Table 8.5 summarizes the values in those six buildings. In 4 of these buildings, all sample sites recorded above the detection limit, while only 36 of the 126 total indoor sample sites were above the detection limit. Although these readings were lower than the 8-h EPA standard and ASHRAE

Table 8.4 Summary Statistics for CO Concentrations (ppm) Reported in the HBI Study[a]

Situation	Range	Mean	Std Dev[b]	Number of Observations
Offices				
Smoking	1.4-8.7	3..6	1.3	118
Control[c]	2.0-6.6	3.3	1.0	70
Other Workplaces				
Smoking	1.7-6.7	3.2	0.9	142
Control[c]	2.0-6.4	3.0	0.8	135
Public Buildings				
Smoking	2.2-6.4	3.8	1.3	21
Control[c]	2.0-7.9	3.4	1.2	22
Restaurants and Cafeterias				
Smoking	2.0-7.9	3.4	1.2	49
Control[c]	2.0-3.0	2.9	0.5	11
Bars and Taverns				
Smoking	3.0-3.2	3.1	0.1	2

[a]Healthy Building International, 1990 and Turner et al. 1992.
[b]Standard deviation of sample data.
[c]Control situation considered where no cigarettes were observed to have been smoked and nicotine levels were below limits of detection.

Table 8.5 Summary of Elevated CO Concentrations from the Pacific Northwest Building Study

Building No.	Underground Parking?	City/Season	CO Concentration, ppm Range Indoor	Outdoor	Background
6	Y	Portland/W	3.3–7.0	NM	2[a]
7	N	Portland/W	3.0–10	NM	5[a]
27	Y	Spokane/S	2.5–3.0	5.0	5[a]
34	N	Spokane/W	2.2–6.0	BD	3[a]
35	N	Spokane/W	2.1–3.5	NM	3[a]
37	Y	Portland/W	2.7–3.0	NM	3[a]

[a]Fixed site monitoring by outside agency.
NM = not measured.
BD = below 2-ppm detection limit.

guidelines, short-term concentrations could have been much higher since the sampling method averaged over an 8-h period. Buildings #7 and #35 show outdoor CO concentrations equal to or slightly exceeding building averages. Since neither building had an attached parking garage, and other indoor sources were not identified, the elevated indoor readings were probably due to the outdoor air, presumably from locally heavy vehicular traffic.

Elevated CO concentrations in Buildings #6, #27, and #37 were probably the result of heavy automobile exhaust emissions in their underground parking areas. During the test of Building #6, the local air pollution monitoring agency reported only 2 ppm average outdoor CO concentration, whereas this value was exceeded by 6 of the 7 interior readings taken in this building. Only in Building #34, where a CO level of 6 ppm was measured in a smoky lunch room, was tobacco smoking suspected to be the dominant source. Outdoor levels were below the detection limit at the building and were 3 ppm as measured by the Spokane County Air Pollution Control Authority approximately 0.3 km from the building. Another indication of smoking in this area is the respirable suspended particulate (RSP) concentration of 116 μg/m^3, well above average. Nitrogen dioxide was 29 ppb (for an 80-h test period), the highest recorded for this building. Formaldehyde, which might also be expected to occur at high levels where heavy smoking occurs, was below detection at this site. While not unequivocal, it seems justified to conclude that heavy smoking in this lunchroom is the dominant CO source.

Only 2 site concentrations approached the EPA-NAAQS outdoor standards and ASHRAE guidelines for CO concentration of 9 ppm for 8-h exposure. The Building #6 reading of 7 ppm occurred near an underground parking garage and exposures to persons working in this area may rise to higher levels under circumstances of heavier parking load, different ventilation regimes, or higher personal travel in the garage space. The Building #34 local reading of 6 ppm seemed to be due to heavy smoking. During high use periods, i.e., lunch hours, the concentration may rise much higher.

Tobacco Industry ETS Restaurant Database (Six-Cities Study)[2]. One of the most extensive databases for ETS has been developed by Lorillard and the R. J. Reynolds Tobacco Company, which compared smoking environments in six cities, viz. Winston-Salem, NC (two studies performed); Greensboro, NC; Dallas, TX; Washington, D.C.; New York City, NY; and Hong Kong. The study results included the number of smokers present, room area and volume, and the concentrations of species of interest, including nicotine, RSP, and CO. Background concentrations for CO were provided for Washington, D.C., Hong Kong, and the second Winston-Salem, NC, study. Table 8.6 provides statistical descriptive measurements are given in Table 8.6.

Table 8.6 Summary of CO Concentrations Taken from the Tobacco Industry Six-Cities Restaurant Study[a]

Parameter	Indoors	Outdoors
Hong Kong, Washington, and Winston-Salem (2nd study)		
Range	-1.5–42.3	-0.3–13.7
Arithmetic mean, ppm	4.2	2.5
Arithmetic standard deviation, ppm	2.73	2.1
Geometric mean, ppm	1.7	1.3
Number of observations	99	99
All Six Cities		
Range	-1.5–42.3	
Arithmetic mean, ppm	3.7	—[b]
Arithmetic standard deviation, ppm	3.7	—[b]
Geometric mean, ppm	1.6	—[b]
Number of observations	184	—[b]

[a]Data from Oldaker et al. 1990; Oldaker et al. 1990a, 1991; Crouse and Oldaker 1990; Crouse and Carson 1989.
[b]Not available for all six cities.

The following are observations from the authors' review of the database.

1.   The range of CO background concentrations varied among the test locations. The upper bound for less-densely-populated Winston-Salem, NC, 3.5 ppm, compared to heavily urbanized Hong Kong, 12.9 ppm. Background data for Washington, D.C. were intermediate, with an upper bound of 9.2 ppm. In all three cases, the lowest value was 0.2 ppm (possibly the detection limit).

2.   In all, 99 background measurements were performed, breaking down to 0–1 ppm (18), 1–2 ppm. (34), 2–3 ppm (22), and >3 ppm (25). Only 1 value exceeded 10 ppm, viz. Hong Kong, with 12.9 ppm. The arithmetic and geometric means of the background data were 2.54 and 1.32 ppm, respectively, with an arithmetic standard deviation of 2.05 ppm.

---

[2]As part of a large, integrated investigation, detailed studies of nicotine and other component levels in the restaurants and offices of individual cities have been reported in the open literature (Oldaker et al. 1990, 1990a, 1991; Crouse and Oldaker 1990; Crouse and Carson 1989; Oldaker and McBride 1988). Additional, unpublished data from these studies was obtained from the investigators through the Center for Indoor Air Research, and evaluated independently by the authors of this monograph.

3.   The 99 indoor CO measurements associated with the background values reported above exhibited arithmetic and geometric means of 4.20 and 1.70 ppm, respectively, with an arithmetic standard deviation of 2.73 ppm. A Student's t-test performed between the background and indoor CO measurements from Winston-Salem; Washington D.C.; and Hong Kong showed no statistically significant difference at the 95% confidence level.

4.   Data from all six cities showed that CO concentrations do not reflect either the number of smokers or cigarettes smoked in a given room of known volume. As a typical example, in the Hong Kong data, an indoor concentration of 2.5 ppm was achieved by 4 smokers smoking a total of 5 cigarettes in restaurants with a volume of either 79 or 7647 m^3. Approximately the same concentration, 3.3 ppm, was observed in a Greensboro, NC, restaurant with a room volume of 510 m^3, where 11 smokers smoked 11 cigarettes.

5.   The measured concentrations of CO do not correlate well, if at all, with those of other measured species, viz. UV-PM (see Chap. 6), RSP, and nicotine. In the specific case of CO levels equal to 1.7 ppm, for example, the RSP concentrations ranged almost twentyfold, between 19.0 and 231.6 $\mu g/m^3$. A similar situation was observed for CO levels at 2.9 ppm, where the RSP concentrations differed tenfold, between 16.0 and 165.1 $\mu g/m^3$. The RSP concentration exceeded 400 $\mu g/m^3$ in only 3 measurements over the CO range of 0–10 ppm. The relationship was equally poor between CO and nicotine. Again, at 1.7 ppm CO the measured concentrations of nicotine ranged between <1.6–26.3 $\mu g/m^3$. A plot of UVPM vs. indoor concentration shows essentially random scatter. It was not possible to make a smoking/nonsmoking comparison because of the very limited number of situations in the latter category.

6.   Correcting the indoor CO concentration for outdoor background frequently yielded a negative final value. Such a problem would result if the two values were essentially identical, as demonstrated from the Student's-t test data described above.

7.   The distribution of CO levels, as shown in Fig. 8.1 exhibits a mode value of approximately 2 ppm and appears log-normally distributed with values extending out to 14 ppm.

Tobacco Industry ETS Office Database (Five-Cities Study)[2]. The Five Cities Study considered normal smoking-related parameters from offices in New York City; Ottawa, Winston-Salem, Dallas, and Washington, D.C. The study measured indoor CO concentrations for all cities, but only outdoor background data for Ottawa and Washington (see Table 8.7 for the descriptive statistics). The following are observations from the authors' review of the database.

1.   The data were statistically indistinguishable using the mean values (2.3 vs. 2.6 ppm), arithmetic standard deviations (2.0 vs. 2.3 ppm) and ranges (0.1– 10.5 vs. NA–10.4 ppm) for indoors vs. outdoors, respectively, in Ottawa and Washington.

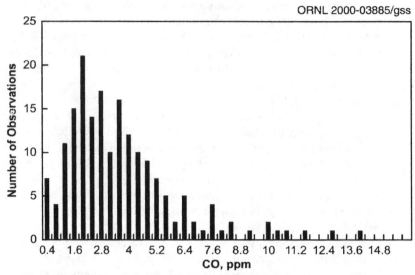

Fig. 8.1. Distribution of carbon monoxide (CO) concentrations for restaurants in six cities. (Data combined from several studies.)

Table 8.7 Summary of CO Data Taken from Tobacco Industry Five-City Office Study[a]

Parameter	Indoors	Outdoors
Ottawa and Washington, D.C.		
Range	0.1–10.5	NA–10.4
Arithmetic mean, ppm	2.3	2.6
Arithmetic standard deviation, ppm	2.0	2.3
Number of observations	66	57
All Five Cities		
Range	-1.1–10.5	_[b]
Arithmetic mean, ppm	2.1	_[b]
Arithmetic standard deviation, ppm	1.5	_[b]
Number of observations	144	_[b]

[a]Data from Oldaker et al. 1990a; Oldaker et al. 1991; Crouse and Oldaker 1990; Crouse and Carson 1989, Oldaker and McBride 1988.
[b]Background data not available for New York City, NY; Winston-Salem, NC; and Dallas, TX.

2.   The range, arithmetic mean, and arithmetic standard deviation of the outdoor concentrations from all five cities agree well with those found in Ottawa and Washington.

3.   There was no correlation between CO and the number of smokers found in a given room. Also, it was not possible to make a smoking/nonsmoking comparison, because of the very limited number of situations in the latter category.

4.   A distinct clustering of data was observed when the indoor CO concentration was plotted against a number of other cigarette-related parameters. Generally, most of the independent values appeared when the CO concentration ranged between 0–4 ppm. This range accounted for the bulk of the nicotine (0–10 $\mu g/m^3$), RSP (0–300 $\mu g/m^3$), and UVPM (0–50 $\mu g/m^3$) concentrations, as well as number of cigarettes (0–5), room volume (0–500 m^3), and number of smokers (2–4). This clustering was in marked contrast to the industry restaurant data (above), where the scatter of the data was fairly random and uniform. This suggests "typical" values for all the parameters mentioned.

5.   A specific relationship between the indoor CO concentration and the various parameters described above was not observed, with the exception of the clustering.

The distribution of CO concentrations observed in the Five-Cities Office Study is shown in Fig. 8.2. The distribution is very similar to that observed in the Six-Cities Restaurant Study described earlier (Fig. 8.1) in that the mode value is approximately 2 ppm. In general, the data also appear normally distributed with values extending to ~11 ppm.

It is only logical that cigarette smoking contributes to the indoor air burden of CO because cigarette smoking releases significant (in the sense of measurable) quantities of CO. The significance (in the sense of percentage contribution to the total CO burden) of the contribution is generally difficult to quantify in natural indoor environments where smoking occurs normally. Circumstances or environments involving heavy smoking, especially coupled with poor ventilation, result in or are accompanied by readily detectable elevated CO concentrations. The relative physiological importance of long-term exposure to barely detectable elevated CO concentrations and short-term exposure to readily detectable elevated concentrations is unknown. The small contribution of ETS to ambient indoor CO concentrations and the variability due to other sources of CO suggest that CO uptake is not a reliable measure of exposure to ETS.

SUMMARY

Although CO is a major component of sidestream smoke, it is diluted rapidly in indoor air. In spite of the ready availability of acceptable analytical methods and many careful independent studies, it has been difficult to discern the effect of CO from ETS in any situation except extreme conditions of experimentally designed exposure. CO has been measured repeatedly in rooms where there is adequate ventilation, with and without cigarette smoking. In

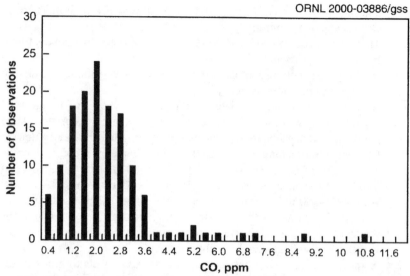

Fig. 8.2. Distribution of carbon monoxide (CO) concentrations for smoking offices in five cities. (Data combined from several studies.)

many cases, the difference in values is small, and is easily masked by either normal variation in the data or the precision of the analytical measurement. Studies that measure the concentration of COHb in both ETS-exposed and non-exposed subjects is consistent with environmental CO measurements, and have demonstrated no consistent significant differences in most indoor environments. The chief difficulty is interference from other sources of CO, particularly cooking, heating, and vehicle exhaust. One author (Eatough 1988) states that about 90% of all CO arises from sources other than ETS.

Taken together, the public probably encounters CO concentrations not higher than 4 ppm CO at any given time, assuming a reasonable air exchange rate in a given room. This expected value may vary slightly, dropping for offices and workplaces (often 2–3 ppm is typical) and increasing for public functions and restaurants (4–5 ppm frequently reported). CO concentrations exceeding the ASHRAE standard, which is 9 ppm, are comparatively rare, and may be observed under conditions where air exchange is limited, the number of occupants is large, and combustion sources such as cooking or transportation, are nearby.

REFERENCES

Akland, G. G., Hartwell, T. D., Johnson, T. R., & Whitmore, R. W. (1985) Measuring human exposure to CO in Washington, D.C., and Denver, Colorado, during the winter of 1982–1983. *Environ. Sci. Technol.*, 19, 911–918.

ASHRAE 62-1989 (1989) *ASHRAE Standard.—Ventilation for Acceptable Indoor Air Quality, Atlanta, GA*, American Society of Heating, Refrigerating, and Air Conditioning Engineers, Inc., pp. 1–26.

Aviado, D. M. (1988) Suspected pulmonary carcinogens in environmental tobacco smoke. *Environ. Technol. Letts.*, 9, 539–544.

Badre, R., Guillerm, R., Abran, N., Bourdin, M. & Dumas, C. (1978) Atmospheric pollution by smoking. *Annales Pharmaceutiques Francaises*, 36 (9–10), 443–452.

Baek, S. O., Kim, U. S., & Perry, R. (1997) Indoor air quality in homes, offices, and restaurants in Korean urban areas—indoor/outdoor relationships. *Atmos. Environ.*, 31, 529–544.

Baker, R. R., Case, P. D., & Warren, N. D (1988) The build-up and decay of environmental tobacco smoke constituents as a function of room conditions. In: Perry, R. & Kirk, P. W. W., eds., *Indoor and Ambient Air Quality*, London, Selper, Ltd., pp. 121–130.

Bridge, D. P. & Corn, M. (1972) Contribution to the assessment of nonsmokers to air pollution from cigarette and cigar smoke in occupied spaces. *Environ. Res.*, 5, 192–209.

Colome, S. D., Lambert, W. E., & Castaneda, N. (1987) Determinants of CO exposure in residences of ischemic heart disease patients. In: Seifert, B., Esdorn, H., Fischer, M., Ruden, H., & Wegner, J., eds., *Proceedings of the 4th International Conference on Indoor Air Quality and Climate, Vol. 1*, pp. 283–287.

Crouse, W. E. & Carson, J. R. (1989) Surveys of environmental tobacco smoke (ETS) in Washington, D.C. offices and restaurants. *43rd Tobacco Chemists' Research Conference*, Richmond, VA.

Crouse, W. E. & Oldaker, G. B. (1990) Comparison of area and personal sampling methods for determining nicotine in environmental tobacco smoke. 1990 *EPA/AWMA Conference on Toxic and Related Air Pollutants*, Raleigh, NC.

(DOT) Department of Transportation (1971). *Health Aspects of Smoking in Transport Aircraft*, Washington, D.C., U.S. National Technical Information Service.

(DOT) Department of Transportation (1973). *CO as an Indicator of Cigarette-caused Pollution Levels in Intercity Buses*.

Eatough, D. J., Hansen, L. D., & Lewis, E. A. (1988) Assessing exposure to environmental tobacco smoke. In: Perry, R. & Kirk, P. W. W., eds., *Indoor and Ambient Air Quality*, London, Selper, Ltd., pp. 131–140.

Elliot, L. P. & Rowe, D. R. *(1975)* Air quality during public gatherings. *JAPCA*, 25(6), 635–636.

First, M. W. (1983) Environmental tobacco smoke measurements: retrospect and prospect. In: Rylander, R., Peterson, Y., & Snella, M.-C., eds., ETS-*Environmental Tobacco Smoke. Report from a Workshop on Effects and Exposure Levels,* pp. 916.

Fugas, M., Sega, K., & Sisovic, A. (1988) Indoor air quality—a relevant factor in human exposure. In: Perry, R. & Kirk, P. W., eds., *Indoor and Ambient Air Quality,* London, Selper Ltd., pp. 287–292.

Haley, N. J., Sepkovic, D. W., Brunnemann, K. D., & Hoffmann, D. (1988) Biomarkers for assessing environmental tobacco smoke uptake. *Proceedings, Air Pollution Control Association,* Niagara Falls, New York.

Harke, H. P. & Peters, H. (1974) The problem of passive smoking. 111. The influence of smoking on the CO concentration in driving automobiles. *Int. Arch. Arbeitsmed.,* 33, 221–229.

Harmsden, H. & Effenberger, E. (1957) Tabakrauch in verkehrsmitteln, wohn-und arbeitsraumen [Tobacco smoke in transportation vehicles, living and working rooms. *Archiv fur Hygiene und Bakertiologie,* 141(5), 383–400.

Healthy Buildings International (1990) *Measurements of environmental tobacco smoke in general office areas.* Fairfax, VA.

Hedge, A., Erickson, W. A., & Rubin, G. (1994). The effects of alternative smoking policies on indoor air quality in 27 smoking office buildings. *Ann. Occup. Hyg.,* 38, 265–278.

Hoffmann, D. & Hoffmann, 1., (1987) Significance of exposure to sidestream tobacco smoke, In: O'Neill, 1. K., Brunnemann, K. D., Dodet, B., & Hoffmann, D., eds., *Environmental Carcinogens—Methods of Analysis and Exposure Measurement. Vol. 9, Passive Smoking (1ARC Scientific Publications No. 81),* International Agency for Research on Cancer, Lyon, France, pp. 3–10.

Houck, J. E., Simmons, C. A., & Snow, G. C. (1988) The impact of residential wood combustion on ambient wintertime carbon monoxide concentrations in residential areas in six northwestern cities. *Proceedings, 1988 EPA/APCA International Symposium on Measurement of Toxic and Related Air Pollutants,* Research Triangle Park, NC, pp. 664–669.

Jackson, M. D., Clayton, R. K., & Stephenson, E. E., Jr. (1988) EPA's indoor air quality test house. 2. Kerosene heater studies. *Proceedings, 1988 EPA/APCA International Symposium on Measurement of Toxic and Related Air Pollutants,* Research Triangle Park, NC, pp. 715–719.

Jarvis, M. J., Russell, M. A. H., & Feyerabend, C. (1983) Absorption of nicotine and carbon monoxide from passive smoking under natural conditions of exposure. *Thorax,* 21, 994–999.

Jarvis, M., Tunstall-Pedoe, H., Feyerabend, C., Vesey, C., & Salloojee, Y. (1984) Biochemical markers of smoke absorption and self reported exposure to passive smoking. *J. Epidemiol. and Community Health,* 38, 335–339.

Jenkins, R. A. & Guerin, M. R. (1984) Analytical chemical methods for the detection of environmental tobacco smoke constituents. *European J. Resp. Dis.* 65 (Supplementum No. 133), 33–46.

Jenkins, R. A., Holmberg, R. W., Wike, J. S., Moneyhun, J. H., & Brazell, R. S. (1984) *Chemical Characterization and Toxicologic Evaluation of Airborne Mixtures—Chemical and Physical Characterization of Diesel Fuel Smoke.* ORNL/TM-9196, Oak Ridge National Laboratory, Oak Ridge, TN, June 1984.

Kandarjian, L. (1989) Federal policy options for indoor air pollution from combustion appliances. In: Harper, J. P., ed., *Combustion Processes and the Quality of the Indoor Environment,* Air & Waste Management Association, Pittsburgh, PA, pp. 310–321.

Kirk, P. W. W., Hunter, M., Baek, S. O., Lester, J. N. & Perry, R. (1988) Environmental tobacco smoke in indoor air. In: Perry, R. & Kirk, P. W. W., eds., *Indoor and Ambient Air Quality,* London, Selper, Ltd., pp. 99–112.

Klus, H., Begutter, H., Ball, M., & Intorp, M. (1987) Environmental tobacco smoke in real life situations. *Proceedings of the 4ᵗʰ International Conference on Indoor Air Quality and Climate, Vol.* 2. Berlin, pp. 137–141.

Lambert, J. L., Liaw, Y.-L., Paukstelis, J. V., & Chiang, Y. C. (1987) Palladium (II) acetamide complex as a solid monitoring reagent for CO. *Environ. Sci. Technol.,* 21, 500–503.

Lebret, E., Noy, D., Boleij, J., & Brunekreef, B. (1987) Real-time concentration measurements of CO and N02 in twelve homes. In: Seifert, B., Esdorn, H., Fischer, M., Ruden, H., & Wegner, J., eds., *Proceedings of the 4ᵗʰ International Conference on Indoor Air Quality and Climate, Vol.* 1, Berlin, pp. 435–439.

Lodge, J. P., Jr., ed. (1990) *Methods of Air Sampling and Analysis, Third Edition,* Lewis Publishers, Inc., Chelsea, MI, pp. 296–306.

Maugh, T. H., II (1972) Carbon monoxide: Natural sources dwarf man's output. *Science,* 177, 338–339.

Nagda, N. L., Koontz, M. D., Konheim, A. G., & Hammond, S. K. (1992) Measurement of cabin air quality aboard commercial airlines, *Atmos. Environ.,* 26A, 2203–2310.

Nguyen, V. H. & Goyer, N. (1988) A global approach to investigate indoor air quality and performance of the ventilation system. In: Perry, R. & Kirk, P. W., eds., *Indoor and Ambient Air Quality,* London, Selper Ltd., pp. 327–332.

Nguyen, V. H. & Martel, J.-G. (1987) A field study in three office towers in Quebec, Canada. In: Seifert, B., Esdorn, H., Fischer, M., Ruden, H., & Wegner, J., eds., *Proceedings of the 4ᵗʰ International Conference on Indoor Air Quality and Climate, Vol.* 2, Berlin, pp. 512–514.

Oldaker, G. B. & McBride, R. L. (1988) Portable air sampling system for surveying levels of environmental tobacco smoke in public places. *Symposium on Environment and Heritage. World Environment Day, Hong Kong, 1988,* Hong Kong University, Hong Kong, June 6, 1988.

Oldaker, G. B., Perfetti, P. F., Conrad, F. C., Jr., Conner, J. M., & McBride, R. L. (1990) Results of surveys of environmental tobacco smoke in offices and restaurants. *Int. Arch. Occup. Environ. Health,* 99–104.

Oldaker, G. B., Ogden, M. W., Maiolo, K. C., Conner, J. M., Conrad, F. C., Jr., Stancill, M. W., & DeLuca, P. O. (1990a) Results from surveys of environmental tobacco smoke in restaurants in Winston-Salem, North Carolina. *Proceedings of the 5th International Conference on Indoor Air Quality and Climate, Vol. 2*, pp. 281–285.

Oldaker, B. G., Stancill, M. W., Conrad, F. W., Jr., Morgan, W. T., Collie, B. B., Fenner, R. A., Lephardt, J. 0., Baker, P. G., Lyons-Hart, J., & Parrish, M. E. (1991) Results from a survey of environmental tobacco smoke in Hong Kong restaurants. *Personal communication.*

Pennanen, A. S., Salonen, R. O., Alm, S., Jantunen, M. J., & Pasanen, P. (1997) Characterization of air quality problems in five Finnish indoor ice arenas. *J. Air & Waste Manage. Assoc.*, 47, 1079–1086.

Proctor, C. J. (1989) A comparison of the volatile organic compounds present in the air of real-world environments with and without environmental tobacco smoke. *82nd Annual Meeting of the Air and Waste Management Association*, Anaheim, CA.

Repace, J. L. (1987) Indoor concentrations of environmental tobacco smoke: models dealing with effects of ventilation and room size. In: O'Neill, I. K., Brunnemann, K. D., Dodet, B., & Hoffmann, D., eds., *Environmental Carcinogens. Methods of Analysis and Exposure Measurement. Vol. 9, Passive Smoking, (IARC Scientific Publications No. 81)*, International Agency for Research on Cancer, Lyon, France, pp. 25–41.

Rowe, D. R., Al-Dhowalia, K. H, & Mansour, M. E. (1989) Indoor-outdoor carbon monoxide concentrations at four sites in Riyadh, Saudi Arabia. *JAPCA*, 39(8), 1100–1102.

Scherer, G., Westphal, K., Sorsa, M., & Adlkofer, F. (1988) Quantitative and qualitative differences in tobacco smoke uptake between active and passive smoking. In: Perry, R. & Kirk, P. W., eds., *Indoor and Ambient Air Quality*, London, Selper Ltd., pp. 189–194.

Schievelbein, H. & Richter, F. (1984) The influence of passive smoking on the cardiovascular system. *Preven. Med.*, 13, 626–644.

Sebben, J., Pimm, P., & Shepard, R. J. (1977) Cigarette smoke in enclosed public facilities. *Arch. of Env. Health*, 32(2), 52–58.

Spengler, J. D. & Sexton, K. (1983) Indoor air pollution: a public health perspective. *Science,* 221(4605), 9–17.

Sterling, E. M., Collett, C. W., Kleven, S., & Arundel, A. (1988) Typical pollutant concentrations in public buildings. In: Perry, R. & Kirk, P. W. W., eds., *Indoor and Ambient Air Quality*, London, Selper, Ltd., pp. 399–404.

Sterling, T. D., Collett, C. W., & Sterling, E. M. (1987) Environmental tobacco smoke and indoor air quality in modern office work environments. *J. Occupat. Med.*, 29(l), 57–62.

Traynor, G. W., Apte, M. G., Carruthers, A. R., Dillworth, J. F., Grimsrud, D. T., & Gundel, L. A. (1987) Indoor air pollution due to emissions from wood-burning stoves. *Environ. Sci. Technol.*, 21, 691–697.

Turk, A. (1963) Measurements of odorous vapors in test chambers, theoretical. *ASHRAE J.*, 5(10), 55–58.

Turk, B. H., Brown, J. T., Geisling-Sobotka, K., Froehlich, D. A., Grimsrud, D. T., Harrison, J., Koonce, J. F., Prill, R. J., & Rezvan, K. L. (1987) *Indoor Air Quality and Ventilation Measurements in 38 Pacific Northwest Commercial Buildings, Vol. I—Measurement Results and Interpretation (LBL~22315 112)*, Lawrence Berkeley Laboratory, Berkeley, CA.

Turner, S., Cyr, L., & Gross, A. J. (1992) The measurement of environmental tobacco smoke in 585 office environments. *Environ. Int.*, 18, 19–28.

Valerio, F., Pala, M., Lazzarotto, A., & Balducci, D. (1997) Preliminary evaluation, using passive tubes, of carbon monoxide concentrations in outdoor and indoor air at street level shops in Genoa (Italy) *Atmos. Environ.*, 31, 2871–2876.

Weber A. & Fischer, T. (1980) Passive smoking at work. *Int. Arch. Occup. Environ. Health*, 47, 209–221.

Weinstock, B. & Niki, H. (1972). Carbon monoxide balance in nature. *Science*, 176, 290–292.

INTRODUCTION

The oxides of nitrogen, commonly abbreviated NO_x, are principally formed by the combustion of nitrogen-containing constituents of fuels and by the high temperature oxidation of nitrogen. The majority of the NO_x, (NO, N_2O, and NO_2) in tobacco smoke arises by oxidation of nitrogen bound in a variety of nitrogen-containing tobacco components such as nicotine alkaloids, amino acids, and proteins and by the thermal decomposition of nitrate. NO plus NO_2 comprise more than 90% of the oxides of nitrogen in cigarette smoke. As such, NO, is taken to mean NO plus NO_2 in describing oxides of nitrogen produced by tobacco smoking.

The Kentucky reference 1R4F filter cigarette delivers approximately 230 μg of NO in its mainstream smoke and 900 μg of NO in its sidestream (Chap. 3, Table 3.1). Commercial nonfilter cigarettes are reported to deliver from 100–600 μg/cigarette of NO in their mainstream smoke and to exhibit sidestream-to-mainstream ratios ranging from 4–10 (Chap. 3, Table 3.2). More than 95% by weight of the NO_x in both mainstream (Norman et al. 1983, Jenkins and Gill 1980) and sidestream (Norman et al. 1983) smoke is present as nitric oxide (NO). Martin et al. (1997) have completed a detailed analysis of NO and NO_2 emissions as ETS for the 50 top-selling brand styles of cigarettes in the United States. Average NO and NO_2 emissions for all brand styles were 1645 and 198 μg/cigarette, respectively. Note that in ETS chamber measurements, NO_2 represents ~11% of the NO_x, reflective the conversion of NO to NO_2 through reaction with ambient oxygen.

The acute and chronic health effects of NO include bronchitis, irritation of the nose and throat, increased susceptibility to respiratory infections, blocking of oxygen transfer in blood, and irritation of the skin and eyes at concentrations below the federal outdoor air standard of 53 ppb (ASHRAE 1989). NO_x has been considered an important constituent of mainstream cigarette smoke because of these effects and because of its possible role in the formation of N-nitrosamines.

The quantity of NO_x in cigarette smoke is much smaller than the quantities of CO, nicotine, and particulate matter. These lesser emissions, combined with significant emissions from many competing sources, make it more difficult to determine ETS-specific contributions to ambient NO_x concentrations.

SAMPLING AND ANALYSIS

Many methods have been employed for the determination of NO, NO_2, and NOx in mainstream cigarette smoke and in the outdoor atmosphere. The most commonly employed methods for the analysis of indoor air environments are summarized in the following.

Chemiluminescence Monitoring of NO_2 or NO. The most common and accepted procedure for measuring either NO or NO_2 involves chemiluminescence monitoring (Jenkins and Guerin 1984), which is extremely sensitive (direct measurement down to the part-per-billion range is possible), and, in most cases, very accurate. The analytical procedure, which normally includes "active" sampling of a test environment with a built-in sampling pump, is relatively straightforward. The weight and size of the NO_x analyzer preclude single-handed portability. In addition, the system requires tanks of gas standards for calibration. However, the equipment can be mounted in a moveable rack. The accuracy and sensitivity of the chemiluminescent analysis make it the method of choice for determination of NO_x concentrations.

In the NO analysis, ozone produced within the analyzer reacts with sample NO to produce NO_2. Approximately 10% of the NO_2 molecules are converted to an excited state, followed by immediate de-excitation and emission of photons. These impinge on a photo multiplier detector, generating current that is amplified to provide a readable output. The determination of NO_x (combined NO and NO_2) is identical with that for NO except that, before entry into the reaction chamber, the sample is routed through a converter where native NO_2 is catalytically reduced to form NO, according to $2\,NO_2 \rightarrow 2\,NO + O_2$. Instrument response is proportional to total NO in the converted sample, i.e., the sum of the NO present in the original sample plus the NO produced by the dissociation of NO_2.

A number of potential interferences can confound the interpretation of chemiluminescence results. Matthews et al. (1977) described quenching caused either by general "third body" quenching or by specific species. Underestimation errors caused by the former can range between 2–30%, depending on the exact converter conditions. Carbon dioxide and water, for example, can quench the chemiluminescent reaction, resulting in a value-biased low, but this is usually not a problem at ambient levels. Hydrogen cyanide, methylamine, and ammonia are converted into NO or NO_2 by the stainless steel reducing catalysts used in some of the converter assemblies at 68, 82, and 86% efficiency, respectively, and thus yield a positive interference if present. In addition, hydrogen cyanide has been shown to be a positive interference when analyzing concentrated smokes with carbon-based converter assemblies. The extent of all these interferences at ambient smoke levels is unknown; however, ambient concentrations of these species in a typical indoor atmosphere are not expected to cause significant bias. Low-biasing of results is conceivable in test chambers where the concentrations of smoke and related constituents are artificially high.

Electrochemical Analysis. A number of manufacturers sell instrumentation based on electrochemical detection of NO or NO_2. These units function by drawing air over a membrane-encased (usually) electrode, which is held at an applied potential necessary to oxidize or reduce the target analyte. The amount of current flowing through the electrode when the analyte is present is proportional to its concentration in the air. Potential interferences are numerous: any species that is oxidized or reduced at the same potential will provide a positive response. It should be noted that most of these systems are intended for industrial applications, where NO_x levels are expected to be 1 or 2 orders of magnitude higher than in environments typically encountered by the public, and NO_x is likely to be the primary constituent of interest.

Active Sampling of NO_x With Gas Sampling Tubes. Several manufacturers offer small gas sampling tubes packed with a solid reagent material that is selective for NO, NO_2, or NO_x. The glass tubes are precalibrated such that the concentration of desired analyte may be read directly from the tube body when a specified volume of gas (often pumped through the tube with a hand-activated pump) passes through the tube. One analytical method for detecting NO_2 involves reaction of the gas with o-toluidine and subsequent reduction to form yellowish-orange nitroso-o-toluidine. NO can be measured jointly with NO_2 with a two-stage sampling tube in which the former is oxidized to the latter with $CrO3$ and sulfuric acid, and the determination of the resulting NO_2 is identical with that described above. The ranges and detection limits, respectively, for each component are 10–300 ppm and 1 ppm for NO and 1–40 ppm and 0.5 ppm for NO_2. While these limits and ranges are considerably elevated compared to what is normally required for measurements involving indoor air, these values may be lowered easily by simply sampling more air. The only known interferences include gaseous halogens and their oxides, which are not normally present in indoor air and would not be produced by ETS.

An alternative strategy involves reaction of NO_2 with N,N'-diphenylbenzene to yield a characteristic bluish-grey reaction product. NO + NO_2 would be measured by first oxidizing NO to NO_2 with $CrO3$, then reacting the resulting NO_2 with N,N'-diphenylbenzene as described. The detection limits are similar to those cited earlier. The most significant interference to the second scheme is ozone, which can be a problem even at low ppm levels. Because 1 ppm ozone would yield approximately the same response as 0.5 ppm NO_2, and because low concentrations of ozone are possible in indoor air, particularly when large volumes are sampled, this latter strategy is discouraged as a method for estimating NO_2 or NO_x concentrations in this situation.

Passive Sampler Measurements of NO₂. The great sensitivity and selectivity of the chemiluminescence procedure described above are partially offset by the weight and bulk of both the instrument itself and the associated tanks of gas standards. While such equipment is entirely appropriate for monitoring NO/NO_2 levels in a room, for example, they are clearly unsuitable for personal sampling. Personal exposure monitoring is frequently carried out with a "passive" sampler, which is worn for a specified period by an individual and later analyzed. Palmes et al. (1976) tested a passive sampler for measuring NO_2 in underground mines. Triethanolamine (TEA) was coated onto stainless steel screens, which in turn were mounted into a small acrylic tube. The TEA-NO_2 complex formed was sufficiently stable that the trapped NO_2 could be stored for a considerable period before analysis. Final quantification was performed by reaction with N-1-naphthylethylenediamine-dihydrochloride to form a colored product, followed by an absorbance measurement of that product. The authors report that SO_2 is a potential interference; however, this material is typically not a major component of either indoor air or ETS.

A high-efficiency passive sampler capable of monitoring both NO_2 and formaldehyde has been recently described (Mulik et al. 1989) and tested extensively by the U. S. Environmental Protection Agency (EPA) and is now a standard OSHA method when used in an active sampling format. Glass fiber filters or molecular sieves are treated with TEA and used for the collection of NO_2. Subsequent analysis features the determination of nitrite by ion chromatography, thereby adding additional selectivity and sensitivity not possible with the method employed above. As with passive sampling in general (Chap. 5), the effective sampling rate is calculated from Fick's first law of diffusion, and was noted (Mulik et al. 1989) as 154 mL/min. Concentration measurements of NO_2 in ambient outdoor air with either the passive sampler or a tunable diode laser were virtually indistinguishable.

Lambert et al. (1987) have demonstrated that paper coated with 1-methylperimidine combined with calcium chloride is an effective trapping medium for NO_2. A visible-ultraviolet spectrophotometer modified for reflection measurements was used to complete the quantitation following passive sampling. In this instrument, the incident beam from the light source is deflected onto the reagent paper surface, then back to its original path to the photo multiplier tube. The loss of incident radiant energy is read as absorbance in the usual manner.

Method Intercomparison. The concentration of NO_2 in homes has been measured with both Palmes diffusion tubes and a NO_x chemiluminescence monitor as a reference (Boleij et al. 1986). The experimental design challenged the Palmes diffusion tubes with the indoor air in 9 homes (kitchen, living room, and bedroom) versus an outdoor blank for each with periods ranging between 3 to 12 days, and comparing the resulting data to those obtained with the chemiluminescence monitor. In most cases, the values obtained from both methods in samples of the outdoors, bedroom, and living room were comparable and typically ranged between ca. 21–48 ppb for all three sites. The greatest disparity in both the methods and in values obtained occurred in the kitchen. In kitchens where the NO_2 concentrations were below ~ 32 ppb, the agreement between the methods was quite reasonable. In kitchens where the concentrations exceeded 53 ppb, the methods disagreed by up to a factor of 2. The authors attempted to explain the difficulties by citing the following possible factors: (1) water vapor quenching of the chemiluminescence response; (2) inhomogeneous mixing of air; or (3) temperature, humidity, or "starvation" effects on the diffusion tubes. The authors concluded that the performance of the diffusion tubes, while certainly promising, required additional testing and evaluation of key parameters before being accepted for general use. In any event, the accuracy figure of ±10% for NO_2 measurements quoted from the studies of Apling et al. (1979) and Cadoff et al. (1979) seems too optimistic for the use of the tubes in homes.

FIELD STUDIES

Appendix 4 and Table 9.1 present background data (also see Chap. 2) from outdoor and indoor situations where no smoking was observed. In general, the means of the outdoor background concentration for NO_2 range between 11 and 51 ppb. Those for the indoor background concentrations where either the entire dwelling was sampled or a particular location was not specified are somewhat smaller, and range between 29–46 ppb. These values serve as benchmark data against which indoor NO_2 concentrations may be compared when smoking is present or suspected.

These typical indoor and outdoor background values are summarized from several selected major studies. Bouhuys et al. (1978), for example, calculated similar values for the outdoor background using repeated measurements from an urban and rural setting in Connecticut. In other studies, notably those by Baker et al. (1987) and Drye et al. (1989), indoor background concentrations of NO_2 were established from large numbers of samples that were similar to those calculated for outdoor background levels. Colome et al. (1987) evaluated NO_2 levels in the homes of heart patients where combustion sources tend to be restricted. The data closely match those of Baker et al. (1987) and Drye et al. (1989).

Drye et al. (1989) observed that the NO_2 concentrations are weakly dependent upon the seasons. The authors evaluated ~ 1000 ambient air measurements from 5 studies conducted in 4 distinct geographical areas (Boston Standard Metropolitan Statistical Area; Southern California; Portage, WI; and St. Louis, MO), and calculated summer and winter concentrations (mean -± standard deviation) of 29 ± 16 and 34 ± 22 ppb, respectively. In contrast, Kulkarni and Patil (1998) demonstrated major (factor of two) summer/winter fluctuations in personal exposure NO_2 levels.

Table 9.1 Background Concentrations of Nitrogen Dioxide (NO$_2$)

Situation	Concentration of NO$_2$, ppb	
	Range of Data	Means
Outdoors, background	5 – 64	11 – 51
Indoors, background	6 – 13	29 – 46
Indoors, with electric stoves		
Activity rooms	1 – 23	7 – 8
Bedrooms	1 – 33	6 – 8
Living rooms	8 – 51	22 – 24
Kitchens	9 – 65	6 – 26
Indoors, with gas stoves		
Activity rooms	0 – 204	35 – 41
Bedrooms	4 – 167	16 – 33
Kitchens	6 – 353	4 – 97
Average indoor	–	44 – 45

Values taken from Drye et al. 1989, Marbury et al. 1988, Petreas et al. 1988, Kulkarni and Patil 1998 and Appendix 4.

Two additional and more important contributors to indoor NO$_2$ background levels are interrelated, viz. the presence/absence of gas-fired appliances and the exact room of the dwelling (proximity to the appliance) where the sample is taken. At least four studies demonstrate this dependence clearly. Data from Drye et al. (1989), for example, show similar means for NO$_2$ measurements taken in bedrooms of private homes, but somewhat elevated concentrations for kitchens. Marbury et al. (1988) and Petreas et al. (1988) observed very little difference in NO$_2$ concentrations among kitchens, living rooms, activity rooms, and bedrooms in homes that used electric cooking; a typical value was 10 ppb. Sega and Fugas (1988) observed somewhat higher typical values, up to 28 ppb. The maximum and, to some degree, the mean, NO$_2$ levels were substantially elevated in homes where gas-fired stoves were present, compared to those without these devices. Sega and Fugas (1988), for example, observed a range of 6–204 ppb NO$_2$ in the living rooms of homes equipped with a gas-fired stove, but 8–33 ppb when electric stoves were present at the same time of year. These results agreed with those of Petreas et al. (1988), who reported NO$_2$ concentrations (mean ± standard deviation) of 20 ± 13 ppb and 8 ± 7 ppb NO$_2$ in the bedrooms of mobile homes equipped with gas-fired and electric stoves, respectively. The numbers of samples involved (230 vs. 46, respectively) allow a comparison of means with a Student's t-test, from which one concludes that the means are statistically different even at a 0.001 confidence level.

The maximum indoor background concentration of NO$_x$ normally does not exceed ~ 60 ppb. Variations are common, however, and normally relate to the season (winter), proximity to a primary source, type of heater and/or stove present (gas-fired), and whether those appliances are properly vented. Chapter 2 discusses other factors influencing NO$_x$ background concentrations.

The concentrations of NO_x measured in a variety of common smoking and control situations can vary between "not detected" (ND) and almost 600 ppb for NO_x and from ND–350 ppb for NO_2. The studies detailed in Appendix 4 and Table 9.2 show that the mean NO_x concentration typically ranges between 5–100 ppb, with the sole exception of bars or taverns. Similar work summarized in Appendix 4 and Table 9.3 shows that the mean NO_2 concentration typically does not exceed ~ 80 ppb. In some of these studies, control data have not been presented. The data ranges for NO_x and NO_2 in smoking and control situations frequently overlap, as shown in both Tables 9.2 and 9.3. Further, a comparison of Tables 9.1 and 9.3 shows that the NO_2 concentrations observed in smoking and nonsmoking situations frequently overlap those observed in both outdoor and indoor background situations, where no smoking was reported.

Table 9.2 Summary of Field Measurements for Total Oxides of Nitrogen (NO_x[a])

Situation	NO_x Values, ppb	
	Means	Range
Offices		
Smoking	42–51	-
Control	–	-
Other Workplaces		
Smoking	ND–82	ND–500
Control	27	ND–570
Restaurants and Cafeterias		
Smoking	5–120	-
Control[b]	4–115	2–218
Transportation		
Airports		
Smoking	–	150–330[c]
Nonsmoking	–	–
Bars		
Smoking	195	66–414
Control[b]	4–115	–
Residences		
Smoking	4–18	4–18
Nonsmoking	5	27

[a]Denotes either $NO + NO_2$ or NO alone.
[b]Outdoor air background. Triebig and Zober (1984).
[c]Measured in airport gate area; smoking not specifically excluded.
Taken from Thurston (1987).
ND = not detected.
Values drawn from Appendix 4.

Typically, the concentration of either NO_x or NO_2 is a few tens of ppb (nominal value 50 ppb), regardless of whether or not smoking is present. Such an estimate is supported by the extensive work of Weber and Fischer (1980), in which ~500 measurements were performed in 44 workrooms. The concentrations of NO and NO_2 observed in this study after correction for blank

Table 9.3 Summary of Field Measurements for Nitrogen Dioxide (NO₂)

	NO₂ Values, ppb	
	Means	Range
Offices		
Smoking	27 ± 11[a]	6 – 40
Control	14 ± 6[b]	6 – 21
Other Workplaces		
Smoking	2 – 60	ND – 200
Control	7	–
Restaurants and Cafeterias		
Smoking	24 – 76	15 – 105
Control[c]	34 – 63	–
Transportation (Airports)		
Smoking	–	218 – 350[d]
Control	–	–
Bars		
Smoking	21	1 – 61
Control[c]	34 – 63	–

ND = Not Detected

[a]10 buildings, open smoking policy, mean ± standard deviation, Turk et al. (1987).
[b]5 buildings, restricted smoking policy, mean ± standard deviation, Turk et al. (1987).
[c]Outdoor air background. Triebig and Zober (1984).
[d]Measured in airport gate area; smoking not specifically excluded.
Taken from Thurston (1987).
Values drawn from Appendix 4.

(unoccupied room) were 32 ± 60 and 24 ± 22 ppb, respectively. Eatough et al. (1988) reported NO_x concentrations for a hair salon, library, lunchroom, and general "offices" ranging between 7 and 53 ppb.

Excursions from the typical values listed above frequently occur when a given situation includes (a) frequent cooking, (b) close proximity to a transportation source, or (c) both, where the common source is high-temperature combustion of nitrogen to form NO_x. These considerations may explain the elevated NO levels observed in restaurants (2–218 ppb) and a bar (195 ppb) (Triebig and Zober 1984); smoking may be a further contributing source. An extreme example showing the importance of vehicular contribution alone to NO_x concentrations is a study in which NO_x was measured both inside and outside a terminal building at Newark airport (Thurston 1987). The elevated concentrations reported almost certainly reflect the contribution of NO_x from jet exhaust and the local custom of opening jet way doors to permit "fresh" air to enter the gate area. Excursions from "typical" values can be both frequent and extreme. In one study involving "worst" workplace environments (not defined further), as selected by the workers, the NO and NO_2 concentrations averaged 500 and 50 ppb, respectively, with the former a clear deviation from values commonly observed (Nguyen and Goyer 1988).

A particularly informative study in which the effects of smoking vs. nonsmoking and rural vs. urban settings were compared for their effect upon several parameters, including NO_2, was reported by Turk et al. (1987). This work employed extensive measurements of pollutants in 38 commercial buildings in the Pacific Northwest. This included NO_2 measurements from the background, building-wide averages, and specific sites of buildings.

NO_2 was sampled with the Palmes' passive sampler, followed by the spectrophotometric quantitation of analyte, described previously. The tubes were originally developed to sample continuously for a 7-day period, and upon analysis, provide a measure of the average pollutant concentration. This traditional method of continuous exposure was modified since the buildings in this study were generally occupied during only a portion of the monitoring period. Each day for 10 working days, samplers were uncapped at the start of occupancy and capped when the building emptied for a total exposure ranging from 75–100 hours. Thus, the passive sampler data represent time-weighted average (TWA) pollutant concentrations for hours when the buildings were occupied and ventilated. All of Building #3 (a suburban office in Portland, OR) and various departments and sections within other buildings were occupied 24 h/day. At these locations, passive samplers were continuously open for 75–100 hours. Passive samplers were included at outdoor sampling sites as well. The location varied from below street level to rooftop, depending on building design.

Of the 40 building measurement periods, 33 included sampling for NO_2 in multiple sites for a total of 245 sites. Outdoor samples were taken at 13 buildings. Potential indoor sources of NO_2 were largely limited to unvented combustion appliances, which were commonly not found in commercial buildings. For all sites, the geometric mean was 18.3 ppb with a geometric standard deviation of 1.7 ppb. The concentration observed in the great majority of buildings and sites fell well below the established EPA guideline level of a 50-ppb annual average for NO_2, in general agreement with the values observed by other investigators and described above.

Table 9.4 summarizes results obtained in the Pacific Northwest Building Study for 15 buildings. The average outdoor background concentration is contrasted with that from nonsmoking and smoking locations within a given building, as well as the overall building mean. Of these 10 buildings, (9 urban sites, 1 suburban site) employed a nonrestricted smoking policy; the remaining 5 (4 suburban, 1 urban) employed a restricted smoking policy. The overall building mean, which includes data from both smoking and nonsmoking sites within a given structure, is useful for assessing potential exposures to individuals who move freely throughout the building.

The arithmetic mean and standard deviation values for the buildings with a nonrestricted smoking policy depicted in Table 9.4 appear indistinguishable, regardless of whether the background, smoking, nonsmoking, or building means are considered. The same observations apply to the buildings listed in

Table 9.4 Nitrogen Dioxide Indoor-Outdoor Concentrations Pacific Northwest Commercial Buildings Study

Building Smoking Policy	Outdoor Background Concentration, ppb	Indoor Concentration, ppb		
		Nonsmoking	Smoking[a]	Overall Building Mean[b]
Open	21	23	31	27
Open	20	14	18	16
Open	23	24	25	24
Open	6	10	11	10
Open	26	18	20	18
Open	25	19	22	22
Open	40	31	30	31
Open	37	28	24	27
Open	37	34	32	33
Open	37	32	31	31
Summary Statistics, Buildings with Open Smoking Policy				
AM	27	23	24	24
ASD	11	8	7	7
Restr.	21	15	19	16
Restr.	15	13	15	13
Restr.	10	13	20	14
Restr.	6	5	7	5
Restr.	17	18	19	18
Summary Statistics, Buildings with Restricted Smoking Policy				
AM	14	13	16	12
ASD	6	5	5	4
Summary Statistics, All Buildings				
AM	23	20	22	20
ASD	11	9	7	8
GM	20	18	20	18
GSD	2	2	2	2

[a]Smoking within 30-ft radius of site.
[b]Arithmetic average of all sites in building.
Open=open smoking policy (no restrictions).
Rest.=restricted smoking policy.
AM=arithmetic mean.
ASD=arithmetic standard deviation.
GM=geometric mean.
GSD=geometric standard deviation.
Table data taken from Turk et al. (1987).

Table 9.4 with a restricted smoking policy; however, the concentrations are clearly smaller in the restricted group compared to the nonrestricted group.

This difference may be caused by the location of the buildings in each sample. Of the 10 buildings with a nonrestricted policy, 9 were located in an urban area; of the 5 with a restricted policy, only 1 was so located. Since the primary source of NO_2 is usually located outside the building shell, there should be a substantial correlation between the indoor and outdoor concentrations, with the outdoor concentration being the larger of the two. The linear regression data from the comparison of the concentrations of outdoor vs. indoor building average NO_2 (both in ppb) illustrate that 83% (R^2) of the variation of building average of indoor NO_2 concentrations is related to changes in outdoor NO_2 concentrations. If only indoor nonsmoking area concentrations are regressed against outdoor levels, the coefficient of determination, R^2, increases slightly to 87%. It would seem that the presence of smoking has a small (2 ppb), but measurable, effect upon NO_2 concentrations. Even better correlation could be expected if indoor source terms (e.g., parking garages and local combustion sources) and removal mechanisms (e.g., ventilation and chemical reactions) are taken into account. Only two sites were found where potential hazardous levels of NO_2 occur. One of these is a building with an underground parking garage. It is surrounded by heavy congested traffic as well and also has an elevated CO level.

SUMMARY

Indoor air concentrations of NO, NO_2, or NO_x can be somewhat elevated when compared to the outdoor background, but show very little dependence upon the presence of ETS. The presence of gas-fired appliances and of elevated outdoor concentrations appear to contribute more significantly to indoor NO_2 concentrations than ETS. Eatough et al. (1988) discussed the utility, or lack thereof, of NO_x as a "marker" for ETS. These authors provided a graphical summary which shows that ETS contributes not more than ~ 25–30% of the NO_x present in indoor air, and that the rest arises from other, unspecified, sources.

REFERENCES

Apling, A. J., Stevenson, J. J., Goldstein, B. D., Melia, R. J. W. & Atkins, D. H. G. *(1979). Air Pollution in Homes-2. Validation of Diffusion Tube Measurements,* Warren Spring Laboratory, Report LR 311 (AP), Stevenage, U. K.

ASHRAE *62-1989 (1989) ASHRAE Standard. Ventilation for Acceptable Indoor Air Quality,* Atlanta, GA, American Society of Heating, Refrigerating, and Air Conditioning Engineers, Inc., pp. 1–26.

Baker, P. E., Cunningham, S. J., Becker, E. W., Colome, S. D., & Wilson, A. L. (1987) Evaluation of housing and appliance characteristics associated with elevated indoor levels of nitrogen dioxide. In: Seifert, B., Esdorn, H., Fischer, M., Ruden, H., & Wegner, J., eds., *Proceedings of the 4th International Conference on Indoor Air Quality and Climate, Vol. 1,* Berlin, pp. 390–395.

Boleij, J. S. M., Lebret, E., Hoek, F., Noy, D., & Brunekreef, B. (1986) The use of Palmes diffusion tubes for measuring NO_2 in homes. *Atmos. Environ.,* 20 (3), 597–600.

Bouhuys, A., Beck, G. J., & Schoenberg, J. B. (1978) Do present levels of air pollution outdoors affect respiratory health? *Nature,* 276, 466–471.

Cadoff, B. C., Knox, S. F., & Hodgeson, J. A. *(1979)* Personal exposure samplers for nitrogen dioxide. National Bureau of Standards.

Colome, S. D., Lambert, W. E., & Casteneda, N. (1987) Determinants of carbon monoxide exposures in residences of ischemic heart disease patients. In: Seifert, B., Esdorn, H., Fischer, M., Ruden, & Wegner, J., eds., *Proceedings of the 4ᵗʰ International Conference on Indoor Air Quality and Climate,, Vol. 1,* Berlin, pp. 283–287.

Drye, E. E., Ozkaynak, H., Burbank, B., Billick, I. H., Baker, P. E., Spengler, J. D., Ryan, P. B., & Colome, S. D. (1989) Development of models for predicting the distribution of nitrogen dioxide concentrations. *JAPCA,* 39(9), 1169–1177.

Eatough, D. J., Hansen, L. D. & Lewis, E. A. (1988) Assessing exposure to environmental tobacco smoke. In: Perry, R. & Kirk, P. W. W., eds., *Indoor and Ambient Air Quality,* London, Selper Ltd., pp. 131–140.

Eatough, D. J., Lewis, L., Lamb, J. D., Crawford, J., Lewis, E. A., & Hansen, L. D. (1988) Nitric and nitrous acids in environmental tobacco smoke. In: *Proceedings of the 1988 EPA/APCA International Symposium on Measurement of Toxic and Related Air Pollutants,* Research Triangle Park, NC, pp. 104–112.

Harrison, R. M., Colbeck, I., & Simmons, A. (1988) Comparative evaluation of indoor and outdoor air quality chemical considerations. In: Perry, R. & Kirk, P. W. W., eds., *Indoor and Ambient Air Quality,* London, Selper, Ltd., pp. 3–12.

Jenkins, R. A. & Gill, B. E. (1980). Determination of oxides of nitrogen (NO_x) in cigarette smoke by chemiluminescent analysis. *Anal. Chem.,* 52, 925–928.

Jenkins, R. A. & Guerin, M. R. (1984) Analytical chemical methods for the detection of environmental tobacco smoke constituents. *European J. Resp. Dis., 65* (Supplementum No. 133), 33–46.

Kandarjian, L. (1989) Federal policy options for indoor air pollution from combustion appliances. In: Harper, J. P., ed., *Combustion Processes and the Quality of the Indoor Environment,* Pittsburgh, PA, Air & Waste Management Association, pp. 310–321.

Klus, H., Begutter, H., Ball, M., & Intorp, M. (1987) Environmental tobacco smoke in real life situations. *Proceedings of the 4ᵗʰ International Conference on Indoor Air Quality and Climate, Vol. 2,* Berlin, pp. 137–141.

Kulkarni, M. M. & Patil, R. S. (1998) Factors influencing personal exposure to nitrogen dioxide in an Indian metropolitan region. *Indoor Built Environ.*, 7, 319-332.

Lambert, J. L., Trump, E. L;, & Paukstelis, J. V. *(1987)* I-Methylperimidine as a solid monitoring reagent for nitrogen dioxide. *Environ. Sci. Technol.,* 21, 497–500.

Marbury, M. C., Harlos, D. P., Samet, J. M., & Spengler, J. D. *(1988)* Indoor residential NO_2 concentrations in Albuquerque, New Mexico. *JAPCA,* 38, 392–398.

Martin, P., Heavner, D. L., Nelson, P. R., Maiolo, K. C., Risner, C. H., Simmons, P. S., Morgan, W. T., & Ogden, M. W. (1997) Environmental tobacco smoke (ETS): a market cigarette study. *Environ. Int.,* 23, 75–90.

Matthews, R. D., Sawyer, R. F., & Schefer, R. W. *(1977)* Interferences in chemiluminescent measurement of NO and NO_2 emissions from combustion systems. *Environ. Sci. Technol.,* 11(12), 1092–1096.

Mulik, J. D., Lewis, R. G., & McClenny, W. A. *(1989)* Modification of a high efficiency passive sampler to determine nitrogen dioxide or formaldehyde in air. *Anal. Chem.,* 61, 187–189.

Nguyen, V. H. & Goyer, N. *(1988)* A global approach to investigate indoor air quality and performance of the ventilation system. In: Perry, R. & Kirk, P. W. W., eds., *Indoor and Ambient Air Quality,* London, Selper Ltd., pp. 327–332.

Norman, V., Ihrig, A. M., Larsen, T. M., and Moss, B. L. (1983) The effect of some nitrogeneous blend components on NO/NO, and HCN levels in mainstream and sidestream smoke. *Beitr. Tabakforsch. Int.,* 12, 55–62.

Palmes, E. D., Gunnison, A. F., Dimattio, J., & Tomczyk, C. (1976) Personal sampler for nitrogen dioxide. *Am. Ind. Hyg. Assoc. J.,* 37, 570–577.

Petreas, M., Liu, K-S., Chang, B-H., Hayward, S.. B., & Sexton, K. (1988) A survey of nitrogen dioxide levels measured inside mobile homes. *JAPCA,* 38(5), 647–651.

Sega, K. & Fugas, M. (1988) Nitrogen dioxide concentrations in residences. In: Perry, R. & Kirk, P. W. W., eds., *Indoor and Ambient Air Quality,* London, Selper Ltd., pp. 493–496.

Spicer, C. W., Coutant, R. W., Ward, G. F., & Joseph, D. W. (1987) Rates and mechanisms of NO_2 removal from indoor air by residential materials. In: Seifert, B., Esdorn, H., Fischer, M., Ruden, H., & Wegner, J., eds., *Proceedings of the 4th International Conference on Air Quality and Climate, Vol. 1,* Berlin, pp. 371–375.

Thurston, G. D. (1987) A field study of nitrogen oxide levels inside and outside an international airport terminal. In: Seifert, B., Esdorn, H., Fischer, M., Ruden, H., & Wegner, J., eds., *Proceedings of the 4th International Conference on Air Quality and Climate, Vol. 1,* Berlin, pp. 451–455.

Triebig, G. & Zober, M. A. (1984) Indoor air pollution by smoke constituents—a survey. *Prev. Med.,* 13, 570–581.

Turk, B. H., Brown, J. T., Geisling-Sobotka, K., Froehlich, D. A., Grimsrud, D. T., Harrison, J., Koonce, J. F., Prill, R. J., & Rezvan, K. L. (1987) *Indoor Air Quality and Ventilation Measurements in 38 Pacific Northwest Commercial Buildings,* Final Report to the Bonneville Power Administration, Indoor Environment Program, Applied Science Division, Lawrence Berkeley Laboratory, Berkeley, CA. LBL-22315.

Weber, A. & Fischer, T. (1980). Passive smoking at work. *Int. Arch. Occup. Environ. Health,* 47, 209–221.

Yanagisawa, Y., Fasano, A. M., Spengler, J. D, & Ryan, P. B. (1987) Removal of nitrogen dioxide by interior materials. In: Seifert, B., Esdorn, H., Fischer, M., Ruden, H., & Wegner. J., eds., *Proceedings of the 4th International Conference on Air Quality and Climate, Vol. 1,* Berlin, pp. 376–380.

Field Studies—
Formaldehyde

INTRODUCTION

Formaldehyde is an important contaminant of indoor air because it is considered a Class 2A Probable Human Carcinogen by the International Agency for Research on Cancer (IARC) and because it is commonly found at higher concentrations in indoor air than in outdoor air. Formaldehyde is further important because even short-term exposure to concentrations near or below federally accepted standards can result in discomfort. Acute effects can include eye, nose, and throat irritation; headaches; and dizziness (Kandarjian 1989). Formaldehyde is among the agents frequently sought in response to complaints of illness or discomfort that appear to be caused by indoor environment.

Upon initial examination, it might be suspected that environmental tobacco smoke (ETS) would be an important contributor to indoor air concentrations of formaldehyde because formaldehyde is known to be a constituent of cigarette smoke. Popular commercial cigarettes deliver approximately 20–90 μg of formaldehyde in their mainstream smoke and 1–2 mg of formaldehyde in their sidestream smoke (Chap. 3). While this contribution may at first appear highly significant, it has generally been found to be very minor when compared with other sources.

The commonly observed excess of formaldehyde in indoor air as compared to outdoor air is most frequently found to result from formaldehyde being emitted by building materials, furnishings, and consumer products (Chap. 2). Major sources include products made from urea-formaldehyde resins, such as urea formaldehyde foam insulation (UFFI) and pressed-wood products such as interior plywood, medium-density fiberboard, and particle board. Formaldehyde is also used in the manufacture of common household items and personal products such as cleaning supplies, permanent press clothing, shampoos, grocery bags, varnishes, and adhesives. Combustion sources also contribute to indoor formaldehyde concentrations. Gas ranges and space heaters, for example, emit 20–40 mg/h of formaldehyde when operating. The variety and generally overwhelming magnitude of other sources have resulted in less attention being given to cigarette smoking as a source of formaldehyde than has been given to other constituents of ETS.

SAMPLING AND ANALYSIS

Measuring formaldehyde is more difficult than measuring constituents such as particulate matter, nicotine, and carbon monoxide because of the former's reactivity. The measurement of formaldehyde is also much more subject to error resulting from interfering species with similar chemical properties. Current sampling and analysis methods rely heavily on chemical reactions between reagents and formaldehyde. Because of this, sampling and analysis methods are discussed in greater detail here than in other chapters.

Active Sampling. Formaldehyde samplers may be either "active" or "passive." Active samplers draw a known volume of air through a liquid or solid trapping medium using either positive or negative pressure. Examples include classical glass impingers (Lodge 1990) or smoke collection flasks (Manning et al. 1983). The sampling times usually range from a few minutes to not more than a few hours. Glass impingers with a liquid trapping medium, while rapid and potentially quantitative in their analyte collection, are fragile, are likely to contain a corrosive and/or otherwise hazardous reagent medium, and must be operated with an external mechanical pump that may be heavy and cumbersome. For these reasons, the impingers are rapidly disappearing from field sampling. Field-acceptable active-sampling alternatives to the glass impingers include small trapping tubes containing solid sorbents, which permit small volumes of air to be sampled by battery-operated pumps. Color reactions in the tube (described below) permit direct on-site measurement of formaldehyde. Such tubes are available from a variety of manufacturers. Small cartridges containing silica coated with acidified 2,4-dinitrophenylhydrazine (2,4DNPH) (Stray and Oehme 1987, Tejada 1986) are commercially available, as are tubes filled with 2-(hydroxymethyl) piperidine coated XAD-2 resin. The air sampling would be conducted in a manner similar to the trapping tubes, with final quantification being performed off-line in the laboratory.

One report (Arnts and Tejada 1989) describes an ozone (O_3) interference with solid samplers, but not with impingers with liquid trapping solutions when 2,4-DNPH was the reagent used. The deviations of the solid sampler results from those of the impingers became more severe with increasing concentrations of O_3. Free radicals generated by O_3 attack both reagent and hydrazone product. For these reasons, using 2,4-DNPH-silica gel cartridges for ambient monitoring is discouraged unless a carbonyl-passive ozone scrubber is used. This problem is present, but usually not significant, when the glass impingers are used because the latter contain a greater mass of 2,4-DNPH reagent than do the solid samplers. Hence, degradation of reagent and product may occur in the impinger, yet produce only small low-biases. An alternative to the 2,4-DNPH cartridges described above is 2(hydroxymethyl) piperidine (2-HMP) coated on XAD-2 sorbent (OSHA 1985). The procedure employs active atmospheric sampling with a small cartridge followed by off-line gas chromatographic determination of the formaldehyde-oxazolidine derivative.

Passive Samplers. Passive sampling badges are useful devices for quantifying formaldehyde because they allow integrated measurements over long periods and do not require mechanical pumps (Chap. 5). The two types of passive samplers, diffusion and permeation, consist of two main components: the barrier to diffusion and the trapping or adsorbent medium. The adsorbent may consist of almost anything with a strong affinity for the analyte, including water and adsorbent paper or resin. In the diffusion sampler, the barrier is distance; one end of the tube is open. This permits the analyte to diffuse toward the adsorbent medium at some rate which depends, in part, on the size of the opening, the distance to the adsorbent, and the size of the molecule. The actual sampling rate, in cm^3/min, is calculated from Fick's first law of diffusion (Mulik et al. 1989). In the permeation sampler, the primary barrier to mass transport is a permeable membrane. Many permeation samplers now employ polydimethylsiloxane membranes, whose permeability does not depend on analyte concentrations (Jenkins and Guerin 1984). The membrane has the additional effect of improving the selectivity of the sampler. The chief driving force in both of these systems is the depletion of the constituent in question at the surface of the adsorbent. Each system has some minor drawbacks. Diffusional samplers are more affected by ambient humidity, whereas membrane samplers require calibration in standard atmospheres of the analyte in question.

Typically, passive samplers have been used at ambient concentrations that are greater than those usually observed in smoking environments. However, recent improvements in the final analysis procedures, coupled with longer sampling times, suggest that passive samplers could offer significant utility for monitoring at ETS constituent levels. The evaluation of analytical precision, accuracy, and performance with time are all obvious, necessary, and fruitful research areas.

Passive samplers used for formaldehyde have been as simple as a few milliliters of water in a small cup covered with a polydimethylsiloxane membrane (Hawthorne et al. 1983), or may employ organic binder-free glass fiber filter treated with sodium bisulfite (Geisling et al. 1982). A third type, developed and tested by EPA (Mulik et al. 1989), permits sampling of nitrogen dioxide and formaldehyde simultaneously. Collection of the latter is achieved with a glass fiber filter soaked with an acidic solution of 2,4-DNPH.

At least two types of passive formaldehyde dosimeter sampling badges are now commercially available. One type employs a filter coated with 2,4-DNPH, which permits the formation of a stable hydrazone in the presence of formaldehyde. The final quantification is performed off-line in the laboratory. One disadvantage of such a sampler is that large quantities of carbonyl compounds, e.g. acetone, may consume the reagent, thereby reducing the uptake of formaldehyde. This can be a problem when quantifying formaldehyde in indoor air because acetone and acetaldehyde are major carbonyl constituents of sidestream smoke (Hoffmann et al. 1987, Repace

1987) and exist in chamber -confined ETS at comparable levels (Martin et al. 1997). Another type employs a proprietary color reagent that changes from dark yellow to reddish black. This badge, which does not require off-line analysis, does not respond to phenols, alcohols, ketones, or other organic solvents. Both badges are relatively free of effects from temperature, humidity, and wind velocity.

Dye-Based (Colorimetric) Measurements. The most frequently cited dye-based analytical procedures for formaldehyde in air employ either chromotropic acid (NIOSH 1984) or pararosaniline (Miksch et al. 198 1). Both procedures presume three common steps: First, an air sample is bubbled through a glass impinger containing an appropriate trapping solution. Second, a dye and acid are combined with an aliquot of trapping medium and heated to form a colored product. Third, the absorbency of the colored product is measured and related back to the concentration of formaldehyde in air.

The chief drawback of any quantitation procedure employing a dye is its comparative lack of specificity for formaldehyde. This may be of little consequence when monitoring industrial atmospheres, where a single compound may predominate. However, indoor air and ETS comprise many constituents, some or all of which may either compete with formaldehyde for the reagent, or may otherwise interfere with the measurement at $\mu g/m^3$ levels.

The chromotropic acid procedure (NIOSH 1984) has been developed to the point where specific interferences and their effect upon the final measurement have been considered in detail. Phenols, when present in eightfold excess over formaldehyde, produce a -10 to -20% bias. Ethanol and higher molecular weight alcohols, olefins, aromatic hydrocarbons, and cyclohexanone also produce small negative interferences. Little interference is seen from other aldehydes. The pararosaniline procedure is also subject to some interferences, although perhaps less so than chromotropic acid. Only low molecular weight aldehydes gave positive interferences, and then only when present in large excess over formaldehyde. Negative interferences could be caused by compounds that react with either pararosaniline or formaldehyde. ETS, by its very nature, contains all of the interferences specifically listed in the NIOSH method; hence, chromotropic acid is not recommended as a reference procedure for determining formaldehyde in ETS.

Gas and Liquid Chromatographic-Based Procedures Employing 2,4Dinitropheny/hydrazine (2,4-DNPH). The use of 2,4-DNPH, the classical derivatizing reagent for aldehydes and ketones, is the key to several procedures allowing more selectivity than those described above (Tejada 1986, Manning et al. 1983, Arnts and Tejada 1989). DNPH will react with virtually all aldehydic and ketonic carbonyls to form the related 2,4-dinitrophenylhydrazone derivative. The individual species may then be separated and quantified by either gas or high-performance liquid chromatographic (HPLC) procedures. The chief disadvantage with this approach is that more sophisticated analytical instrumentation is required.

Nevertheless, given the potential interferences in cigarette smoke, DNPH derivatization and analysis would appear to be the method of choice when accurate quantification is required, and has been employed in modern studies of formaldehyde associated with ETS (Hedge et al. 1994).

Gas Chromatographic-Based Procedures Employing 2-Hydroxymethyl Piperidine (2-HMP). The use of 2-HMP has become more popular as a reagent for the derivatization of formaldehyde since the procedure has been validated by OSHA. Briefly, air is sampled through sorbent traps containing 2-HMP-coated XAD-2. The resulting oxazolidine derivative is eluted from the trap with toluene, and analyzed using gas chromatography with nitrogen selective detection. The chief advantage of this approach is that more laboratories typically have gas chromatographs than have the HPLC systems usually employed with the DNPH derivatization procedure (above).

Gas Detector Tubes. Gas detector tubes are usually made of glass, and contain a solid packing material designed to give a selective color reaction for the analyte of interest. The tubes are precalibrated by the manufacturer so that the analyte concentration can be read directly from the tube barrel, assuming a specified volume of air has been sampled. Several manufacturers offer gas sampling tubes that can detect formaldehyde at concentrations as low as 60 $\mu g/m^3$, a level approaching the range for utility for indoor air characterization. By increasing the volume of air sampled beyond that recommended by the manufacturer, the detection limit can be reduced further by at least an order of magnitude. Several types feature a reaction whereby formaldehyde reacts with xylene (isomer not specified) to form a dimer, which is in turn converted to a pink quinoid compound via reaction with sulfuric acid. This quinoid product may be dehydrated further with sulfuric acid for additional selectivity. Alternatively, formaldehyde may react with hydroxylamine phosphate to liberate phosphorous acid, which discolors a pH indicator to reddish brown from its normal yellow.

Because they are both simple and rugged, gas detector tubes are ideal for active sampling applications. However, the tubes described are not specific for formaldehyde. The concentrations measured must be interpreted as approximate (possibly maximum) values only because there is no inherent mechanism present to isolate a product which is truly specific for formaldehyde. Some interferences described by the manufacturers are constituents of ETS and should be expected to yield a positive interference. These include other aldehydes (e.g., acetaldehyde and acrolein); styrene; ketones, esters, and ethers in concentrations exceeding 0.1% that of formaldehyde; as well as acidic gases.

Fourier-Transform Infrared Spectroscopy (FT-IR). At least one study (Pitts et al. 1989) describes a wholly instrumental method capable of detecting formaldehyde, gaseous nitric acid, methanol, and formic acid in room air. The FT-IR spectrometer employed was a Mattson Instruments, Inc., Sirius 100, fitted by the investigator with external transfer optics and an external liquid nitrogen-cooled HgCdTe detector. The pathlength and resolution were 380 m and 0.13 cm^{-1}, respectively. Under these conditions, the detection limits for formaldehyde, formic acid, methanol, and gaseous nitric acid were 8, 3, 8, and 12 ppb, respectively (10, 4, 10, and 15 μg/m^3), suggesting that such technology may be suitable for the quantitation of formaldehyde from ETS. The absorption peaks employed for identification of formaldehyde were 2778.5 and 2781.0 cm^{-1}; the latter was employed for quantitative measurements.

FIELD STUDIES

The importance of the ETS contribution to indoor air concentrations of formaldehyde with respect to other possible sources may be evaluated from the summary data presented in Table 10.1 and Appendix 5. In most cases, background concentrations are on the order of a few tens of μg/m^3 in an older building; a typical value would be 30 μg/m^3. Some of the situations described include stationary homes, both with and without urea formaldehyde foam insulation (UFFI), stationary homes with wood-burning stoves, offices/workplaces, and mobile homes. UFFI was used extensively in the past in the latter two situations. In offices, in addition, furniture and medium density fiberboard containing formaldehyde-based resins or fabrics with formaldehyde-based adhesives is frequently used. As such, they cause incremental excursions from the typical background concentrations, reaching as high as 5200 μg/m^3. The problems with UFFI are compounded in mobile homes because the air exchange rate is much smaller and the loading factor of pressed wood products is higher than that for a conventional stationary dwelling.

Perhaps the most important difference between formaldehyde contributed to indoor air from ETS and that from other sources is that the former is generated very quickly, as a pulse, and that the concentrations in the immediate vicinity of the cigarette may be substantial. Formaldehyde arising from UFFI, furniture, etc., tends to create a "standing" or "steady state" concentration in a room that decreases slowly with time, aging over a timeframe of several years. Non-vented combustion sources such as kerosene-fueled space heaters tend to produce an intermediate-term pulse of formaldehyde release, further complicating assessments of ETS-specific contributions to ambient indoor concentrations.

There are very limited data available describing the ETS contribution to indoor air levels of formaldehyde in a variety of natural situations. Those data reported frequently appear to result from brief surveys that accompany studies of other issues rather than studies designed to specifically address ETS-derived formaldehyde. An example is presented in Table 10.2 [taken from Klus et al. (1987)]. This survey, albeit crude and without extensive controls, suggests that formaldehyde concentrations in the presence of ETS are comparable to background levels and normally do not exceed approximately 40 μg/m^3. While

**Table 10.1 Summary of Field Measurements for Formaldehyde
in Natural Situations**
(Ranges of Formaldehyde Concentrations, μg/m3)

Situation	Formaldehyde Concentrations, μg/m^3	
	Means	All Data (Range)
Restaurant		
Smoking	-	15–40
Control	-	8
Pubs and Taverns		
Smoking	89,104	7–104
Control	–	–
Railway Compartment		
Smoking	35	–
Control	12	–
Automobile		
Smoking	27	–
Control	12	–
Public Buildings		
Smoking	20[1] –32	ND–740
Control	ND[2] –34	ND–270
Outdoors	1[3] –38	ND–58
Stationary Homes		
Observed[4]	27–50	2–280
Control	7	2–14
Stationary Homes w/UFFI		
Observed[4]	61–98	25–250
Control	–	–
Mobile Homes		
Observed	75–1100	< 12–5200
Control	–	–
Offices/Workplace		
Observed	50–540	12–1300
Control	–	25–26

[1]Median, not mean, of 200 individual smoking situations.
[2]Median, not mean, of 7 individual nonsmoking situations.
[3]Median, not mean, of 24 individual outdoor situations.
[4]Data reported without stating presence/absence of smoking. Data summarizes Appendix 5.

in general the formaldehyde concentrations appear elevated when smoking is occurring, the highest level, 53 μg/m^3, was actually observed in a room identified as a nonsmoking environment.

Sterling et al. (1987) have compiled a database for 230 situations, including total outdoor, total indoor, indoors/ smoking permitted and indoors/ smoking restricted. Table 10.3 presents the number of measurements, median, and range of concentrations for each of the four cases. The ranges for smoking-restricted (ND–270 μg/m^3) and permitted (ND–740 μg/m^3) environments not only are very wide but also clearly overlap each other as well as that for outdoor environments (ND–38 μg/m^3). The degree of skew in the data is evident from the median values, which are less than 20 μg/m^3 for the ranges noted. The small number of samples obtained for indoor/smoking

Table 10.2 Formaldehyde Concentrations Determined in Natural Situations; Smoker and Nonsmoker Areas

Smoker/ Nonsmoker Area	No. of Cigarettes	No. of Persons	Sampling Time (h)	Formaldehyde (μg/m³)
Living Room, Four Person Household				
Nonsmoker	–	4	4	53
Smoker	10	4	4	50
Smoker	11	3	3.5	42
Smoker	40	6	4.5	23
Restaurant				
Nonsmoker	–	12	4	8
Smoker	5	6	3	15
Smoker	20	20	4	40
Pub				
Smoker	>100	50	4	7
Smoker	>100	38	4	38
Railway Compartment				
Nonsmoker	–	4	3	12
Smoker	3	8	2	35
Personal Automobile				
Nonsmoker	-	2	4	12
Smoker	8	2	4	27
Office Buildings Where Smoking Restricted to Rooms with No Additional Air Treatment				
Nonsmoking (13 locations)	-	-	3	63[a]
Smoking (23 locations)	-	-	3	75[a]
Office Buildings Where Smoking Restricted to Rooms with Local Air Filtration				
Nonsmoking (26 locations)	-	-	3	25[a]
Smoking (22 locations)	-	-	3	50[a]

*Taken from Klus et al. 1987 and Hedge et al. 1994. No attempt made to influence smoking and working habits of people. No controls exercised.
[a] Mean Value

Table 10.3 Comparison of Formaldehyde Concentrations from 230 Situations Contained in Building Performance Database

Situation	Number of Cases	Formaldehyde Concentration, μg/m³	
		Median	Range
Total indoor	207	16	ND –740
Total outdoor	24	1	ND –38
Smoking permitted	200	20	ND –740
Smoking restricted	7	ND	ND –270

Detection limit not specified.
Taken from Sterling et al. (1987)

restricted and outdoor background environments yields inconclusive comparisons between these two situations and indoors/smoking permitted.

Turk et al. (1987) reports one of the more extensive studies of common pollutants found in commercial buildings. Formaldehyde was sampled in approximately 40 buildings in the Pacific Northwest with the Geisling (1982) passive sampler. The investigators report summary statistics for 401 sampling sites taken from 38 buildings (2 were sampled twice). The geometric mean and standard deviation were 22 and 2 $\mu g/m^3$ (18 and 1.7 ppb), respectively; the corresponding arithmetic mean and standard deviation were 26 and 20 $\mu g/m^3$ (21 and 16 ppb). The reported method detection limit for formaldehyde was 25 $\mu g/m^3$ (20 ppb). The geometric mean and standard deviation for all building averages (i.e., the average of all results taken in a particular building) were 26 and 2 $\mu g/m^3$ (21 and 1.6 ppb), respectively; the arithmetic mean and standard deviation were 29 and 12 $\mu g/m^3$ (23 and 10 ppb) (same convention). With a minimum detection level of 25 $\mu g/m^3$ (20 ppb), only 21 (53 %) of the buildings tested had a building mean concentration above the detection limit. None of the buildings exhibited whole-building means that approached the Department of Housing and Urban Development guideline of 492 $\mu g/m^3$ (400 ppb) for manufactured homes (the only existing federal standard), and none of the 401 sampling sites exceeded this value (ASHRAE 1989). Further information concerning the effect of ETS upon the formaldehyde concentrations may be gleaned by grouping the building means according to the characteristics of the building itself, i.e., by smoking policy (restricted vs. open) or by location (rural, suburban, urban), as shown in Table 10.4. This table shows that most (22) of the buildings in this study were located in either urban areas with an open smoking policy or in suburban areas with a restricted smoking (10). In both cases, the arithmetic means and standard deviations were virtually indistinguishable. Because these two situations contain the greatest number of data values, the resulting comparisons may be more meaningful than those where less than five measurements were made. Except for one situation (restricted smoking policy, rural sampling area) where there was exactly one set of measurements performed, none of the cases were significantly different from the background. In short, there were no differences in the mean formaldehyde concentration when open or restricted smoking policies were compared, and the average formaldehyde concentration typically observed did not exceed 35 $\mu g/m^3$. Outdoor background concentrations exhibited considerable fluctuation, as shown in Table 10.4. The unusual precision of the suburban background, 26 ± 1 $\mu g/m^3$, taken from four data points, is caused by two values just above the detection limit (26 $\mu g/m^3$) and two at the detection limit (< 25 $\mu g/m^3$, with 25 $\mu g/m^3$ used for calculation purposes).

Other studies confirm the difficulty of establishing a clear relationship between ETS and indoor air concentrations of formaldehyde. Sheldon et al. (1988) examined public buildings including both smoking and nonsmoking environments and rarely detected formaldehyde concentrations greater than the detection limit (25 $\mu g/m^3$) of their method. Turner and Binnie (1990) have reported a survey of 26 office buildings in Switzerland included measurements of formaldehyde along with nicotine, and several other parameters. Mean formaldehyde concentrations range from 12 –244 $\mu g/m^3$ (10 –200 ppb) across the buildings but were completely independent of the nicotine concentrations.

Table 10.4 Summary of Formaldehyde Concentrations Observed In the Pacific Northwest Building Study

Smoking Policy	Bldg. Location	No. of Values	Formaldehyde Concentration μg/m^3	
			Mean ± std dev	Range
Open	All	26	30 ± 6	< 25–93
Restricted	All	14	34 ± 14	<25–236
Background Data	All	13	34 ± 12	< 25–58
Open	Suburban	4	32 ± 9	< 25–60
Restricted	Suburban	10	33 ± 14	< 25–92
Background Data	Suburban	4	26 ± 1	< 25–26
Open	Urban	22	29 ± 6	< 25–93
Restricted	Urban	3	34 ± 16	<25–236
Background Data	Urban	9	38 ± 13	< 25–58
Restricted	Rural	1	47	37–54

Method detection limit is 25 μg/m^3. For calculation of statistical summary, values below detection limit taken as 25 μg/m^3
Data taken from Turk et al. (1987).

Formaldehyde concentrations can be significantly enhanced in environments where heavy smoking occurs and no other major source is present. This is illustrated by the work of Klus et al. (1987) who monitored the air in an unventilated office during heavy smoking. In this study, 11 samplings were performed for 2-h periods in which 8–18 cigarettes (and an occasional small number of cigars and pipes) were smoked during each period. Maximum increases over the background formaldehyde concentrations ranged from 15–99 μg/m^3. The correlation between the number of cigarettes (and cigars and pipes) smoked and the maximum excess concentration of formaldehyde was found to be very poor, but the experimental design demands that the excess be due to cigarette smoking. Hedge et al. (1994) have demonstrated differences in formaldehyde levels for modern offices in which smoking was restricted to specific locations. Mean formaldehyde concentrations were ~25 μg/m^3 in nonsmoking areas, and ~75 μg/m^3 in smoking offices with no air treatment.

Contributions of ETS to formaldehyde concentrations of the magnitude found in the office studies are highly unlikely under normal conditions of smoking, ventilation, and the presence of other sources. The contribution of ETS to background levels of formaldehyde in environments of principal concern (e.g., residences, offices, other locations associated with chronic exposure) has not yet been determined but would be expected to be at or near the detection limits of many of the technologies employed for monitoring.

SUMMARY

There can be little doubt that formaldehyde is an important component of sidestream smoke, a major precursor of ETS. However, formaldehyde is a relatively reactive species, and there are a number of consumer products and building materials which emit formaldehyde at rates comparable to or exceeding those from smoldering cigarettes. Many commonly encountered levels of formaldehyde, whether or not measured in the presence of ETS, are at or near the detection limits of the sampling and analysis systems employed in many field studies. As such, it has been difficult for many studies to demonstrate consistently elevated levels of formaldehyde due to ETS. It is true that in some situations, where smoking rates are high and ventilation is minimal, there is a clear contribution to formaldehyde from ETS on the order of a few tens of $\mu g/m^3$. To determine the exact contribution of formaldehyde from ETS in most commonly encountered situations, it will likely be necessary to make careful background measurements in the target area, using sampling and analytical systems which are approximately an order of magnitude more sensitive than most of those in use today.

REFERENCES

Arnts, R. R. & Tejada, S. B. (1989) 2,4-Dinitrophenylhydrazine-coated silica gel cartridge method for determination of formaldehyde in air: identification of an ozone interference. *Environ. Sci. Technol.,* 23, 1428–1430.

ASHRAE 62-1989 (1989) *ASHRAE Standard, Ventilation for Acceptable Indoor Air Quality,* Atlanta, GA, American Society of Heating, Refrigerating, and Air Conditioning Engineers, Inc., pp. 1–26.

Beck, S. W. & Stock, T. H. (1990) An evaluation of the effect of source and concentration on three methods for the measurement of formaldehyde in indoor air. *Am. Ind. Hyg. Assoc. J.,* 51(1), 14–22.

DeBortoli, M., Knoppel, H. Pecchio, E., Pcil, A., Rogora, L., Schauenburg, H., Scholitt, H., & Vissers, H. (1985) *Measurement of indoor air quality and comparison with ambient air—A study on 15 homes in northern Italy,* EUR 9656 EN.

Geisling, K. L., Tashima, M. K., Girman, J. R., Miksch, R. R., & Rappaport, S. M. (1982) A passive sampling device for determining formaldehyde in indoor air. *Environ. Int.,* 8, 153–158.

GMD Systems, Inc. (1989) 570 Series formaldehyde dosimeter badge. GMD Systems, Inc., Hendersonville, PA.

Griest, W. H., Guerin, M. R., Quincy, R. B., Jenkins, R. A., & Kubota, H. (1980) Chemical characterization of experimental cigarettes and cigarette smoke condensates in the fourth cigarette experiment. In: Gori, G. B., ed., *Toward Less Hazardous Cigarettes.* In *Fourth Set of Experimental Cigarettes,* U.S. Department of Health, Education, and Welfare, Public Health Service, National Institutes of Health, National Cancer Institute, Smoking and Health Program, Report No. 4.

Hawthorne, A. R., Gammage, R. B., Dudney, C. S., Womack, D. R., Morris, S. A., Westley, R. R., & Gupta, K. C. (1983) Preliminary results of a forty-home indoor air pollutant monitoring study. *Speciality Conference on Measurement and Monitoring of Non-Criteria (Toxic) Contaminants,* Air Pollution Control Association, Chicago, IL.

Hedge, A., Erickson, W. A., & Rubin, G. (1994) The effects of alternative smoking policies on indoor air quality in 27 smoking office buildings. *Ann. Occup. Hyg.* , 38, 265-278.

Hoffmann, D., Adams, J. D., & Brunnemann, K. D. (1987) A critical look at N-nitrosamines in environmental tobacco smoke. *Toxicol. Letts.,* 35, 1–8.

Jenkins, R. A. & Guerin, M. R. (1984) Analytical chemical methods for the detection of environmental tobacco smoke constituents. *Eur. J. Res. Dis.,* 65(Supplementum No. 133), 33–46.

Kandarjian, L. (1989) Federal policy options for indoor air pollution from combustion appliances. In: Harper, J. P., ed., *Combustion Processes and the Quality of the Indoor Environment,* Air & Waste Management Association, Pittsburgh, PA, pp. 310–321.

Klus, H., Begutter, H., Ball, M., & Intorp, M. (1987) Environmental tobacco smoke in real life situations. *Proceedings of the 4th International Conference on Indoor Air Quality and Climate, Vol. 2,* Berlin, pp. 137–141.

Lodge, J. P., Jr., ed. (1990) *Methods of Air Sampling and Analysis, Third Edition,* Lewis Publishers, Inc., Boca Raton, FL.

Manning, D. L., Maskarinec, M. P., Jenkins, R. A., & Marshall, A. H. (1983) High performance liquid chromatographic determination of selected gas phase carbonyls in tobacco smoke. *JAOAC,* 66(1), 8–12.

Martin P., Heavner, D. L., Nelson, P. R., Maiolo, K. C., Risner, C. H., Simmons, P. S., Morgan, W. T., & Ogden, M. W. (1997) Environmental tobacco smoke (ETS): a market cigarette study. *Environ. Int.,* 23, 75–90.

Matthews, T. G., Hawthorne, A. R., Howell, T. C., Metcalfe, C. E., & Gammage, R. B. (1982). Evaluation of selected monitoring for formaldehyde in domestic environments. *Environ. Int.,* 8, 143–151.

Miksch, R. R., Anthon, D. W., Fanning, L. Z., Hollowell, C. D., Revzan, K., & Glanville, J. (1981) Modified pararosaniline method for the determination of formaldehyde in air. *Anal. Chem.,* 53, 2118–2123.

Mulik, J. D., Lewis, R. G., & McClenny, W. A. (1989) Modification of a high efficiency passive sampler to determine nitrogen dioxide or formaldehyde in air. *Anal. Chem.*, 61, 187–189.

NIOSH (1984) *Formaldehyde. Method 3500; issued 2115184.* NIOSH Manual of Analytical Methods, Third Edition, pp. 3500-1–3501-3.

OSHA Analytical Laboratory (1985) *Acrolein and/or Formaldehyde, Method 52.* OSHA Analytical Methods Manual, pp. 52-1–52-32.

Pitts, J. N., Jr., Biermann, H. W., Tuazon, E. C., Green, M., Long, W. D., & Winer, A. M. (1989) Time-resolved identification and measurement of indoor air pollutants by spectroscopic techniques: gaseous nitrous acid, methanol, formaldehyde and formic acid. *JAPCA*, 39, 1344–1347.

Repace, J. L. (1987) Indoor concentrations of environmental tobacco smoke: field surveys. In O'Neill, I. K., Brunnemann, K. D., Dodet, B., & Hoffmann, D., eds., *Environmental Carcinogens—Methods of Analysis and Exposure Measurement. Vol. 9, Passive Smoking* (IARC *Scientific Publications No. 81)*, International Agency for Research on Cancer, Lyon, France, pp. 141–162.

Sexton, K., Liu, K.-S., & Petreas, M. X. (1986) Formaldehyde concentrations inside private residences: A mail-out approach to indoor air monitoring. *JAPCA*, 36, 698–704.

Sheldon, L. S., Handy, R. W., Hartwell, T. D., Whitmore, R. W., Zelon, H. S., & Pellizzari, E. D. (1988) *Indoor Air Quality in Public Buildings, Vol. I,* EPA/ 600/6-88009a.

Sheldon, L., Zelon, H., Sickles, J., Eaton, C., & Hartwell, T. (1988) *Indoor Air Quality in Public Buildings, Vol. 11*, EPA/600/6-88/009b.

Sterling, D. (1985) Volatile organic compounds in indoor air: an overview of sources, concentrations, and health effects. In: Gammage, R. B. & Kaye, S. V., eds., *Indoor Air and Human Health,* Lewis Publishers, Inc., Boca Raton, FL, pp. 387–402.

Sterling, T. D., Collett, C. W., & Sterling, E. M. (1987) Environmental tobacco smoke and indoor air quality in modern office work environments. *J. Occupat. Med.*, 29(1), 57–62.

Sterling, T. D. & Sterling, E. M. (1984) Environmental tobacco smoke. 1.2. Investigations on the effect of regulating smoking levels of indoor pollution and on the perception of health and comfort of office workers. *Eur. J. of Resp. Dis.*, 65,(Supplementum 133), 17–32.

Stray, H. & Oehme, M. (1987) Convenient method for determination of sub-ppb to ppm levels of C_1-C_5 aldehydes in air by solid adsorbent sampling. *Proceedings of the 4th International Conference on Indoor Air Quality and Climate,* pp. 705–708.

Surgeon General (1986) *The Health Consequences of Involuntary Smoking.* Rockville, MD, U.S. Department of Health and Human Services.

Tejada, S. B. (1986) Evaluation of silica gel cartridges coated in situ with acidified 2,4dinitrophenyl hydrazine for sampling aldehydes and ketones in air. *Int. J. Environ. Anal. Chem.,* 26, 167-485.

Turk, B. H., Brown, J. T., Geisling-Sobotka, K., Froehlich, D. A., Grimsrud, D. T., Harrison, J., Koonce, J. F., Prill, R. J., & Rezvan, K. L. (1987) *Indoor Air Quality and Ventilation Measurements in 38 Pacific Northwest Commercial Buildings.* Final Report to the Bonneville Power Administration. Indoor Environment Program, Applied Science Division, Lawrence Berkeley Laboratory, Berkeley, CA. LBL-22315.

Turner, S. & Binnie, P. W. H. (1990) An indoor air quality survey of twenty-six Swiss office buildings. *Proceedings of the 5th International Conference on Indoor Air Quality and Climate, Vol. 4,* Toronto, Canada, pp. 27–32.

INTRODUCTION

Volatile organic compounds (VOCs) are those chemicals that are common major constituents of the gas phase of cigarette smoke (Chap. 3) and are commonly reported as volatile organic contaminants of indoor air. These are primarily aliphatic, olefinic, and aromatic hydrocarbons and include benzene and its alkyl-substituted derivatives. Chlorine-substituted hydrocarbons (not expected to be ETS-related), simple aldehydes and ketones, and nitrogen-substituted organics are also considered. N-Nitrosamines, some of which are gas-phase constituents of cigarette smoke, are discussed in a separate chapter (Chap. 12) because they are present at generally significantly lower concentrations in indoor air. Formaldehyde (Chap. 10) is treated as a special case because of its suspected carcinogenicity and its association with ETS, common building construction materials, and consumer products.

VOCs are important constituents of indoor air because they are ubiquitous and because several (e.g., formaldehyde and benzene) of their class are suspect human carcinogens. That they are commonly detected in both smoking and nonsmoking environments demonstrates that they result from multiple sources common to many indoor situations. Discomfort and illness associated with "Sick Building Syndrome" are often assumed to be associated with exposure to such chemicals because they are frequently detected in the suspect environments.

Many of the studies considered here involve the use of multiple-component analytical methods, i.e., those methods that allow the simultaneous measurement of a number of individual chemicals. It is characteristic of such studies that the constituents reported are often a function of the analytical method used rather than a function of the issue being addressed. Data are often reported merely because they are readily acquired. As such, the appearance of a given chemical on a list of reported analytes does not imply that the chemical is related to ETS. Conversely, given the large number of chemicals identified as cigarette smoke components (see Chap. 3), it is possible that indoor air contaminants listed but not now associated with ETS will be found to be ETS-related as studies progress. It is also likely that many chemicals known to be components of the gas phase of cigarette smoke but not yet studied in indoor air will be detected in indoor air when they are specifically sought.

Research to date clearly shows that cigarette smoking contributes to the VOC burden of indoor air. The magnitude of the contribution, however, is often difficult to determine because of the variety and variability of other sources of VOCs. Toluene and other simple alkyl derivatives of benzene, for example, as well as a wide range of n-paraffins, are known to be contributed by smoking but are also often major products of out-gassing from building materials, furnishings, and office activities. Gasoline is an important source of benzene and other aromatic species in indoor environments. Background source strengths depend on ventilation, penetration of outside air (e.g., Baek et al. 1997); the building's age (e.g., Sheldon et al. 1988a, b); work activity (e.g., photocopying) (Hodgson and Daisey 1989), and the presence or recent use of many consumer products (e.g., Tichenor and Mason 1988). The concentration of a background constituent can remain relatively constant over a long period or be present at a very high concentration for a short period depending on its source. Given the wide variety of activities taking place at any given time and the number of micro-environments in, for example, office buildings and residences, concentrations of highly source-specific contaminants would be expected to vary spatially as well as temporally.

The presence of multiple sources combined with the very large number of chemicals present (e.g., Table 11.1) has probably contributed to the fact that most ETS-related studies of VOCs reported to date have been of a survey nature. Exceptions to this are the detailed study (e.g., Wallace et al. 1987) of benzene, ethylbenzene, styrene, and xylenes as part of EPA's Total Exposure Assessment Methodology (TEAM) program, and personal exposure studies reported by Heavner et al. (1995, 1996). This chapter summarizes observations to date. Appendix 6 tabulates results and literature references from work reviewed here. The reader is also referred to a published (Moschandreas 1991) review of VOC analysis and air quality studies and a detailed review of indoor and outdoor VOCs by Shah and Singh (1988).

SAMPLING AND ANALYSIS METHODS

Analyses designed to determine hydrocarbons, chlorinated hydrocarbons, simple ketones, and other volatile and relatively nonreactive constituents are most commonly based on thermal desorption gas chromatography (e.g., Table 11.2). The sample is collected on a suitable solid sorbent and the constituents are introduced into the chromatograph by heating the trap with a concurrent carrier gas flow. Aldehydes, e.g., acetaldehyde and acrolein, can be determined similarly with appropriate choice of chromatographic column but also by methods such as those described in Chap. 10 for the determination of formaldehyde. These methods typically involve collecting a sample on specially designed solid sorbents, converting the aldehydes to more stable and

Table 11.1 Example Sampling and Analysis Conditions for VOC Studies

	TEAM Study[a]	Proctor 1989	Guerin 1991	Persson et al. 1988	Heavner et al. 1996	Lebret et al. 1984
Target analytes	Hydrocarbons, chlorocarbons	Hydrocarbons, chlorocarbons	Hydrocarbons, nitriles, amines	Ethene, propene, gaseous hydrocarbons	Hydrocarbons, chlorocarbons	Hydrocarbons, chlorocarbons
Sampling						
Solid sorbent	Tenax GC	Tenax TA	Multisorbent[b]	Molecular sieve 13X	Dual cartridge[c]	Charcoal
Sampling rate	30 mL/min	50 mL/min	200 mL/min	40 mL/min	1700 mL/min	100 mL/min
Sampling time	12 h	1 h	1 h	2-min increments "continuous"	14 hr/8 h (2 pumps)	5–7 days
Total collected	~20 L	~3 L	~12 L		2.2 m^3	variable
Analysis						
Gas chromatography	Capillary	Capillary 30 m DB5	Capillary 60-m DB5	Packed, alumina	Capillary 60 m DB-Wax/capillary 27.5-m Poraplot Q	Capillary
Detection	Mass spectrometry	Mass spectrometry	Flame ionization, nitrogen thermionic	Flame ionization	Mass spectrometry	Flame ionization, mass spectrometry

[a]e.g., Wallace et al. 1987.
[b]Tenax-CarboTrap-Ambersorb XE340 in series.
[c]Tenax/CarboTrap in front trap, Carboxen 569 in rear trap.

Table 11.2 Chemicals Considered in Field VOC Studies with ETS as a Variable

Saturated Hydrocarbons

Chemical	Study Number
Ethane[a]	6
Propane	6, 7
n-Butane	7
Isobutane	7
n-Hexane[b]	4
n-Heptane[b]	4
n-Octane[b]	1, 2, 3, 4
n-Nonane[b]	4, 9
n-Decane[b]	2, 3, 4, 9
n-Undecane[b]	2, 3, 4, 9
n-Dodecane[a]	2, 3, 4, 9
n-Tridecane	4, 9
n-Tetradecane	4

Aromatic Hydrocarbons

Chemical	Study Number
Benzene	1, 2, 3, 4, 5, 8, 10
Toluene[b]	3, 4, 5, 8, 10
o-Xylene[b]	1, 2, 3, 4, 5, 9
m + p-Xylene[b]	1, 2, 3, 4, 5, 9
Ethylbenzene[b]	1, 2, 3, 4, 5, 9
Styrene[b]	1, 2, 3, 5, 9
n-Propylbenzene[b]	4, 9
i-Propylbenzene[b]	4, 9
o-Methylethylbenzene[b]	4, 9
m-Methylethylbenzene[b]	4, 9
p-Methylethylbenzene[a]	4, 9
1,2,3-Trimethylbenzene	4, 9
1,2,4-Trimethylbenzene[b]	4

Compound	1	2	3	4	5	6	7	8	9	10
n-Pentadecane				4					9	
n-Hexadecane				4					9	
3-Methylpentane				4						
2-Methylhexane				4						
3-Methylhexane				4						
Cyclohexane				4				8		
Methylcyclohexanes				4						
Dimethylcyclopentanes				4						
1,3,5-Trimethylbenzene[b]				4					9	
n-Butylbenzene				4					9	
p-Methyl-i-propylbenzene				4						
Naphthalene[b]				4	5					
1-Methylnaphthalene				4						
Chlorocarbons[b]										
Vinylidenechloride	1								9	
Chloroform	1	2								
1,2-Dichloroethane	1	2	3							
1,1,1-Trichloroethane	1	2								
Carbon tetrachloride	1	2		4						
Trichloroethylene	1	2	3	4						
1,1,1,2-Tetrachloroethane		2								
Unsaturated Hydrocarbons										
Ethene						6	7			
Ethyne							7			
Propene						6	7		9	
1,3-Butadiene						6			9	10

Table 11.2 continued

Chemical	Study Number[a]									
	1	2	3	4	5	6	7	8	9	10
Isoprene					5	6			9	10
α-Pinene[b]		2	3							
Limonene			3	4	5				9	
Chlorocarbons[b] (continued)										
Bromoform	1									
Dibromodichloromethane	1									
1,2-Dibromoethane		2								
Dibromochloropropane	1									
Tetrachloroethylene	1	2	3	4					9	
Chlorobenzene	1	2	3	4						
m-Dichlorobenzene	1	2		4						
o-Dichlorobenzene	1	2	3							

Chemical	Study Number[a]							
	1	2	3	4	5	6	7	8
1,1,2,2-Tetrachloroethane		2						
Bromodichloromethane	1							
Other Oxygenated								
2-Methylfuran					5			
2,5-Dimethylfuran					5			
1,4-Dioxane		2						
Ethanol								8
Isopropanol								8
Nitriles								

Compound	1	2	3	4	5	6	7	8	9
p-Dichlorobenzene	1	2		4					
1,2,3-Trichlorobenzene				4					
1,2,4-Trichlorobenzene				4					
1,3,5-Trichlorobenzene				4					
Carbonyls									
Formaldehyde[b]								8	
Acetaldehyde								8	
Propionaldehyde								8	
Acrolein								8	
Acetone[b]					5			8	9
2-Butanone[b]					5			8	

Compound	1	2	3	4	5	6	7	8	9
Acetonitrile					5				
Acrylonitrile					5				
Propionitrile					5				
Butyronitrile									
Isobutyronitrile					5				
Valeronitrile					5				
Nitrogen Heterocyclics									
Pyridine					5				9
Vinylpyridine			3		5				9
Picoline					5				9
Pyrrole					5				

Table 11.2 continued

Chemical	Study Number[c]	Chemical	Study Number[c]
Ethylacetate	8		
2-Pentanone	8		

[a]Italics indicate suspect as ETS being significant contributor.

[b]Major contribution by building materials and/or consumer products (Molhave, 1982, Miksch et al. 1982, Tichenor and Mason, 1988).

[c]Studies.

1. TEAM Study, New Jersey. Wallace et al. 1987 (See also Wallace et al. 1987, Wallace and Pellizzari, 1986).
2. TEAM Study, California. Wallace et al. 1984 (See also Wallace et al. 1987, Wallace and Pellizzari, 1986).
3. Proctor, 1989, Proctor et al. 1989.
4. Lebret at al., 1984.
5. Guerin, 1991.
6. Löfroth et al. 1989.
7. Persson et al. 1988.
8. Badre et al. 1978.
9. Heavner et al. 1996
10. Phillips, Howard, and Bentley, 1998

readily detectable derivatives, and final determination by gas or liquid chromatography. A brief description of thermal desorption sampling and analysis methods follows. The reader is also referred to a review by Pellizzari (1991).

Sampling. Samples for thermal desorption analysis are typically taken by drawing the air through cartridges packed with a solid sorbent (see Table 11.1). The cartridges are generally glass or stainless steel tubes with typical dimensions being 10 cm long by 1.4 cm internal diam (Sheldon et al. 1988a, b) or 90 × 6 mm outside diam (e.g., Heavner et al. 1996). Personal sampling pumps or diaphragm pumps are commonly employed to provide sampling rates of 10–1700 mL/min. Sampling times range from a few minutes to several days depending on the nature of the study, but most frequently range from 1–16 h. Sample sizes ranging from 5–20 L are generally adequate to collect a sufficient quantity of common VOCs for analysis, although greater quantities collected can mean lower volumetric detection limits.

Sampling flow rate is an important consideration. Each solid sorbent material has a finite affinity and capacity for each individual VOC. Excessive flow rates result in poor collection efficiencies and can serve to strip the sorbent of previously collected analyte, especially when long sampling periods are used. For most of the sorbents and configurations given in Table 11.1, flow rates of approximately 50 mL/min yielded good collection efficiencies. In the study reported in Guerin (1991), a specially designed multisorbent trap was used to obtain good collection efficiencies at up to 200 mL/min, but noticeable breakthrough was observed at 500 mL/min. In the Heavner et al. (1996) study, multiple traps in series were employed, permitting much higher sampling rates.

One of the most commonly used sorbents has been Tenax [poly(p-2,6-diphenylphenylene) oxide]. Tenax is available from a variety of chromatography and sampling supply houses, is an effective medium for collecting mid- to high-boiling-range volatiles, and may be thermally desorbed at up to almost 300°C without breakdown. Tenax is less efficient at collecting low-boiling-range VOCs such as isoprene and is incapable of collecting gaseous hydrocarbons such as ethane and ethene. Heavner et al. (1996) have employed Tenax backed by CarboTrap in a dual sorbent trap, which was in turn backed by a second trap filled with Carboxen 569. Higgins et al. (1987) employed a trap containing Tenax followed by CarboTrap followed by Ambersorb XE340 in series to retain the advantages of Tenax while increasing collection efficiencies for the more volatile constituents of interest. Multisorbent traps consisting of glass beads, Tenax-TA, Ambersorb XE340, and activated charcoal (Hodgson and Girman 1989) and consisting of glass beads, Tenax, and Ambersorb XE340 (Tsuchiya and Stewart 1990) have also been employed for indoor air studies.

The collection of gaseous hydrocarbons or of low-boiling-range hydrocarbons over a long period requires the use of more active sorbents. Molecular Sieve 13X (Persson et al. 1988) and activated charcoal (e.g., Lebret et al. 1984) have been used to address such needs. Sorbents such as activated charcoal are especially efficient for collecting VOCs in air but cannot be used for the determination of mid- and higher-boiling compounds because they are irreversibly adsorbed.

It is important to note that stringent procedures are necessary to prepare sorbents and in handling sampling cartridges to ensure reproducible and interference-free sampling. In studies of VOCs in public buildings, for example, Sheldon et al. (1988b) prepared their Tenax adsorbent by 24-h Soxhlet extractions with methanol and n-pentane, drying under nitrogen for 24 h, and then drying at 100°C for 24 h in a vacuum (28 in. of water) oven. The treated Tenax was then sieved to 35/60 mesh, packed into cartridges, and the cartridges were desorbed at 270°C for 5 h with an inert gas purge. This treatment was necessary because the investigators recycled Tenax from previous sampling trips. Commercially obtained cartridges (Tenax, other polymers for thermal desorption, or carbon-based sorbents such as CarboTrap) or those prepared in-house from bulk sorbents should at least be subjected to thermal treatment similar to that described above, the temperature of which will depend on the sorbent. Once packed and thermally conditioned, the cartridges must be carefully sealed until use to avoid laboratory or field contamination. It is prudent to analyze at least 5% of the cartridges prior to field sampling and to carry some cartridges to the field as field blanks to ensure that results are not due to sorbent decomposition or contamination.

Analysis. Collected sample is most commonly analyzed by thermal desorption gas chromatography (e.g., Table 11.1). In this procedure, the cartridge is quickly heated (e.g., to 250–270°C for Tenax, 350°C for materials such as CarboTrap) while being purged with chromatographic carrier gas (most commonly helium or nitrogen) and the analytes are collected in a small volume cold trap in line with the chromatographic column or at the head of the chromatographic column itself. The cold-trap and/or column is then rapidly heated to introduce the analytes into the column for chromatographic separation. The constituents are detected most commonly with a flame ionization detector or a mass spectrometer.

Gas chromatography/mass spectrometry (GCMS) is generally considered the best analytical system because it provides both quantitative and qualitative information. The ability to acquire mass spectra of each detected constituent during the analysis increases confidence that the constituent is properly identified and provides information concerning the purity of the chromatographic peak. Hydrogen flame ionization detectors (FID) are commonly used because they are very much less expensive, can be operated by less highly trained personnel, and provide a more uniform relative response to common indoor VOCs. A useful practice is to analyze a few samples by GCMS and GCFID to confirm identities and establish chromatographic patterns. Subsequent analyses can be performed by GCFID; changes in the chromatographic pattern indicate a need for further analysis by GCMS.

Specialty detectors are also available for specific applications. As examples, sulfur-containing compounds can be sought with a flame photometric detector (FPD), halogen-containing and other electronegative compounds can be sought with electron capture detection (ECD), nitrogen-containing compounds can be detected with the nitrogen-phosphorus thermionic detector (NPD), and N-nitrosamines can be sought with the pyrolysis chemiluminescence-based detectors. Of these, only the NPD has been used (Guerin 1991) for field studies of ETS VOCs considered in this chapter. In this case, the gas chromatographic effluent was split between an FID and an NPD to provide a simultaneous measurement of hydrocarbon and nitrogenous constituents.

Thermal desorption gas chromatographic analysis is becoming a common organic analytical method. Detailed descriptions of analytical and quality assurance procedures for VOCs in indoor air are available in Heavner et al. (1996) and EPA reports on its TEAM study (e.g., Wallace et al. 1987b) and its public buildings study (Sheldon et al. 1988a, b). General descriptions of various methods are included in many of the references to this chapter.

FIELD STUDIES

More than 90 chemicals (Table 11.2) have been reported in 1 or more of 9 studies of VOCs in indoor air where ETS is a variable. (Note that more studies have been performed. However, those in Table 11.2 are considered to be key investigations and have clearly delineated smoking and nonsmoking environments.) Of these, about half are thought to be associated with ETS. Others are likely to be identified as more detailed investigations are conducted. Benzene and its simple alkyl derivatives have received the greatest attention possibly because benzene is classed as a Group 1 human carcinogen, and it and its derivatives are commonly detected in indoor air. Chlorine-containing compounds are also frequently reported, but these are not expected to be associated with ETS. Cigarette smoke has been reported (Chap. 3) to contain trace quantities of simple chlorocarbons, but most of those detected in indoor air are not present in cigarette smoke. Saturated and unsaturated aliphatic hydrocarbons have also received some attention. n-Decane, n-undecane, and n-dodecane have been studied because they are present in cigarette smoke and selected n-paraffins may possess co-carcinogenic activity. Simple alkanes such as ethene and propene have been studied (e.g., Persson et al. 1988) because of their possible in-vivo conversion to potentially carcinogenic epoxides.

One study reported in Guerin (1991) has specifically targeted organonitriles and simple nitrogen heterocyclics. The chemicals were selected for study as possible ETS-specific markers of exposure and not based on expected health hazard. An earlier study (Jermini et al. 1976) identified nitriles as major constituents in chamber studies. A small number of aldehydes, ketones, and other oxygenated compounds have also been investigated in a few studies of ETS contamination. An observation (Gordon 1990) that 2,5-dimethylfuran is uniquely present in the breath of smokers may increase attention to furans as ETS-specific VOCs. A characteristic of most chemicals given in Table 11.2 is that they derive from many sources. Building materials (e.g., Sheldon et al. 1988a,b) and consumer products (e.g., Tichenor and Mason 1988) are major sources of aromatic and aliphatic hydrocarbons and the overwhelming source of chlorine-containing organics.

Results of VOC analyses are tabulated in Appendix 6 for reference. Table 11.3 summarizes concentration ranges for representative VOCs by location type based on the data tabulated in Appendix 6. The purpose of Table 11.3 is to illustrate concentration ranges reported in such a way as to allow comparison across location types, between smoking and nonsmoking areas of the same location, and between chemicals. However, such a presentation must be studied with caution. As with all other chemicals treated in this volume, the distribution of individual results is highly skewed toward the low end of the concentration range. The low concentration of the range is more representative of the typical situation than is the high concentration. High concentration results are important, however, because they illustrate conditions that have been encountered in the field. These tend to be rare occurrences due to proximity to some specific source at the time of sampling. Some border on the artificial, e.g., the value of 1040 μg/m^3 of acetaldehyde reported by Badre et al. (1978) corresponds to two smokers actively smoking in an automobile with windows closed and with no mechanical ventilation.

Most VOCs are commonly present at concentrations ranging from approximately 1–20 μg/m^3. Acetaldehyde and acrolein appear to be commonly present at concentrations of 50–200 μg/m^3 but the data available are very limited and center on environments with high smoking and other activity. Ranges of concentrations reported (Table 11.3) for smoking areas versus nonsmoking areas overlap to a large extent but with a clear trend toward smoking areas exhibiting higher concentrations than the nonsmoking areas for some components. With few exceptions [e.g., toluene and xylenes in a train compartment (Proctor et al. 1989) and residences (Heavner et al. 1996)], the high end of the concentration range is often greater for smoking conditions than for nonsmoking conditions. The differences between the high range concentrations is often small, however.

Except possibly in the case of visibly extreme cigarette smoke contamination, unusually high concentrations (e.g., in Table 11.3, 570 μg/m^3 toluene in a smoking workplace or 328 μg/m^3 m + p-xylene in an office) of a VOC probably indicate an unsuspected major source was encountered. Proximity to a newly furnished or painted area or to copying machines or printing areas are examples. Without careful apportionment studies using ETS-specific markers (e.g., Heavner et al. 1995; see also Chap. 4), results from long-term personal sampling with subjects proceeding through normal daily

Table 11.3 Summary of Concentration ($\mu g/m^3$) Ranges of Representative ETS-Related Volatile Organic Compounds in Indoor Air[a]

Constituent	Offices/Other Workplaces		Residences		Restaurants		Taverns		Transportation	
	S	NS	S	NS	S	NS	S	NS	S	NS
Isoprene	1–61	0–20	3–83	1–17.3	16.6–90	NR	85–150	NR		
n-Octane	1–530	1–15	1.3–4.7[b,c]	1.6–3.1[b,c]					0.2–2.9	0.8–2.0
α-Pinene	1–11	2–8							0.6–6.3	0.7–18.1
Limonene	1–31	0.4–8	0–629	0–124	52–239	NR			0.1–19.4	<0.1–8
Benzene	1–101	0.4–31	1–17	4.4–11[c]	5.6–13.8	NR	21–27	NR	0.9–28.6	2.9–29.3
Toluene	5–570	2–110	10–74	6–102	16.2–215	NR	41–80	NR	3.3–88	7.8–105
Ethylbenzene	0.8–23.4	0.2–66	0.4–10	1.1–26	2.1–5.2	NR	9–32[d]	4–6[d]	0.3–9.4	1.4–19.7
Styrene	2–59	0–79	0.2–4.9	0.1–9.9	1.8–4.4	NR	4–11[d]	4–6[d]	1.4–26.4	1.1–63.8
o-Xylene	0.65–68	0.4–94	0–6.6	0.9–35	1.5–6.7	NR			0.4–18.4	1.3–59.7
m + p-Xylene	14–328	23–170	6.3–25[c]	7.2–21[c]	5.4–20.6	NR			4.1–99.6	2.1–300
Acetaldehyde					170–630	NR	183–204	NR	65–1040	NR
Acrolein			3,4,17.3	ND	30–100	NR	21–24	NR	20–120	NR
Acetonitrile			0.6,6.5	ND–0.2	2.4–48.9	NR				
Pyridine	0–5.3	0–7.6	0.6,10.6	ND	0.8–15.7	NR				
Pyrrole					<0.1–12.1	NR				

[a]See Appendix 6 for data and references.
[b]From Wallace et al. 1987.
[c]Range of geometric means.
[d]Betting shops.
ND=not detected.
NR=not reported or outdoor air as control.

activities can be further subject to overestimates of ETS-related VOC exposure. This is because the subjects can encounter unusually high concentration environments. Pumping gasoline, for example, is reported to result in breathing zone concentrations of benzene of 3000 $\mu g/m^3$ (Wallace 1989).

Identifying absolutely ETS-free background concentrations of VOCs from the existing database is difficult. There is a very great spatial and temporal variability in VOC concentration depending on the specific environment and human activity involved, and no reported study provides evidence of an absolutely ETS-free environment being used as a control. Several data sets may, however, indicate what constitutes background concentrations. These include data from outdoor median concentrations reported (Michael et al. 1990) as part of the TEAM study and ambient roof concentrations reported (Hodgson and Daisey 1989) as part of a study of indoor VOCs in a new office building. Results for indoor air in a hospital (Sheldon et al. 1988a) and for personal sampling of indoor air in nonsmoking residences (Wallace et al. 1987b; Heavner et al. 1995, 1996) are also likely to be free of or at least minimally contributed to by ETS. In the latter studies, levels of the ETS-specific marker 3-ethenyl pyridine (3-EP; see Chap. 4) were used to quantitatively assess contribution from ETS in both smoking and nonsmoking environments. These data suggest that minimum background concentrations of benzene, ethylbenzene, and o-xylene are typically 2–5 $\mu g/m^3$ with excursions to approximately 10 $\mu g/m^3$, concentrations of toluene and m+ p-xylene are typically 2–25 $\mu g/m^3$ with excursions to 40 $\mu g/m^3$, and that C_{10} to C_{12} n-paraffins are commonly present at concentrations of <1 to 4 $\mu g/m^3$ with occasional excursions to 10 $\mu g/m^3$. As shown by the data reported for nonsmoking environments by various investigators, much higher concentrations can be encountered without ETS being an obvious contributor (Appendix 6 and Table 11.3).

The EPA TEAM study deserves special mention. The TEAM study has measured up to 26 VOCs in residences, backyard outdoor air, and drinking water in 7 cities with an experimental design which is expected to represent the exposure of 700,000 individuals. ETS exposure has been studied by measuring approximately 20 VOCs in the breathing zone and in the exhaled breath of smokers and of nonsmokers exposed and not exposed to ETS. The study design (Wallace et al. 1987) included approximately 200 smokers and 322 nonsmokers and included taking sequential 12-h personal exposure air samples during the day (6 a.m.–6 p.m.) and overnight (6 p.m.– 6 a.m.). Overnight exposure is viewed as predominantly residential exposure, and the study compares the exposure of nonsmokers in a nonsmoking residence against those in a smoking residence.

Results (Wallace et al. 1987) of overnight exposure to expectedly ETS-related aromatic hydrocarbons are given in Table 11.4 and compared with more modern residential personal exposure data obtained by Heavner et al.

Table 11.4 Overnight Indoor Air Concentrations ($\mu g/m^3$) in Homes With and Without Smokers

Study Site	No. Subjects		Benzene		Styrene		Ethylbenzene		o-Xylene		m + p-Xylene	
	S	NS	S	NS	S	NS	S	NS	S	NS	S	NS
NJ, fall[a]	252	94	16[b]	8.4	1.9[b]	1.0	7.9[b]	4.5	6.3[b]		3.8	19[b]
NJ, summer[a]	111	44	–	–	1.4	1.2	3.9	4.8	4.6	5.5	11	13
New Jersey, winter[a]	37	12	–	–	1.5	1.0	5.6	7.5	6.1	7.2	17	21
Los Angeles, CA, winter[a]	56	58	17[b]	11	3.4[b]	1.7	9.1[b]	5.6	11[b]	7.5	25[b]	17
Los Angeles, CA, spring[a]	23	28	4.8	4.5	0.4	0.6	2.8	3.1	2.8	2.7	11	11
Antioch, CA, spring[a]	35	33	4.9	4.4	0.6	0.5	1.8	2.1	2.2	2.8	6.3	7.2
Columbus, OH winter[c]	25	24	4.03[b]	2.42	1.92[b]	1.38	2.21	2.12	2.57	2.43	8.0	6.5
Philadelphia, PA/Camden, NJ fall[d]	32	61	3.87	2.97	1.05	0.98	3.40	3.65	2.25	2.27	7.2	6.15

[a]TEAM Study, 6 p.m.–6 a.m., personal sampler, weighted geometric mean, homes (s) with one or more current smokers versus homes (NS) with no current smokers. Wallace et al. 1987.
[b]Elevation of smoking vs. nonsmoking homes significant at $p < 0.05$.
[c]Heavner et al. 1995 15 h away-from-work personal sampler, medians
[d]Heavner et al. 1996 15 h away-from-work personal sampler, medians

(1995, 1996). In the TEAM study, investigators found statistically significant (p < 0.05) increases in overnight exposure to aromatic hydrocarbons in homes with smokers as compared to nonsmokers' homes in the fall and winter but commonly not in the spring or summer. Excess concentrations in homes with smokers were computed to be 3.6 ± 1.8 $\mu g/m^3$ benzene, 1.5 ± 2.1 $\mu g/m^3$ m + p-xylene, 0.58 ± 0.8 $\mu g/m^3$ o-xylene, and 0.53 ± 0.3 $\mu g/m^3$ styrene. n-Octane was also considered as part of the study, but there was no statistically significant distinction between smoking and nonsmoking residences. In contrast, Heavner et al. only observed statistically significant differences for benzene and styrene in the Columbus, OH study. However, they observed statistically significant differences among nitrogen-containing compounds not measured in the TEAM study.

Concentrations of benzene in smoking and nonsmoking residences have also been considered in a study (Krause et al. 1987) of 500 homes in Germany. The results are in excellent agreement with those of the TEAM study; 6.5 vs. 7.0 $\mu g/m^3$ benzene in nonsmoking residences and 11 vs. 10.5 $\mu g/m^{3'}$ benzene in smoking residences. Similar results have been reported (Proctor et al. 1991) in a study comparing residential and workplace exposures. However, in a comparison of rural vs. urban subjects living in smoking and nonsmoking homes, Scherer et al. (1995) have reported that urban vs. rural factors are more important in determining benzene exposure that the presence or absence of ETS. Wallace (1996) has recently reviewed environmental benzene exposure.

A small study reported in Guerin (1991) deserves mention because, with the exception of the two studies by Heavner et al. (1995 1996), it is the only study to address a wider variety of nitrogen-containing constituents of cigarette smoke in indoor air, especially nitriles. Both hydrocarbons and the nitrogenous compounds (aliphatic nitriles, pyridine, pyrrole, picoline, and 3-ethenylpyridine, or 3-EP) studied were found elevated over pre-smoking levels in a private residence (Table 11.5) . An important characteristic of the nitriles and pyridines, however, is that they were not detected in the pre-smoking environment. Indeed, subsequent to this report, 3-EP has become a widely accepted tracer for the vapor phase of ETS (see Chap. 4). Nitrogenous compounds have not been commonly considered in general studies of indoor air so it is not known whether important background sources exist. Specific amines, for example, have been reported (Edgerton et al. 1989) in indoor air as resulting from steam humidifiers. Acrylonitrile has recently been reported (van Faassen and Borm 1990) as a significant emission from water-based paint.

While studies of private residences demonstrate a measurable contribution of ETS to indoor air aromatic hydrocarbon content, studies of other environments have exhibited mixed results. Bayer and Black (1987) were unable to distinguish between a smoking and nonsmoking office based on concentrations of aromatic hydrocarbons. Proctor (Proctor 1989, Proctor et al. 1989) determined 17 VOCs in smoking and nonsmoking offices and found

Table 11.5 Distribution of ETS Volatile Organics in a Residence[a]

| | Concentration in $\mu g/m^3$ | | | |
| | Family Room[b] | | Upstairs Bedroom | |
Constituent	Before Smoking	During Smoking[c]	Before Smoking	During Smoking[c]
Isoprene	2.3	83.3	1.8	18.6
2-Butanone	9.5	18.5	6.8	12.4
2-Methylfuran	1.4	12.1	ND	3.8
Benzene	3.6	17.6	2.8	6.9
2,5-Dimethylfuran	ND	4.9	ND	1.3
Toluene	20.9	51.2	16.0	25.0
Ethylbenzene	3.9	8.0	3.1	2.5
m + p-Xylene	11.3	22.4	7.7	12.0
Styrene	2.0	7.3	1.6	3.0
o-Xylene	5.4	7.1	3.7	4.9
Limonene	12.2	22.0	8.9	11.6
Acetonitrile	–	17.3	–	3.4
Acrylonitrile	–	0.8	–	0.6
Propionitrile	–	3.6	–	0.9
Isobutyronitrile	–	1.2	–	0.3
Butyronitrile	–	1.1	–	0.3
Pyridine	0.2	6.5	·	0.3
Pyrrole	10.6	–	0.6	0.6
Valeronitrile	–	1.2	–	–
Picoline[d]	–	4.0	–	0.5
3-Vinylpyridine[d]	–	8.4	–	–
Nicotine	–	51.8	–	–

[a] Data reported in Guerin, 1991, from a study performed by Higgins et al. 1990.
[b] Smoking source. 3 cigarettes/h for 2 h.
[c] Smoking in downstairs family room. Bedroom sampled for the period 30–60 min after first cigarette lit in family room.
[d] 3- and 4-isomers coelute.
ND=not detected.

detectable increases in smoking offices only for ethylbenzene, limonene, and n-octane. A similar study of train compartments by the same investigators resulted in a similar difficulty in distinguishing between smoking and nonsmoking environments based on VOC content. However, in a study of personal exposure to ETS in smoking vs. nonsmoking workplaces, Heavner et al. (1996) reported differences in 20 VOCs, including benzene, alkyl-substituted benzenes, isoprene, acetone, and 1,3-butadiene.

It is important to note that studies of ETS and of indoor air in general have addressed only a small fraction of the VOCs likely to be present. A review (Wallace et al. 1990) of data from 2500 samples taken as part of the TEAM study estimates a total indoor air burden of VOCs of approximately 8 mg/m^3 during the day and 4 mg/m^3 overnight. A major study (Hodgson and Daisey 1989) of an office building that involved measuring 27 VOCs accounted for from less than 20% up to 50% of the total VOCs present as measured by Total Organic Carbon analyses. Similarly, only a fraction of the known gas-phase constituents (Chap. 3) of cigarette smoke have been specifically sought in indoor air.

SUMMARY

Concentrations of aliphatic and monocyclic aromatic hydrocarbons in environments where ETS is expected to be absent range from < 1–10 μg/m^3. Common indoor air concentrations where ETS may or may not be a contributor generally range from 2–20 μg/m^3. Environments where heavy smoking and/or poor ventilation contribute to visibly obvious high ETS contamination can contain VOCs at concentrations of 50–500 μg/m^3. The highest concentrations of VOCs, 100 μg/m^3 and greater, are generally associated with circumstances other than ETS exposure except possibly under extreme smoking conditions. Detailed study of residential exposures suggests that nonsmoking households experience lower levels of benzene than smoking households. However, degree of urbanization may be more important to overall exposure. Excess levels of styrene and xylenes may range from 0.5–1.5 μg/m^3 in smoking households. Studies of offices, restaurants, train compartments, and public buildings suggest that ETS contributes to the indoor air burden of VOCs but that other sources predominate. Major sources include building materials, furnishings, cleaning products, office machines, gasoline, and combustion sources associated with cooling, heating, and transportation. Newly constructed or decorated buildings typically exhibit the highest background concentration of VOCs. Background decreases with time as the building materials and furnishings off-gas and the VOCs are removed by ventilation. Of the VOCs studied to date, only the major chlorocarbons are considered to be completely unrelated to ETS. Most constituents of the gas phase of cigarette smoke have not yet been specifically sought in indoor air.

REFERENCES

Badre, R., Guillerm, R., Abran, N., Bourdin, M., & Dumas, C. (1978) Atmospheric pollution by smoking. *Annales Pharmaceutiques Francaises,* 36 (9–10), 443–452.

Baek, S. O., Kim, U. S., & Perry, R. (1997) Indoor air quality in homes, offices, and restaurants in Korean urban areas—indoor/outdoor relationships. *Atmos. Environ.*, 31, 529–544.

Bayer, C. W. & Black, M. S. (1987) Thermal desorption/gas chromatographic/mass spectrometric analysis of volatile organic compounds in the offices of smokers and nonsmokers. *Biomed. Environ. Mass Spectrom.,* 14, 363–367.

Brunnemann, K. D., Kagan, M. R., Cox, J. E., & Hoffmann, D. (1990) Analysis of 1,3-butadiene and other selected gas phase components in cigarette mainstream and sidestream smoke by gas chromatography-mass selective detection. *Carcinogenesis,* 11, 1863–1868.

Eatough, D. J., Caka, F. M., Wall, K., Crawford, J., Hansen, L. D., & Lewis, E. A. (1989) An automated sampling system for the collection of environmental tobacco smoke constituents in commercial aircraft. *Proceedings of the 1989 EPA/ AWMA International Symposium Measurement of Toxic and Related Air Pollutants,* Raleigh, NC, pp. 564–575.

Edgerton, S. A., Kenny, D. V. & Joseph, D. W. (1989) Determination of amines in indoor air from steam humidification. *Environ. Sci. Technol.,* 23(4), 484–488.

Fischer, T., Weber, A., & Grandjean, E. (1978) Air pollution due to tobacco smoke in restaurants. *Int. Arch. Occup. Environ. Health,* 41, 267–280.

Gordon, S. M. (1990) Identification of unique volatile organic compounds in exhaled breath of smokers. *Proceedings of the 5th International Conference on Indoor Air Quality and Climate, Vol. 2,* Toronto, Canada, pp. 177–182.

Guerin, M.R. (1991) Environmental tobacco smoke. In: Hansen, L. D. & Eatoughc D. J., eds., *Organic Chemistry of the Atmosphere,* CRC Press, Boca Raton, FL. pp. 79-119.

Heavner, D. L., Morgan, W. T., & Ogden, M. W. (1995) Determination of volatile organic compounds and ETS apportionment in 49 homes. *Environ. Int.,* 21, 3–21.

Heavner, D. L., Morgan, W. T., & Ogden, M. W. (1996) Determination of volatile organic compounds and respirable suspended particulate matter in New Jersey and Pennsylvania homes and workplaces. *Environ. Int.,* 22, 159–183.

Higgins, C. E., Jenkins, R. A., & Guerin, M. R. (1987) Organic vapor phase composition of sidestream and environmental tobacco smoke from cigarettes. *Proceedings of the 1987 EPA/APCA Symposium on Measurement of Toxic and Related Air Pollutants,* pp. 140–151.

Higgins, C. E., Thompson, C. V., Ilgner, R. H., Jenkins, R. A., & Guerin, M. R. (1990) Determination of vapor phase hydrocarbons and nitrogen-containing constituents in environmental tobacco smoke. Personal communication, Analytical Chemistry Division, Oak Ridge National Laboratory, Oak Ridge, TN 37831-6120.

Hodgson, A. T. & Daisey, J. M. (1989) Source strengths and sources of volatile organic compounds in a new office building. In: *Session 80—Evaluating Indoor Sources of Gas Phase Organic Compounds,* AWMA 82nd Annual Meeting, June 25–30, Anaheim, CA.

Hodgson, A. T., & Girman, J. R. (1989) Application of a multisorbent sampling technique for investigations of volatile organic compounds in buildings, design and protocols for monitoring indoor air quality, *ASTM STP 1002,* N. L. Nagda and J. P. Harper, Eds., ASTM, Philadelphia, PA, 1989, pp. 244–256.

Jermini, C., Weber, A., & Grandjean, E. (1976). Quantitative determination of various gas phase components of the sidestrearn smoke of cigarettes in room air. *Int. Arch. Occup. Environ. Health,* 36, 169–181.

Krause, C., Mailahn, W., Nagel, R., Schulz, C., Seifert, B., & Ullrich, D. (1987). Occurrence of volatile organic compounds in the air of 500 homes in the Federal Republic of Germany. *Proceedings of the 4th International Conference on Indoor Air Quality and Climate, Vol. 1,* West Berlin: Institute for Soil, Water, and Air Hygiene, pp. 102–106.

Lebret, E., van de Wiel, H. J., Bos, H. P., Noij, D., & Boleij, J. S. M. (1984) Volatile hydrocarbons in Dutch homes. *3rd International Conference on Indoor Air Quality and Climate,* pp. 169–174.

Löfroth, G., Burton, R. M., Forehand, L., Hammond, K. S., Seila, R. L., Zweidinger, R. B., & Lewtas, J. (1989). Characterization of environmental tobacco smoke. *Environ. Sci. Technol.,* 23, 610–614.

Michael, L. C., Pellizzari, E. D., Perritt, R. L., Hartwell, T. D., Westerdahl, D., & Nelson, W. C. (1990) Comparison of indoor, backyard, and centralized air monitoring strategies for assessing personal exposure to volatile organic compounds. *Environ. Sci. Technol.,* 24, 996–1003.

Miksch, R. R., Hollowell, C. D., & Schmidt, H. E. (1982) Trace organic chemical contaminants in office spaces. *Environ. Int.,* 8, 129–137.

Molhave, L. (1982) Indoor air pollution due to organic gases and vapours; of solvents in building materials. *Environ. Int.,* 8, 117–127.

Moschandreas, D. J. & Gordon, S. M. (1991) Volatile organic compounds in the indoor environment: review of characterization methods and indoor air quality studies. In: *Organic Chemistry of the Atmosphere*, Hansen, L. D. & Eatough, D. J., eds., CRC Press, Boca Raton, FL, pp. 121–154.

Pellizzari, E. D. (1991) Sampling gas-phase compounds with sorbent beds. In: *Organic Chemistry of the Atmosphere,* Hansen, L. D. & Eatough, D. J., eds., CRC Press, Boca Raton, FL, pp. 1–51.

Persson, K.-A., Berg, S., Tornqvist, M., Scalia-Tomba, G.-P., & Ehrenberg, L. (1988) Note on ethene and other low-molecular weight hydrocarbons in environmental tobacco smoke. *Acta Chemica Scandinavica,* B42, 690–696.

Phillips, K., Howard, D. A., & Bentley, M. C. (1998) Exposure to tobacco smoke in Sydney, Kuala Lumpur, European, and Chinese cities. *Proceedings of the 1ᵗʰ International Clean Air & Environment Conference, Melbourne, Australia, Oct. 1998,* pp. 347-352.

Proctor, C. J. (1989) A comparison of the volatile organic compounds present in the air of real-world environments with and without environmental tobacco smoke. In: *Session 80-Evaluating Indoor Sources of Gas Phase Organic Compounds,* AWMA 82ⁿᵈ Annual Meeting, June 25–30, Anaheim, CA.

Proctor, C. J., Warren, N. D., & Bevan, M. A. J. (1989) An investigation of the contribution of environmental tobacco smoke to the air in betting shops. *Environ. Technol. Letts.,* 10, 333–338.

Proctor, C. J., Warren, N. D., & Bevan, M. A. J. (1989) Measurements of environmental tobacco smoke in an air-conditioned office building. *Procs., Conference on the Present and Future of Indoor Air Quality,* Brussels.

Proctor, C. J., Warren, N. D., Bevan, M. A. J., & Baker-Rogers, J. (1991) A comparison of methods of assessing exposure to environmental tobacco smoke in nonsmoking British women. *Environ. Int.,* 17, 287–297.

Scherer, G., Ruppert, T., Daube, H., Kossien, I., Riedel, K., Tricker, A. R., et al. (1995) Contribution of tobacco smoke to environmental benzene exposure in Germany. *Environ. Int.,* 21, 779–789.

Shah, J. J. & Singh, H. B. (1988) Distribution of volatile organic chemicals in outdoor and indoor air: a national VOC data base. *Environ. Sci. Technol.,* 22, 1381–1388.

Sheldon, L. S., Handy, R. W., Hartwell, T. D., Whitmore, R. W., Zelon, H. S., & Pellizzari, E. D. (1988a) *Indoor Air Quality in Public Buildings, Vol. I,* EPA/ 660/6-88/009a.

Sheldon, L. S., Felon, H. S., Sickles, J., Eaton, C., & Hartwell, T. D. (1988b) *Indoor Air Quality in Public Buildings, Vol. II,* EPA/600/6-88/009b.

Tichenor, B. A. & Mason., M. A. (1988) Organic emissions from consumer products and building materials to the indoor environment. *JAPCA,* 38, 264–268.

Tsuchiya, Y., & Stewart, J. B. (1990) Volatile organic compounds in the air of Canadian buildings with special reference to wet process photocopying machines. *Proceedings of the 7ᵗʰ International Conference on Indoor Air Quality and Climate, Vol. 2,* Toronto, Canada, pp. 633–638.

van Faassen, A., & Borm, P. J. A. (1990) Indoor air pollution and health hazards by waterborne construction paints. *Proceedings of the 5ᵗʰ International Conference on Indoor Air Quality and Climate, Vol. 3,* Toronto, Canada, pp. 695–700.

Wallace, L. A. (1987) *The Total Exposure Assessment Methodology (TEAM) Study. Summary and Analysis.—Vol. I,* EPA/600/6-97/007a, ORD, USEPA, Washington, D.C. 20460.

Wallace, L. A. (1989) The exposure of the general population to benzene. *Cell Bio. and Toxicol.,* 5(3), 297–314.

Wallace, L. (1989) The total exposure assessment methodology (TEAM) study: an analysis of exposures, sources, and risks associated with four volatile organic chemicals. *J. Am. College of Toxicol.,* 8(5), 883–895.

Wallace, L. A. & Pellizzari, E. D. (1986) Personal air exposures and breath concentrations of benzene and other volatile hydrocarbons for smokers and nonsmokers. *Toxicol. Letts.,* 35, 113–116.

Wallace, L., Pellizzari, E., Hartwell, T. D., Perritt, R. & Ziegenfus, R. (1987) Exposures to benzene and other volatile compounds from active and passive smoking. *Archives of Environ. Health,* 42(5), 272–279.

Wallace, L. A., Pellizzari, E. D, Hartwell, T. D., Sparacino, C., Whitmore, R., Sheldon, L., Zelon, H., & Perritt, R. (1987b) The TEAM study: personal exposures to toxic substances in air, drinking water, and breath of 400 residents of New Jersey, North Carolina, and North Dakota. *Environ. Res.,* 43, 290–307.

Wallace, L. A., Pellizzari, E. D., Hartwell, T. D., Whitmore, R., Zelon, H., Perritt, R., & Sheldon, L. (1988) The California TEAM study: breath concentrations and personal exposures to 26 volatile compounds in air and drinking water of 188 residents of Los Angeles, Antioch, and Pittsburg, CA. *Atmos. Environ.,* 22(10), 2141–2163.

Wallace, L., Pellizzari, E., & Wendel, C. (1990). Total organic concentrations in 2500 personal, indoor, and outdoor air samples collected in the U.S. EPA TEAM studies. *Proceedings of the 5ᵗʰ International Conference on Indoor Air Quality and Climate, Vol. 2,* Toronto, Canada, pp. 639–644.

Wallace, L. (1996) Environmental exposure to benzene: an update. *Environ. Health Perspec.* 104(Suppl. 6), 1129-1136.

Weber, A., Fischer, T., & Grandjean, E. (1979) Passive smoking in experimental and field conditions. *Environ. Res.,* 20(l), 205–216.

INTRODUCTION

For the purposes of this chapter, ETS trace constituents are considered as those constituents of tobacco smoke that are commonly delivered in mainstream and sidestream smoke in ng/cigarette quantities. These constituents, whether ETS-derived or from other sources, are typically found in indoor air at concentrations in the low ng/m^3 (parts per trillion) range. Though present in very small quantities, some of these constituents are carcinogenic and thus remain of public health concern. Some, particularly under conditions involving the presence of major sources and/or poorly ventilated environments, have been detected at μg/m^3 (parts per billion) concentrations for at least brief periods.

As discussed in Chap. 3, an increasingly large number of constituents are detected in cigarette smoke as the quantity being considered is lowered. Constituents present at the μg/cigarette and ng/cigarette ranges probably account for more than 95% of the several thousand individual chemicals in cigarette smoke. The number of constituents produced by other combustion sources is similarly predominated by those present at trace concentrations.

ETS-related trace constituents of currently greatest concern in indoor air are polycyclic aromatic hydrocarbons (PAHs), amino- and nitro-substituted polycyclic aromatics, N-nitrosamines, trace elements, and radioactive species. Except for tobacco-specific nitrosamines, which are clearly ETS-specific, these constituents are contributed to indoor air by various sources. Radioactivity and nitro-polycyclics appear to be overwhelmingly contributed to by sources other than tobacco smoke. For indoor situations, with minimal penetration of unconditioned outside air, volatile N-nitrosamines such as N-nitrosodimethylamine (NDMA) appear to principally result from tobacco smoking, but other sources are possible.

Concentrations and estimated contributions of ETS to trace constituents in indoor air are briefly treated below.

SAMPLING AND ANALYSIS

The sampling and measurement methods required to determine trace constituents in indoor air are tailored to the specific constituent(s) of interest. The methods generally involve combining sampling technologies designed to sample adequate volumes of air with measurement technologies capable of highly sensitive and specific detection. The required sensitivity and selectivity is commonly provided by more sophisticated analytical technologies such as gas chromatography/mass spectrometry (e.g., polycyclic aromatics) and atomic absorption spectrometry (e.g., trace elements). The principal concerns in trace analyses are the avoidance of interferences from sampling materials and methods and the need to confidently distinguish the constituent of interest from similar constituents that are commonly present.

The best descriptions of sampling and analysis methods are generally found in the original papers describing the results of studies of any given constituent of interest. The reader is also referred to the recent summary (Anon 1990) of methods for the determination of indoor environmental carcinogens and the extensive compendium (O'Neill et al. 1987) of methods for the measurement of ETS indoor-air constituents.

POLYCYCLIC AROMATIC HYDROCARBONS

PAHs ranging in size from naphthalene (2 rings) to coronene (6 rings) have been detected in indoor air. The most commonly reported PAHs are naphthalene and its methyl substituted derivatives; anthracene and phenanthrene; fluoranthene; pyrene, chrysene, and benz[a]anthracene; benzofluoranthenes; benzo[a]pyrene and benzo[e]pyrene; dibenz[a,h]anthracene; benzo[ghi]perylene; indeno[1,2,3-cd]pyrene; and coronene. It is not clear whether these chemicals are so frequently reported because they are the only or the predominant PAHs present, or because they are the most confidently detected by commonly employed analytical methods. One would expect that perylene and alkyl-substituted derivatives of many of the above parent compounds would also be frequently present, but they are much less often reported.

Studies that address PAHs in indoor air as a cancer hazard frequently report concentrations of benzo[a]pyrene (BaP) and/or of "total" PAHs. BaP concentrations are important because BaP exhibits a high carcinogenic activity in animal testing and it is considered a suspect (at the least) human carcinogen by many authorities. BaP is also frequently determined to provide a relative measure of PAHs as a whole. Results for total PAH concentrations are also important but must be reviewed carefully to determine the definition of "total." Results summarized in Table 12.1, for example, sum the concentrations of 4- to 6-ring PAHs reported by the investigators as the definition of "total." Including results for naphthalene in computing total PAHs would increase the

Table 12.1 Polycyclic Aromatic Hydrocarbon (PAH) Concentrations

	Concentration (ng/m^3)					
	Pacific Northwest Buildings[a]			Residences[b]	Residences/Offices[c]	
					Arithmetic Mean	
	Arithmetic Mean	Geometric Mean	Range	Arithmetic Mean	Site 1	Site 2
Benzo[a]pyrene						
Outdoors	0.52	0.31	0.04–2.84	0.22	0.09	0.03
Smoking	1.07	0.71	ND–3.60	3.3	0.61	0.38[d]
Nonsmoking	0.39	0.27	0.09–1.35	0.31	–	–
Total PAH						
Outdoors	5.47	3.84	1.53–19.66	4.6	1.36	1.04
Smoking	9.36	7.32	ND–23.74	23.14	4.37	3.07
Nonsmoking	2.40	2.15	ND–5.43	3.57	–	–
Smoking minus Nonsmoking						
Benzo[a]pyrene	+0.68	+0.44	Neg–3.4	+3.0	–	–
Total PAH	+6.96	+5.17	Neg–24	+19.57	–	–

[a]Turk et al. 1987. Total PAH Chrysene, benzo[b]fluoranthene, benzo[k]fluoranthene, benzo[a]pyrene, dibenz[a,h]anthracene, benzo[ghi]perylene, and indeno[1,2,3-cd]pyrene.

[b]Gold et al. 1990. Total PAH = Benz[a]anthracene, chrysene, benzofluoranthenes, benzo[c]pyrene. benzo[a]pyrene. benzo[ghi]perylene, and coronene.

[c]Gundel et al. 1990. Total PAH = benz[a]anthracene, benzo[c]pyrene, benzo[b/k]fluoranthenes, benzo[a]pyrene, dibenz[a,h]anthracene, benza[ghi]perylene, indeno[1,2,3-cd]pyrene, and coronene.

[d]Possible positive interference is noted.

ND=Not detected.

Neg=negative.

concentration of total PAHs in a nonsmoking residential environment (Gold et al. 1990) from 3.57 ng/m³ (Table 12.1) to 1803.57 ng/m³. Naphthalene is frequently present at 2–3 orders of magnitude greater concentrations than are 4- to 6-ring PAHs. For example, Gundel et al. (1995) have determined naphthalene emissions from a reference cigarette to be 150 times greater than those of BaP. Anthracene and phenanthrene (3-ring PAHs) are frequently present at 1 to 2 orders of magnitude greater than are the 4- to 6-ring PAHs, which include most PAHs known or suspected to be environmental carcinogens. As such, summation of data from 4- to 6-ring PAHs appears to be the most relevant definition of "total" PAHs with regard to cancer hazard.

Table 12.1 summarizes results for BaP and for total PAH (4- to 6-ring) in environments defined as smoking and nonsmoking and, where available, corresponding outdoor environments. Table 12.2 details results for BaP concentrations from the study (included in Table 12.1) of buildings in the Pacific Northwest (Turk et al. 1987) to provide the reader with a measure of site-to-site variability.

Outdoor air concentrations of BaP most commonly range between 0.05 and 1.0 ng/m³ and occasionally reach 3 ng/m³ in the U.S. (also see Chap. 2). The high end of the range reflects heavily trafficked urban areas while the low end reflects rural areas. Special circumstances can lead to unusually high BaP concentrations in outdoor air. A most extreme example is the finding (Guenther et al. 1988) of 11.3 ng/m3 of BaP in the outdoor air of Fairbanks, AK. This result was obtained during winter when wood burning as a source of indoor heating was common, and a weather inversion prevented the normal dispersion of ambient air pollutants. Areas where industrial, economic, or cultural practices involve the use of crude fuels commonly experience outdoor BaP concentrations an order of magnitude greater than that commonly encountered in the U.S. In all outdoor environments, the concentration of 4- to 6- ring PAHs is approximately 10 times that of BaP alone.

Elliot and Rowe (1975) were among the first to determine the contribution of ETS to indoor concentrations of BaP. They found concentrations ranging from 7–22 ng/m³ of BaP in arenas where smoking occurred, as contrasted to 0.7 ng/m³ for the arenas before they were occupied. More recent results (Tables 12.1, 12.2) of more commonly encountered environments suggest that background (ETS-free) indoor concentrations of BaP typically range from 0.1–1 ng/m³ while the same or similar environments containing ETS range in concentrations from 0.3–1.5 ng/m³. Total (4- to 6-ring) PAH concentrations are again approximately 8 to 10 times that of BaP alone for both environments.

Tables 12.1 and 12.3 present crude assessments of the contribution of ETS to indoor air PAH concentrations. The data suggest that ETS contributes between 0.5 and 1 ng/m³ of BaP to indoor environments containing readily measurable ETS-contamination. Total PAH contributions are proportionally equivalent to the non-ETS background concentrations and commonly range

Table 12.2 Benzo[a]pyrene Concentrations in Pacific Northwest Buildings Study[a]

Site	Outdoor	Indoor Concentrations, ng/m³			
		Smoking	Nonsmoking	Bldg. Mean	Smkg. minus Nonsmoking
Salem, Oregon					
Urban office 8[b]	0.22	ND	0.18	0.18	<-0.18
Urban office 9	0.27	0.37	0.09	0.31	0.28
Urban office 10	0.04	1.14	0.56	1.00	0.58
Urban office 11	0.09	2.45	0.17	0.93	2.28
Urban office 11[c]	0.40	1.44	0.19	0.69	1.25
Urban office 15	0.34	0.35	0.13	0.20	0.22
Spokane, Washington					
Suburban school 16	0.29	0.57	0.23	0.43	0.34
Suburban library 23	0.11	ND	0.10	0.10	<-0.10
Suburban office 25	0.12	0.25	0.10	0.15	0.15
Urban office 26	0.29	0.57	0.23	0.43	0.34
Urban school 27	0.38	0.26	0.26	0.26	0
Urban multiuse 28	0.23	0.28	0.30	0.29	-0.02
Suburban school 30	2.84	1.44	1.35	1.36	0.09
Suburban multiuse 31	0.58	3.60	0.25	1.36	3.35
Urban office 34	0.73	1.45	0.24	0.72	1.21
Urban office 35	1.25	1.65	1.10	1.21	0.55
Portland, Oregon					
Urban office 37	0.42	0.67	0.98	0.81	-0.31
Urban office 39	0.47	0.12	ND	0.12	>0.12
Arithmetic Mean	0.52	1.07	0.39	0.60	—
Geometric Mean	0.31	0.71	0.27	0.43	—

[a]Turk et al. 1987.
[b]Building site designation.
[c]Repeat of earlier sampling trip.

Table 12.3 Distribution of Polycyclic Aromatic Hydrocarbons in Indoor Air

| | Suburban School[a] | | Urban Office[a] | | Residences[b] | | Residence/Office Study[c] | |
	Smoking	Nonsmoking	Smoking	Nonsmoking	Smoking Occupants	Nonsmoking Occupants	Smoking Site 1	Smoking Site 2
Chemical								
Naphthalene					2200	1800		
Acenaphthylene					120	10		
Phenanthrene					210	59		
Anthracene					1.5	2.0	0.86	0.22
Fluoranthene					23	7.2	1.64	0.82
Pyrene					17	5.6	0.13	0.52
Benzo[a]anthracene	0.93	0.11	0.44	0.09	3.4	0.24	3.58	5.54
Chrysene	1.20	0.12	0.77	0.26	7.2	0.93	0.42	0.58
Benzo[b]fluoranthene	0.33	0.07	0.33	0.06			0.13	0.15
Benzo[k]fluoranthene								
Benzofluoranthenes					5.1	0.78	0.93	0.39
Benzo[e]pyrene	1.49	0.07	0.39	0.09	1.0	0.68	0.61	0.38
Benzo[a]pyrene	0.09	0.09	0.56	0.11	3.3	0.31	0.18	<0.11
Dibenz[a,h]anthracene	0.67	0.25	1.21	0.73			0.88	0.44
Benzo[ghi]peryiene	1.35	0.65	0.31	0.35	2.5	0.32	0.55	0.50
Indeno[1,2,3-cd]pyrene				0.64			0.54	< 0.01
Coronene					0.31			

Concentration in ng/m³

[a]Turk et al. 1987.
[b]Gold et al. 1990.
[c]Gundel et al. 1990.

from 4–9 ng/m^3. Excursions in BaP concentrations due to ETS in such environments reach approximately 2.5 ng/m^3. These are similar to background concentrations (and much less than excursion concentrations) observed (Table 2.9) in environments with wood-burning stoves, wood-burning fireplaces, and unvented kerosene or gas space heaters. The magnitude of the ETS contribution is generally small and often difficult to confidently detect.

The BaP and total PAH content of indoor air are closely related to the particulate matter content of the air. This is because BaP and other 4- to 6-ring PAHs are generally found as constituents of airborne particulate matter or as constituents adsorbed onto airborne particulate matter. Environments exhibiting a high RSP matter content (Chap. 6) due to ETS are accompanied by a correspondingly high PAH content. High RSP environments due to other combustion sources are similarly enriched in PAH content.

Table 12.3 summarizes the concentrations of individual PAHs reported in several studies of commonly encountered environments. One study that cites 2- to 3-ring PAH concentrations is included in the table because its data represent the common observation that 2-ring (naphthalene, alkyl naphthalenes, etc.) and 3-ring (phenanthrene, anthracene) PAHs quantitatively predominate over the higher-ring-system PAHs. Naphthalenes can result from consumer products and are generally a major component of a variety of combustion products. Naphthalene, anthracene, and phenanthrene (i.e., 2- to 3-ring PAHs) are found as vapor-phase constituents of indoor air and thus relate very poorly to the higher-ring, particle-bound PAHs suspected of being important environmental carcinogens.

Concentrations of individual 4- to 6- ring PAHs present in any given environment at any given time differ from one another by approximately an order of magnitude. There is no obvious relationship between the relative quantities of individual PAHs present and the environments sampled. Chapter 2 contains an additional discussion of PAHs.

N-NITROSAMINES

N-Nitrosamines are potentially important constituents of ETS because many are carcinogenic and because tobacco smoking is thus far their only identified source in nonoccupational indoor air. N-nitrosamines of concern (e.g., Hoffmann et al. 1987) fall into two classes, the Volatile Nitrosamines (VNAs) and the Tobacco-Specific Nitrosamines (TSNAs). VNAs of concern include nitrosodimethylamine (NDMA), nitrosodiethylamine (NDEA), and nitrosopyrrolidine (NPYR). TSNAs of concern in ETS exposure are N-nitrosonornicotine (NNN) and 4-(methyl-nitrosamino)-1-(3-pyridyl)-1-butanone (NNK). NNN and NNK are especially potent carcinogens in hamsters, rats, and mice when applied topically, and NNK has been identified as a possible human lung carcinogen.

An additional factor that makes nitrosamines of special interest is that much larger quantities (e.g., Brunnemann et al. 1977; Hoffmann et al. 1987) are released in sidestream smoke than in mainstream smoke. At least 10 times (and up to 50 times) more VNA is released in sidestream than mainstream smoke and at least 2–5 times more TSNAs are released in sidestream than mainstream smoke. Commercial cigarettes, for example yield approximately 400–800 μg/cigarette of NDMA and 80–150 μg/cigarette of NPYR in their sidestream smokes (Mahanama and Daisey 1996). The preference for sidestream release and the quantities released imply that passive inhalation can rival active smoking as the principal route of inhalation exposure. However, reactivity may act to diminish their concentrations in smoking environments.

To the authors' knowledge, there have been only five studies (Brunnemann et al. 1978, Stehlik et al. 1982, Matsushita and Mori 1984, Klus et al. 1987 & 1992, Brunnemann et al. 1992) that report original data on nitrosamines in ETS. Nitrosodimethylamine, nitrosodiethylarnine and/or nitrosopyrrolidine have been determined in artificially highly contaminated environments (Table 12.4) and in natural (but also rather highly contaminated) environments (Table 12.5). Results (Klus et al. 1987a, 1992) have been reported for the tobacco-specific nitrosamines NNN and NNK for an artificially contaminated environment (Table 12.6), and for a few natural environments (Brunnemann et al. 1992) (Table 12.7).

Early work (Brunnemann et al. 1978) found that ambient concentrations of NDMA ranged from 10 ng/m^3 to 240 ng/m^3 in natural environments. Subsequent work by others (e.g., Stehlik et al. 1982; Klus et al. 1987) find concentrations of NDMA to range from < 10 ng/m^3 up to only 70 ng/m^3. Only the most extreme of artificial conditions (e.g., an extremely irritating environment with an accompanying carbon monoxide concentration of 16 ppm) produced NDMA concentrations as great as 150 ng/m^3.

Review of Tables 12.4 and 12.5 suggests that what appears to be an important inconsistency (i.e., that natural environments contain as great or greater quantities of nitrosamines than do artificially contaminated environments) may actually be an artifact of the experimental design. The number of tobacco products consumed in the natural environments for many of the samplings suggest that they were often very nearly as contaminated as were the artificial environments. Considering those relatively few samplings that accompanied reasonable combinations of smoking and room size suggests that the more commonly encountered time weighted average (TWA) concentrations of NDMA range from 1– 20 ng/m^3 rather than the 10–240 ng/m^3 frequently cited in summary tables. For example, if the median ratio of NDMA:nicotine for data reported by Klus et al. (1992) (0.174 ng NDMA/m^3 per μg nicotine/m^3) is maintained at more typically encountered TWA nicotine concentrations, median and 95th percentile TWA level of NDMA in unrestricted smoking workplaces (Jenkins and Counts 1999) would be 0.25 and 4 ng/m^3, respectively. Even if the actual ratio is a factor of 10 greater, due

Table 12.4 Nitrosamine Concentrations in Artificial Environments

Test Environment	Test Conditions	Concentrations (ng/m³)[a]			Carbon Monoxide (ppm)
		NDMA[b]	NDEA[c]	NPYR[d]	
Stehlik et al. 1982					
Unventilated 22-m³ office	35 cigarettes in 2 h by 3 persons	130	<10	—	15
	38 cigarettes in 2 h by 3 persons	150	<10	—	16
	28 cigarettes in 2 h by 3 persons	80	<10	—	—
	12 cigarettes in 2 h by 2 persons	20		—	8
Unventilated 46-m³ office	18 cigarettes 4 pipes in 2 h by 7 persons	80	<10	—	8
Unventilated 43-m³ conference room	64 cigarettes in 2 h by 11 persons	70[e]	<10	—	—
Unventilated 43-m³ conference room	40 cigarettes in 1 h by 10 persons	20[f]	<10	—	—
Klus et al. 1987 and 1992					
Unventilated 84 m³ office	11 cigarettes in 2 h	12.8	nd	6.8	10.4
	18 cigarettes in 2 h	13.2	1.4	4.9	13.0
	11 cigarettes 1 cigar in 2 h	27.1	8.6	11.6	12.5
	11 cigarettes in 2 h	22.0	1.7	13.0	7.7
	11 cigarettes in 2 h	8.2	0.5	3.5	7.6
	13 cigarettes, 1 pipe in 2 h	7.9	0.7	5.2	11.0
	12 cigarettes, 1 pipe in 2 h	13.9	nd	5.9	12.4
	15 cigarettes	9.0	nd	4.5	14.0

[a]Sampled during smoking period unless otherwise noted.
[b]N-nitrosodimethylamine.
[c]N-nitrosodiethylamine.
[d]N-nitrosopyrrolidine.
[e]Sampled during smoking period and 1 h after.
[f]Sampled during smoking period and 2 h after.

Table 12.5   Volatile Nitrosamine Concentrations in ETS-Containing Natural Environments

Environment	Estimated Smoking	Concentrations (ng/m³)	
		NDMA	NDEA
Stehlik et al., 1982			
207 m³ Work room	Continuous for 2 hours	23	<10
301 m³ Conference room	26 cigts, 1 pipe, 6 cigarillos in 2 hrs	24	<10
70 m³ off ice	27 cigts in 2 hrs	30	30
50 m³ Conference room	37 cigts, 4 pipes, 3 cigars in 2 hrs	20	20
120 m³ Restaurant	20-30 cigts, 2 pipes in 2 hrs	<10	<10
160 m³ Restaurant	20 cigts in 1 hr	10	<10
180 m³ Restaurant	25-30 cigts in 1 hr	40	<10
160 m³ Restaurant	15-20 cigts in 1 hr	50	<10
320 m³ Dancing bar	Not determined (but "very smokey")	70	20
Klus et al., 1987			
Five locations[a]	2-40 cigts in 2-4 hrs depending on location	<17	<17
Brunnemann et al., 1978			
Train I (bar car)		130	-
Train II (bar car)		110	-
Bar		240	-
Sports hall		90	-
Betting parlor		50	-
Discotheque		90	-
Large room (bank)		10	-

[a]Residence living room, restaurant, pub, railway compartment, car.

Table 12.6 Tobacco-Specific Nitrosamines in an Artificial Environment[a]

Number of Items Smoked[c]			Concentration Increase[b]			
Cigts	Cigars	Pipes	NNN ng/m³	NNK ng/m³	CO ppm	Nicotine µg/m³
18	0	0	nd	1.6	13.0	110
11	1	0	1.7	nd	12.5	122
8	0	1	2.6	8.0	20.0	87.8
10	0	1	3.8	4.5	13.0	61.9
11	0	0	1.4	2.8	10.4	64.2
11	0	0	3.5	13.5	7.6	69.8
13	0	1	4.9	2.1	11.0	100.4
12	0	1	3.1	3.1	12.4	117.6
15	0	0	3.8	10.8	14.0	84.3
11	0	0	1.6	4.4	7.7	89.8
7	1	1	2.9	4.7	11.4	101.7
6	0	1	2.3	4.9	8.7	110
7	0	1	1.6	3.3	10.7	58.5
7	1	0	6.0	4.2	9.0	93.7

[a]Klus et al. 1987a.and Klus et al. 1992 Unventilated office with heavy smoking.
[b]Maximum values, corrected for background.
[c]2-h period.

Table 12.7 Tobacco-Specific Nitrosamines in ETS-Containing Natural Environments[a] (concentrations, ng/m³)

Locations	Time (h)	Cigts.	NNN	NAT	NNK
Bars (3)	NS	NS	4.3–22.8	3.7–9.2	9.6–23.8
Restaurants (2)[b]	NS	NS	ND–1.8	ND–1.5	1.4–3.3
Automobile (1)[c]	3.3	13	5.7	9.5	29.3
Train cars (2)	NS	NS	ND	ND	4.9–5.2
Office (1)	6,5	25	ND	ND	26.1
Residence	3.5	30	ND	ND	1.9

[a]Data from Brunnemann, et al. 1992.
[b]Smoking sections.
[c]Windows partially open.
NS=not specified.

to nicotine surface adsorption (see Chap. 7), median NDMA concentrations would be 2.5 ng/m³. While concentrations of 100 ng/m³ or more likely occur, they should be considered rare, or as accompanying obviously highly contaminated environments. Concentrations of 1–20 ng/m³ may remain important, given that ETS-free environments have been reported (Brunnemann et al. 1978) to contain < 5 ng/m³ of NDMA. However, Shah and Singh (1988) have reported median outside urban air concentrations of NDMA to be 4.6 ng/m³. Depending on the time spent in typical indoor and outside environments, exposures (the product of concentrations and time) of nonsmokers to NDMA may be greater from nontobacco-related sources.

Tables 12.6 and 12.7 summarize the only known (to the authors) data on ETS-related tobacco specific nitrosamines. Under the conditions of one study in a semi-manipulated environment (unventilated 84-m^3 office area with occasional door opening for egress and the tabulated tobacco product consumption), the investigators reported NNN concentrations to range from not detectable up to 6.0 ng/m^3 for NNN, and from not detectable up to 13.5 ng/m^3 for NNK. The NNN elevation is more frequently found to be 1–4 ng/m^3 and NNK elevation is more frequently 2–5 ng/m^3. It may be worth noting that the correlation between TSNA concentrations and those for carbon monoxide and nicotine are very poor. The other study (Brunnemann et al. 1992) focused on "natural environments with a relatively high incidence of smoking." NNK concentrations tended to be somewhat higher than those reported for the previous study, but within approximately a factor of 2. By today's standards, neither of these studies were directed toward scenarios that would be considered "typical." This may have been due to the need to sample high ETS concentration environments to collect quantifiable levels of these components. "Typical" levels of tobacco-specific nitrosamines are expected to be a factor of 5–100 lower than those reported in these studies, or 0.1–5 ng/m^3 for NNK.

OTHER CONSTITUENTS

PAHs and N-nitrosamines are the primary trace constituents of concern in ETS, but interest also exists in other trace constituents. The interest is based principally on their known or suspected presence in cigarette smoke and the resulting implication that they must also be present in ETS, regardless of whether they have been directly measured. These constituents include heavy metals, radioactive species, polycyclic aromatic amines and polycyclic aromatic nitro compounds. Relatively few data are available to judge the importance of these chemicals in exposures to ETS.

Heavy metals. Elements of particular concern (e.g., Gold et al. 1990) are arsenic, beryllium, cadmium, chromium, nickel, and selenium. Interest exists because either the elemental forms of the element (e.g., hexavalent chromium) or some of their compounds (e.g., nickel carbonyl) are classed as IARC Group I carcinogens (Smith et al. 1997).

Tobacco in cigarettes contains between less than 0.1 μg/g up to approximately 10 μg/g of most trace elements (Table 12.8). Copper and zinc have been reported at concentrations of up to 50 μg/g. Given that a cigarette typically contains 1 g or less of tobacco, the trace element content of the tobacco serves as the upper bound of what is available for release into the environment. A review of the literature coupled with an independent study indicates (Mussalo-Rauhamaa et al. 1990) that transfer percentages to mainstream smoke range from < 0.02% to up to 10%, depending on the metal considered. Cadmium is reported to be transferred to mainstream smoke to the highest degree (2–10%) and to be the most poorly accounted for by analysis

Table 12.8 Trace Element Content[a] of Tobacco and Tobacco Smoke

Element	Tobacco μg/g	Mainstream μg/cigarette	Sidestream μg/cigarette
Chromium	0.4–10	<0.1	-
Nickel	0.4–4.0	<0.1–0.5	0.2–2.5
Arsenic	0.2–4.0	<0.1–0.2	0.015–0.023
Selenium	< 0.1–2.0	<0.1	–
Cadmium	0.2–4.0	< 0.1–0.5	0.4–0.7
Lead	2–10	<0.1–1.0	–
Magnesium	2– 10	–	–
Iron	0.4–1.0	–	–
Zinc	10–50	0.1–0.5	0.02–0.38
Copper	10–20	–	–
Polonium-210	–	<0.1–0.5 pCi	0.4 –1 .6 pCi

[a]Jenkins, 1986; Mussalo-Rauhamma et al. 1986; Bell and Mulchi, 1990; Mahajan and Huber, 1990; Landsberger and Wu, 1995

of the butt and ash. This has led to the suggestion (e.g., Gold et al. 1990) that ETS might be an important contributor to ambient indoor cadmium exposure and that cadmium might serve as a marker of ETS particulate matter contamination (see Chap. 4).

Indoor air concentrations (Table 12.9) of chromium, nickel, arsenic, selenium, and cadmium in indoor air have been reported (e.g., Lebret et al. 1987 in Gold et al. 1990) to range from nondetectable up to 10 ng/m³ depending on the element considered. Of the elements considered, nickel concentrations predominated with a range of 1–11 ng/m³. Chromium, arsenic, selenium, and cadmium were reported as being present at concentrations ranging from nondetectable up to a maximum of 5 ng/m³. A more recent study (Landsberger and Wu 1995) reported that arsenic levels in smoking environments ranged from 0.2–1.0 ng/m³, and cadmium levels ranged from 4–40 ng/m³.

Table 12.9 Indoor Air Concentrations (ng/m³) of Selected Metals

	Indoor Air Survey[a]				Theoretical ETS[b]	
	Median	Minimum	90%	95%	Maximum	Range[c]
Chromium	ND	ND	3.00	4.94	6.04	1.2–8.9
Nickel	5.73	1.02	9.31	10.51	10.57	2.5–7.2
Arsenic	1.24	ND	2.31	2.51	2.52	3.1–3.1
Selenium	0.56	ND	0.93	1.09	1.10	0.2–4.7
Cadmium	1.29	ND	2.87	2.94	2.94	–

[a]20 samples of indoor environments (Lebret et al. 1987 in Gold et al. 1990).
[b]Computed for 30 μg/m³ ETS particulate matter from Benner et al. 1989 (see Table 12.10).
[c]Sample preparation method varied.
ND=not detectable.

A study (Benner et al. 1989) of the fundamental properties of ETS particulate matter provides data allowing an assessment of the contribution of ETS to indoor air trace element concentrations. The investigators introduced sidestream smoke from machine-smoked Kentucky Reference IRI cigarettes into a 30-m^3 Teflon chamber and sampled its contents to determine many constituents including trace elements. Results are reported as micro-moles of element-per-gram of particulate matter for samplings of test environments containing 500 and 1000 μg/m^3 of particulate matter. Two methods of sample preparation (Table 12.10) were evaluated because the measurement method is considered developmental for such applications.

Accepting the results of this study for the trace element content of ETS particulate matter indicates that an environment contaminated with 30 μg/m^3 of ETS particulate matter will contain between 0.1 and 10 ng/m^3 of each trace element due to ETS alone depending upon which trace element is considered. Nickel appears to predominate with a median concentration of approximately 5 ng/m^3. This may be important because nickel carbonyl is carcinogenic and because it has been reported (Huber 1989) that 10–20% of the nickel in mainstream smoke is present as the carbonyl. Nickel carbonyl has not yet been reported as a constituent of ETS, however, and its presence is improbable given its reactivity.

Note, however, that the estimates (based on 30 μg/m3 of ETS RSP) overestimate the actual levels of As measured by Landsberger and Wu (1995) and underestimate Cd concentrations. In contrast, the estimate overestimates the Cd levels reported by Brauer and Mannetje (1998) found in restaurants. This exemplifies the challenges of extrapolating from well controlled chamber studies, with one source, to real world environments, with multiple sources. For the chamber study (Benner et al. 1989), the difference in sample preparation method alone yields particulate matter concentrations of chromium which range from 0.8–5.7 μmole/gm (42–296 μg/gm). Results (e.g., Lebret et al. 1987; Landsberger and Wu 1995) of ambient indoor air sampling are further confounded by the possible contributions of trace elements in particulate matter from outdoor air, from indoor combustion appliances, and from indoor dusts.

The contribution of ETS to indoor air levels of trace elements is yet to be well-defined. It appears that important trace elements such as chromium, nickel, arsenic, and selenium are present in ETS particulate matter at concentrations ranging from 10–100 μg/gm (ppm) depending upon the element considered. Considerable further work is required to better define this range and to determine the importance of ETS relative to other sources of trace elements in indoor air.

Radioactivity. The radioisotope polonium-210 has been identified in cigarette smoke and has been suggested as a possible contributor to cancer associated with active smoking. Po^{210} decays to stable Pb^{206} by emission of a 5.3 MeV alpha particle. 5 MeV alpha radiation penetrates only 0.75 cm in air,

Table 12.10 Trace Element Distribution from Experimental Chamber Study[a]

Element	Molecular Weight, µg/µmole	0.01M HCl Extract, Mean ± Uncertainty		Water Extract, Mean ± Uncertainty	
		µmole/g[a]	ng/m³ [b]	µmole/g[a]	ng/m³ [b]
K	39.098	226 ± 451	265.1 ± 529.0	256 ± 297	300.3 ± 348.4
Ca	40.08	93 ± 117	111.8 ± 140.7	103 ± 196	123.8 ± 235.7
Ti	47.9	9.2 ± 8.3	13.2 ± 11.9	11.7 ± 31	16.8 ± 44.5
Ba	137.33	26.7 ± 28.4	110.0 ± 117.0	7.9 ± 13.7	32.5 ± 564
V	50.9415	9.8 ± 20.2	15.0 ± 30.9	3.3 ± 8.7	5.0 ± 13.3
Cr	51.996	5.7 ± 15.7	8.9 ± 24.5	0.8 ± 2.3	1.2 ± 3.6
Mn	54.938	1.4 ± 2.1	2.3 ± 3.5	< 1	
Fe	55.847	46.3 ± 111	77.6 ± 186.0	13.4 ± 16.2	22.5 ± 27.1
Co	58.933	0.8 ± 1.6	1.4 ± 2.8	0.6 ± 1.5	1.1 ± 27
Ni	58.7	1.4 ± 1.9	2.5 ± 3.3	4.1 ± 5.3	7.2 ± 9.3
Cu	63.546	4.3 ± 9.9	8.2 ± 18.9	0.1 ± 0.3	0.2 ± 0.6
Zn	65.38	2.6 ± 8.2	5.1 ± 16.1	4.5 ± 7.7	8.8 ± 15.1
Pb	207.2	3 ± 5.3	18.6 ± 32.9	1.5 ± 3.1	9.3 ± 19.3
As	74.922	1.4 ± 1.9	3.1 ± 4.3	1.4 ± 2.5	3.1 ± 5.6
Se	78.96	0.1 ± 0.4	0.2 ± 0.9	2 ± 3.9	4.7 ± 9.2
Br	79.904	3.3 ± 5.1	7.9 ± 12.2	1.3 ± 2.4	3.1 ± 5.8

[a]Benner et al. 1989.
[b]Calculated for 30 µg/m³ ETS particulate matter.

and much less in water or physiological media. However, its interactions with biocellular materials are highly carcinogenic. Active smoking yields between 0.02–0.5 pCi per cigarette of polonium-210 (Po^{210}) in mainstream smoke. This quantity is generally considered biologically insignificant in itself, but it has been suggested (Martel 1975 and 1982, Harley et al. 1980) that Po^{210} can be concentrated in and carried by smoke particles derived from tobacco trichomes into the lung where localized high radiation doses occur. Po^{210} is delivered in sidestream smoke at the rate of approximately 0.4 pCi per cigarette (0.015 Bq/cigarette) (e.g., Aviado 1988).

Po^{210} also results as a decay product of radon-222 (Rn^{222}). Rn^{222} emits alpha particles of similar energy to that emitted by Po^{210} as it decays to Po^{218}. Then, through a complex series of transitions, Po^{218} eventually decays to lead-210 (Pb^{210}), which subsequently decays to Po^{210}. Radon-222 is currently recognized (e.g., Gold et al. 1990) as being among the most important of indoor air contaminants and as arising primarily from the soil, building foundation, and certain building materials. The outdoor ground level concentration of Rn^{222} is (e.g., Gold et al. 1990) approximately 0.15–0.30 pC per liter (5–10 Bq/m^3). Mean indoor air concentrations of Rn^{222} are commonly 30–60 Bq/mL, and sites are frequently identified that exhibit concentrations much in excess of 100 Bq/m^3. The contribution of Po^{210} from cigarettes, 0.01 Bq/cigarette or less, appears to be completely insignificant relative to other sources.

It has been speculated that ETS may play a role in indoor-air-radiation-induced cancer although by itself, ETS is a minor source of radionuclides. ETS-derived particulate matter might serve as a vehicle for the airborne scavenging of radionuclides from other sources and the subsequent transport to and deposition deep in the respiratory tract. However, airborne particulate matter from sources other than ETS might function similarly. In many smoking environments, ETS is not the primary source of respirable particles (see Chap. 6). ETS itself is generally not a major source of indoor air radioactivity.

Substituted Polycyclic Aromatics. Many minor (in quantity and/or assumed biological importance) constituents of tobacco smoke have been considered in studies of ETS. Examples include phenols (e.g., Klus et al. 1987, O'Neill et al. 1987) and the large variety of volatile organics treated in Chaps. 2 and 11. Two classes of trace constituents that have not been considered there deserve special mention. These are the polycyclic aromatic primary amines and nitro-substituted polycyclic aromatics. Both the amines and nitro-polycyclics are known to exhibit mutagenic properties. Beta-naphthylamine and 4-aminobiphenyl are also known constituents of mainstream tobacco smoke and are classified as potent human bladder carcinogens.

Beta-naphthylamine and 4-aminobiphenyl are delivered in mainstream smoke at quantities of 4–30 and 2–5 ng/cigarette, respectively. Sidestream deliveries are (Patrianakos and Hoffmann 1979) 30 times greater than are mainstream deliveries. Four- and 5-ring nitro-substituted polycyclic aromatic hydrocarbons have been sought in mainstream cigarette smoke but reported (El Bayoumy et al. 1985) to be undetectable (< 10 ng/cigarette for the method used). As such, the amines and nitro-substituted polycyclic aromatics are commonly present at one-tenth or less the quantity of BaP.

Beta-naphthylamine and 4-aminobiphenyl (4-BAP) have not yet been determined in nonoccupational indoor air. Both, however, have been identified in the fumes of cooking oils (vegetable, sunflower, and refined lard) (Chiang et al. 1999). In addition, hemoglobin adducts of 4-BAP have been identified in pregnant, nonsmoking women (Banner et al 1998) passively exposed to ETS. However, elevations in adduct levels for smoking women were not in proportion to expected tobacco smoke exposure differences. These findings together suggest that while airborne levels of 4-BAP and beta-naphthalamine associated with ETS are yet to be measured, the contribution from ETS, relative to other potential sources, may be quite low.

A number of nitro-substituted polycyclics (e.g., nitro-anthracenes, -phenanthrenes, -fluoranthenes) are known constituents of both outdoor and indoor air. They result from a wide variety of combustion sources (e.g., kerosene space heaters, vehicle exhaust) and have been identified as especially important contributors to the mutagenicity of diesel exhaust particulates and of ambient air particulate matter. Concentrations range from a factor of 10 to a factor of 1000 lower than the corresponding polycyclic aromatic hydrocarbons.

At least one study (Wilson and Chuang 1989 in Gold et al. 1990) provides data implicating ETS as a contributor to indoor air nitro-PAHs. The study reports the concentration of 9-nitroanthracene to be 0.32 ng/m^3 in the living rooms of homes with smokers and 0.13 ng/m^3 in the living rooms of homes without smokers. Differences for 9-nitrophenanthrene (0.25 vs. 0.04 ng/m^3) and 2-nitrofluoranthene (0.14 vs. 0.02 ng/m^3) were more dramatic. Different homes in different locations were compared, and the possible contribution by other sources was not considered. Without careful comparison of other lifestyle and indoor environmental factors, the use of smoking/nonsmoking home designation may be too inexact a tool to assess the contribution of ETS to nontobacco-specific components of indoor air pollution, especially when those components are present at sub-ng/m^3 concentrations. The very small quantities, if any, of polycyclic aromatic nitro compounds in cigarette smoke suggests that other sources predominate as contributors to their indoor air concentrations.

SUMMARY

Trace constituents of ETS that are thought to be potentially hazardous, such as individual PAHs, nitrosamines, and trace elements, are most commonly present at concentrations of <1 up to 10 ng/m^3 in typical environments. Heavily contaminated environments can contain some trace constituents at concentrations of 10–50 ng/m³, and excursions occur up to or slightly exceeding 100 ng/m³. Most studies of the trace constituents of ETS have not thoroughly considered the impact of other sources. ETS is likely to be the primary contributor to tobacco-specific nitrosamines, a partial contributor to polycyclic aromatic hydrocarbons and trace metals, and an unlikely contributor to radioactivity and nitro-polycyclic aromatics.

REFERENCES

Anonymous (1990) *Indoor Air: Assessment. Methods of Analysis for Environmental Carcinogens.* EPA/600/8-90/041, Office of Research and Development, U.S. Environmental Protection Agency, Research Triangle Park, NC 27711, June 1990, 37 pp.

Aviado, D. (1988) Suspected pulmonary carcinogens in environmental tobacco smoke. *Environ. Technol. Letts.,* 9, 539–544.

Bell, P. & Mulchi, C. L. (1990) Heavy metal concentrations in cigarette blends. *Tob. Sci.,* 34, 32–34.

Benner, C. L., Bayona, J. M., Caka, F. M., Hongmao, T., Lewis, L., Crawford, J., Lamb, J. D. , Lee, M. L., Lewis, E. A., Hansen, L. D., & Eatough, D. J. (1989) Chemical composition of tobacco smoke. 2. Particulate-phase compounds. *Environ. Sci. Technol.,* 37, 688–699.

Branner, B., Kutzer, C., Zwickenpflug, W., Scherer, G., Heller, W. D., & Richter, E. (1998) Hemoglobin adducts from aromatic-amines and tobacco specific nitrosamines in pregnant smoking and nonsmoking women. *Biomarkers,* 3, 35–47.

Brauer, M. & 't Mannetje, A. (1998) Restaurant smoking restrictions and environmental tobacco smoke exposure. *Am. J. Public Health,* 88, 1834–1836.

Brunnemann, K. D., Yu, L., & Hoffmann, D. (1977) Assessment of carcinogenic volatile N-nitrosamines in tobacco and in mainstream and sidestream smoke from cigarettes. *Cancer Res.,* 37, 3218–3222.

Brunnemann, K. D., Cox J. E., & Hoffmann, D. (1992) Analysis of tobacco-specific N-nitrosamines in indoor air. *Carcinogenesis,* 13, 2415–2418.

Brunnemann, K. D., Adams, J. D., Ho, D. P. S., & Hoffmann, D. (1978) The influence of tobacco smoke on indoor atmospheres. II. Volatile and tobacco-specific nitrosamines in mainstream and sidestrearn smoke and their contribution to indoor pollution. *Proceedings, 4th Joint Conference on Sensing of Environmental Pollutants,* American Chemical Society, Washington, D.C., pp. 876–880.

Chiang, T. A., Wu, P. F., Ying, L. S., Wang, L. F., & Yo, Y. C. (1999) Mutagenicity and aromatic amine content of fumes from heated cooking oils produced in Taiwan. *Food and Chem. Toxicol.,* 37, 125–134.

El-Bayoumy, K., O'Donnell, M., Hecht, S. S., & Hoffmann, D. (1985) On the analysis of 1-nitronaphthalene, 1-nitropyrene, and 6-nitrochrysene in cigarette smoke. *Carcinogenesis, 6*, 505–507.

Elliott, L. P. & Rowe, D. R. (1975) Air quality during public gatherings. *JAPCA, 25*, 635–636.

Gold, K. W., Naugle, D. F., & Berry, M. A. (1990) *Indoor Concentrations of Environmental Carcinogens.* Research Triangle Park, NC, U.S. Environmental Protection Agency, RTI Report 4479/07-F, issued April, 1990.

Guenther, F. R., Chesler, S. N., Gordon, G. E., & Zoller, W. H. (1988) Residential wood combustion: a source of atmospheric polycyclic aromatic hydrocarbons. *JHRCC&C, 761*–766.

Gundel, L. A., Daisey, J. M., & Offermann, F. J. (1990) Development of an indoor sampling and analysis method for particulate polycyclic aromatic hydrocarbons. *Proceedings of the 5th International Conference on Indoor Air Quality and Climate, Vol. 2,* Toronto, Canada, pp. 299–304.

Gundel L.A., Mahanama K.R.R., and Daisey J.M (1995) Semivolatile and Particulate Polycyclic Aromatic Hydrocarbons in Environmental Tobacco smoke: Cleanup, Speciation, and Emission Factors. *Environ. Sci. Technol.* 29, 1607–1614

Harley, N. H., Cohen, B. S., & Tso, T. C. (1980) Polonium-210: a questionable risk factor in smoking-related carcinogenesis. In: Gori, G. B. & Bock, F. G., eds., *Banbury Report No. 3, A Safe Cigarette,* Cold Spring Harbor Laboratory, Cold Spring Harbor, NY, pp. 93–101.

Hoffmann, D., Adams, J. D., & Brunnemann, K. D. (1987) A critical look at N-nitrosamines in environmental tobacco smoke. *Toxicol. Lett., 35*, 1–8.

Huber, G. L. (1989) Physical, chemical, and biological properties of tobacco, cigarette smoke, and other tobacco products, *Seminars in Resp. Med., 10,* 297–332.

Jenkins, R. A. (1986) Occurrence of selected metals in cigarette tobaccos and smoke. In: O'Neill, I. K., Schuller, P., & Fishbein, L., eds., *Environmental Carcinogens: Selected Methods of Analysis. Vol. 8 —Some Metals. As, Be, Cd, Cr, Ni, Pb, Se, Zn (IARC Publication No. 71),* International Agency for Research on Cancer, Lyon, France, pp. 129–138.

Jenkins, R. A., & Counts, R. W., (1999) Occupational exposure to environmental tobacco smoke, results of two personal exposure studies. *Environ. Health Perspec.,* 107 (Suppl. 2), 341–348.

Klus, H., Begutter, H., Ball, M., & Intorp, M. (1987a) Environmental tobacco smoke in real life situations. In: Siefert, B. et al. eds., *Poster Handout From the 4th International Conference on Indoor Air Quality and Climate.*

Klus, H., Begutter, H., Scherer, G., Tricker, A. R., Adlkofer, F. (1992) Tobacco-Specific and Volatile N-Nitrosamines in Environmental Tobacco Smoke of Offices. *Indoor Environ.* 1, 348–350.

Klus, H., Begutter, H., Ball, M., & Intorp, M. (1987) Environmental tobacco smoke in real life situations. In: Siefert, B. et al. eds., *Proceedings of the 4th International Conference on Indoor Air Quality and Climate*, pp. 137–141.

Landsberger, S. & Wu, D. (1995) The impact of heavy metals from environmental tobacco smoke on indoor air quality as determined by Compton supression neutron activation analysis. *Sci. Total Environ.* 173/174, 323 - 337

Lebret, E., McCarthy, J., Spengler, J. & Chang, B. (1987) Elemental composition of indoor fine particles. In: Seifert, B., Esdorn, H., Fischer, M., Ruden, H. & Wegner, J., eds, Indoor Air '87: Proc. Fourth Int. Conf. on Indoor Air Quality and Climate, Vol. I, Institute for Water, Soil and Air Hygiene, Berlin, pp. 569 - 574.

Mahajan, V. K. & Huber, G. L. (1990) Health effects of involuntary smoking: impact on tobacco use, smoking cessation, and public policy. *Seminars in Resp. Med.*, 11, 87–114.

Mahanama, K. R. R. & Daisey, J. M. (1996) Volatile n-nitrosamines in environmental tobacco smoke: sampling, analysis, emission factors, and indoor air exposures. *Environ. Sci. Technol.*, 30, 1477–1484.

Martel, E. A. (1982) Correspondence. *New Eng. J. Med.*, 307, 309–310.

Martel, E. A. (1975) Tobacco radioactivity and cancer in smokers. *Am. Sci.*, 63, 404–412.

Matsushita, H. & Mori, T. (1984) Nitrogen dioxide and nitrosamine levels in indoor air and sidestream smoke of cigarettes. In: Berglund, B. et al. eds., *-Indoor Air Vol. 2, Radon, Passive Smoking, Particulates and Housing Epidemiology*, Swedish Council for Building Research, Stockholm, pp. 335–340.

Mussalo-Rauhamaa, H., Salmela, S. S., Leppanen, A., & Pyysalo, H. (1986) Cigarettes as a source of some trace and heavy metals and pesticides in man. *Arch. Environ. Health*, 41, 49–55.

O'Neill, I. K., Brunnemann, K. D., Dodet, B., & Hoffmann, D., eds. (1987) *Environmental Carcinogens—Methods of Analysis and Exposure Measurement. Vol. 9—Passive Smoking (JARC Scientific Publication No. 81)*, International Agency for Research on Cancer, Lyon, France, 372 pp.

Patrianakos, C. & Hoffmann, D. (1979) Chemical studies on tobacco smoke. LXIV: on the analysis of aromatic amines in cigarette smoke. *Anal. Toxicol.*, 3, 150–154.

Shah, J. J. & Singh, H. B. (1988) Distribution of volatile organic chemicals in outdoor and indoor air: a national VOC data base. *Environ. Sci. Technol.*, 22, 1381–1388.

Smith, C. J., Livingston, S. D., & Doolittle, D. J. (1997) An international literature survey of IARC Group I carcinogens reported in mainstream cigarette smoke. *Food and Chem. Toxicol.*, 35, 1107–1130.

Stehlik, G., Richter, 0., & Altmann, H. (1982) Concentrations of dimethyinitrosamine in the air of smoke-filled rooms. *Ecotoxicol. and Environ. Safety,* 6, 495–500.

Turk, B. H., Brown, J. T., Geisling-Sobotka, K., Froehlich, D. A., Grimsrud, D. T., Harrison, J., Koonce, J. F., Prill, R. J., & Revzan, K. L. (1987) *Indoor Air Quality and Ventilation Measurements in 38 Pacific Northwest Commercial Buildings, Vol. I—Measurement Results and Interpretation (LBL-22315 112), Vol. 11—Appendices (LBL 22315 212),* Lawrence Berkeley Laboratory, Berkeley, CA 94720.

Wilson, N.K. & Chuang, J.C. (1989) Indoor Air Levels of PHA and Related Compounds in an Eight-Home Pilot Study. In: Proceedings of the Eleventh Intl. Symposium of Polynuclear Aromatic Hydrocarbons, Gaithersburg, MD.

Field Studies—Indoor Particulate Matter Concentrations ($\mu g/m^3$)

Location/Condition	Reference	Smoking		Controls		Measure	Method	
		Mean	Range	Mean	Range		Type	Interval
Residences (Res.)—Area Sampling								
5 homes, wood-burning stoves	Benton et al. 1981			30.3	0–66	RSP	G	17 h
Outdoors	Brunekreef & Boleij 1982			–	41–73	TSP	G	2 mo
1 res./3 smokers	Brunekreef & Boleij 1982	335	–			TSP	G	2 mo
4 res./nonsmoking	Brunekreef & Boleij 1982			55	20–90	TSP	G	2 mo
14 res./2 smokers	Brunekreef & Boleij 1982	152	60–340			TSP	G	2 mo
7 res./1 smoker	Brunekreef & Boleij 1982	125	60–250			TSP	G	2 mo
2 res./kerosene heater	Burton et al. 1990				13.4–84.3	RSP	G	NS
5 homes outdoor/Portage, WI	Colome and Spengler 1981			8.6[b]		RSP	G	24 h
10 homes indoor	Colome et al. 1990			42.5		RSP	G	24 h

Field Studies—Indoor Particulate Matter Concentrations (μg/m³)

Location/Condition	Reference	Smoking		Controls		Measure	Method	
		Mean	Range	Mean	Range		Type	Interval
10 homes outdoor	Colome et al. 1990			60.8		RSP	G	24 h
5 homes indoor/ Steubenville, OH	Colome and Spengler 1981			30.4[b]		RSP	G	24 h
5 homes outdoor/ Steubenville, OH	Colome and Spengler 1981			41.4[b]		RSP	G	24 h
5 homes indoor/ Portage, WI	Colome and Spengler 1981			16.4[b]		RSP	G	24 h
1 home nonsmoking	Conner et al. 1990			4	1–8	RSP	UV	2–3 h
1 home nonsmoking	Conner et al. 1990			58	17–86	RSP	G	2–3 h
4 res./smoking	Eatough, Benner, Tang, et al. 1989	55	14–151	–	–	RSP	P,G	NS
2 res./smoking	Hawthorne et al. 1984	–	96–106			RSP	QCMI	5–15 min
11 Res./nonsmoking	Hawthorne et al. 1984			–	9–40	RSP	QCMI	5–15 min
8 Res./ nonsmoking	Hawthorne et al. 1984			–	12–46	RSP	QCMI	5–15 min

Kitchen/nonsmoking	Kirk et al. 1988			300		RSP	O	30 min
Flat (supposedly nonsmoking)	Kirk et al. 1988			55	50–60	RSP	P	10 h
Home/156 samples/smoking	Kirk et al. 1988	700	70–3150			RSP	O	30 min
Flat (supposedly nonsmoking)	Kirk et al. 1988			69	–	TSP	G	10 h
Kitchen/smoking	Kirk et al. 1988	500		300		RSP	O	30 min
Living room/nonsmoking	Kirk et al. 1988			270	0–2050	RSP	O	30 min
Home/592 samples/nonsmoking	Kirk et al. 1988					RSP	O	30 min
Living room/smoking	Kirk et al. 1988	630				RSP	O	30 min
47 smoking homes	Leaderer and Hammond 1991	44	ca. 10–150			RSP	G	1 week
49 nonsmoking homes	Leaderer and Hammond 1991			15	ca. 5–30	RSP	G	1 week
54 homes, Netherlands	Lebret et al. 1990	191[c]				RSP	G	1 week
98 homes, Netherlands	Lebret et al. 1990			41[c]		RSP	G	1 week
23 homes	Mumford et al. 1989	74.1		28.4		RSP	G	14 h

Field Studies—Indoor Particulate Matter Concentrations ($\mu g/m^3$)

Location/Condition	Reference	Smoking		Controls		Measure	Method	
		Mean	Range	Mean	Range		Type	Interval
Activity rooms of 580 children in smoking homes	Neas et al. 1994	48.5	ca. 5–115			$PM_{2.5}$	G	4 weeks
Activity rooms of 470 children in nonsmoking homes	Neas et al. 1994			17.3	ca. 5–65	$PM_{2.5}$	G	4 weeks
19 res./nonsmoking	Nitschke et al. 1985			26	6–88	RSP	G	168 h
11 res./smoking	Nitschke et al. 1985	59	10–144			RSP	G	168 h
Outdoors	Nitschke et al. 1985			11	1–28	RSP	G	168 h
2 homes	Ogden & Maiolo 1989	200	187–212	–	–	RSP	G	4 h
2 res./1–2 smokers	Parker et al. 1984	–	10–46			TSP	O	24 h
1 res./nonsmoking	Parker et al. 1984			< 10	–	TSP	O	24 h
209 homes, Suffolk City	Perritt et al. 1990	70.1	2.2–284	24.9		RSP	G	NS
224 homes, Onandaga City	Perritt et al. 1990	61.3	0.7–172	18.1		RSP	G	NS

43 res./nonsmoking	Quackenboss et al. 1989		30.3		RSP(PM$_{10}$)	G	NS
18 res./smoking >20 cigarettes/day	Quackenboss et al. 1989	75.0			RSP(PM$_{10}$)	G	NS
27 res./smoking ≤20 cigarettes/day	Quackenboss et al. 1989	46.2			RSP(PM$_{10}$)	G	NS
56 res. N. India	Ramakrishna et al. 1989		3100		TSP	G	46–53 min
60 res. S. India	Ramakrishna et al. 1989		2600		TSP	G	58–68 min
93 res. W. India	Ramakrishna et al. 1989		3900		TSP	G	39–51 min
73 res./nonsmoking	Spengler et al. 1985		28	–	RSP	G	24 h
28 res./nonsmoking	Spengler et al. 1985	74		–	RSP	G	24 h
Outdoors	Spengler et al. 1981		21		RSP	G	24 h
5 res./2 smokers	Spengler et al. 1981	70		–	RSP	G	24 h
15 res./1 smoker	Spengler et al. 1981	36		–	RSP	G	24 h
Outdoors	Spengler et al. 1985		18	–	RSP	G	24 h
35 res./nonsmoking	Spengler et al. 1981		24	–	RSP	G	24 h

Field Studies—Indoor Particulate Matter Concentrations ($\mu g/m^3$)

Location/Condition	Reference	Smoking Mean	Smoking Range	Controls Mean	Controls Range	Measure	Method Type	Method Interval
10 homes/apartments	Van der Wal et al. 1990				<20–370	RSP	O	48 h
5 outdoor/Zigui area	Zhao & YuFeng 1990			80	40–120	RSP	G	8 h
15 res./Zigui area	Zhao & YuFeng 1990			1120	110–2230	RSP	G	8 h
16 res./Wu-Shan area	Zhao & YuFeng 1990			1810	610–4550	RSP	G	8 h
12 res./Qian-Jiang area	Zhao & YuFeng 1990			970	430–2040	RSP	G	8 h
5 outdoor/Wu-Shan area	Zhao & YuFeng 1990			50	20–90	RSP	G	8 h
1 outdoor/Peng-Shui area	Zhao & YuFeng 1990			100	–	RSP	G	8 h
9 res./Peng-Shui area	Zhao & YuFeng 1990			930	480–2390	RSP	G	8 h
4 outdoor/Ba-Dong area	Zhao & YuFeng 1990			440	90–990	RSP	G	8 h
12 res./Ba-Dong area	Zhao & YuFeng 1990			710	310–1260	RSP	G	8 h
3 outdoor/Qian-Jiang area	Zhao & YuFeng 1990			500	420–720	RSP	G	8 h

Res.—Personal Exposure

Subjects	Reference							
58 workers in nonsmoking homes	Heavner et al. 1996			28	8–100	RSP	G	15 h
29 workers in smoking homes	Heavner et al. 1996	89	11.8–825			RSP	G	15 h
306 workers in smoking homes	Jenkins et al. 1996	44.1	125 (95th percentile)			RSP	G	15 h
899 workers in nonsmoking homes	Jenkins et al. 1996			19.7	46.8 (95th percentile)	RSP	G	15 h
56 house persons in nonsmoking homes	Phillips et al. 1998 (Lisbon)			38	54 (90th percentile)	RSP	G	24 h
31 house persons in nonsmoking homes	Phillips et al. 1996 (Sweden)			18	8.2–58	RSP	G	24 h
255 nonsmokers	Phillips et al. 1994 (UK)	179	89–420			RSP	G	24 h
22 house persons in smoking homes	Phillips et al. 1998 (Lisbon)	40	67 (90th percentile)			RSP	G	24 h
25 house persons in smoking homes	Phillips et al. 1999 (Basel)	60	88 (90th percentile)			RSP	G	24 h
58 house persons in nonsmoking homes	Phillips et al. 1999 (Basel)			31	49 (90th percentile)	RSP	G	24 h

Field Studies—Indoor Particulate Matter Concentrations ($\mu g/m^3$)

Location/Condition	Reference	Smoking		Controls		Measure	Method	
		Mean	Range	Mean	Range		Type	Interval
51 house persons in smoking homes	Phillips et al. 1998 (Paris)	71	130 (90th percentile)			RSP	G	24 h
30 workers in smoking homes	Phillips et al. 1998 (Sydney)	37	70 (90th percentile)			RSP	G	24 h
66 workers in nonsmoking homes	Phillips et al. 1999 (Basel)			48	55 (90th percentile)	RSP	G	15 h
38 workers in smoking homes	Phillips et al. 1999 (Basel)	46	85 (90th percentile)			RSP	G	15 h
43 workers in nonsmoking homes	Phillips et al. 1998 (Sydney)			31	41 (90th percentile)	RSP	G	24 h
23 workers in smoking homes	Phillips et al. 1998 (Bremen)	44	89 (90th percentile)			RSP	G	15 h
81 workers in nonsmoking homes	Phillips et al. 1998 (Bremen)			25	39 (90th percentile)	RSP	G	15 h
21 house persons in smoking homes	Phillips et al. 1998 (Bremen)	39	63 (90th percentile)			RSP	G	24 h
58 house persons in nonsmoking homes	Phillips et al. 1998 (Bremen)			27	37 (90th percentile)	RSP	G	24 h

58 workers in smoking homes	Phillips et al. 1998 (Paris)	66	112 (90th percentile)		RSP	G	15 h
40 house persons in nonsmoking homes	Phillips et al. 1997 (Barcelona)			63	90 (90th percentile)	RSP	G	24 h
67 workers in nonsmoking homes	Phillips et al. 1998 (Paris)			36	66 (90th percentile)	RSP	G	15 h
43 house persons in nonsmoking homes	Phillips et al. 1998 (Paris)			44	84 (90th percentile)	RSP	G	24 h
36 house persons in nonsmoking homes	Phillips et al. 1998 (Prague)			35	55 (90th percentile)	RSP	G	24 h
9 house persons in smoking homes	Phillips et al. 1996 (Sweden)	39	15–154		RSP	G	24 h
56 housepersons in smoking homes	Phillips et al. 1998 (Beijing)	123	221 (90th percentile)		RSP	G	24 h
45 house persons in nonsmoking homes	Phillips et al. 1998 (Beijing)			84	161 (90th percentile)	RSP	G	24 h
30 workers in smoking homes	Phillips et al. 1997 (Turin)	76	135 (90th percentile)		RSP	G	15 h
59 workers in smoking homes	Phillips et al. 1998 (Kuala Lumpur)	53	83 (90th percentile)		RSP	G	15 h
88 workers in nonsmoking homes	Phillips et al. 1998 (Kuala Lumpur)			59	87 (90th percentile)	RSP	G	15 h

Field Studies—Indoor Particulate Matter Concentrations ($\mu g/m^3$)

Location/Condition	Reference	Smoking		Controls		Measure	Method	
		Mean	Range	Mean	Range		Type	Interval
40 house persons in smoking homes	Phillips et al. 1998 (Kuala Lumpur)	83	89 (90th percentile)			RSP	G	24 h
51 house persons in nonsmoking homes	Phillips et al. 1998 (Kuala Lumpur)			53	89 (90th percentile)	RSP	G	24 h
47 house persons in nonsmoking homes	Phillips et al. 1997 (Turin)			55	81 (90th percentile)	RSP	G	24 h
50 workers in smoking homes	Phillips et al. 1998 (Hong Kong)	61	96 (90th percentile)			RSP	G	15 h
74 workers in nonsmoking homes	Phillips et al. 1998 (Hong Kong)			54	96 (90th percentile)	RSP	G	15 h
35 house persons in smoking homes	Phillips et al. 1998 (Hong Kong)	49	77 (90th percentile)			RSP	G	24 h
34 house persons in nonsmoking homes	Phillips et al. 1998 (Hong Kong)			47	70 (90th percentile)	RSP	G	24 h
40 workers in nonsmoking homes	Phillips et al. 1997 (Barcelona)			59	105 (90th percentile)	RSP	G	15 h
43 house persons in smoking homes	Phillips et al. 1997 (Barcelona)	82	155 (90th percentile)			RSP	G	24 h

67 workers in nonsmoking homes	Phillips et al. 1998 (Prague)			32	55 (90th percentile)	RSP	G	15 h
28 workers in smoking homes	Phillips et al. 1997 (Barcelona)	95	160 (90th percentile)			RSP	G	15 h
50 house persons in smoking homes	Phillips et al. 1998 (Prague)	61	112 (90th percentile)			RSP	G	24 h
75 workers in nonsmoking homes	Phillips et al. 1997 (Turin)			52	90 (90th percentile)	RSP	G	15 h
70 workers in smoking homes	Phillips et al. 1998 (Prague)	61	105 (90th percentile)			RSP	G	15 h
36 house persons in smoking homes	Phillips et al. 1997 (Turin)	83	140 (90th percentile)			RSP	G	24 h
Workplaces — Personal Exposure								
15 workers in smoking workplaces	Coultas et al. 1990	64	4–146			RSP	G	2–8 h
28 workers in smoking workplaces	Heavner et al. 1996	67	18–217			RSP	G	9 h
52 workers in nonsmoking workplaces	Heavner et al. 1996			30	0–98	RSP	G	9 h

Field Studies—Indoor Particulate Matter Concentrations ($\mu g/m^3$)

Location/Condition	Reference	Smoking		Controls		Measure	Method	
		Mean	Range	Mean	Range		Type	Interval
703 workers in workplaces where smoking was banned	Jenkins and Counts 1999			17.3	47 (95th percentile)	RSP	G	8 h
134 workers in unrestricted smoking workplaces	Jenkins and Counts 1999	62	181 (95th percentile)			RSP	G	8 h
82 workers in nonsmoking workplaces	Philips et al. 1996 (Sweden)			23	9.4–96	RSP	G	8 h
32 workers in nonsmoking workplaces	Philips et al. 1998 (Prague)			93	78 (90th percentile)	RSP	G	8 h
71 workers in smoking workplaces	Philips et al. 1997 (Turin)	92	172 (90th percentile)			RSP	G	8 h
20 workers in smoking workplaces	Philips et al. 1998 (Sydney)	34	51 (90th percentile)			RSP	G	3 8-h shifts > 24 h
39 workers in nonsmoking workplaces	Phillips et al. 1999 (Basil)			43	77 (90th percentile)	RSP	G	8 h
8 workers in nonsmoking offices	Phillips et al. 1997 (Barcelona)			58	90 (90th percentile)	RSP	G	8 h

60 workers in smoking workplaces	Phillips et al. 1997 (Barcelona)	112	197 (90th percentile)			RSP	G	8 h
22 workers in nonsmoking workplaces	Phillips et al. 1998 (Paris)			53	93 (90th percentile)	RSP	G	8 h
33 workers in nonsmoking workplaces	Phillips et al. 1997 (Turin)			73	100	RSP	G	8 h
48 workers in smoking workplaces	Phillips et al. 1998 (Hong Kong)	66	112 (90th percentile)			RSP	G	8 h
68 workers in nonsmoking workplaces	Phillips et al. 1998 (Hong Kong)			42	72 (90th percentile)	RSP	G	8 h
104 workers in smoking workplaces	Phillips et al. 1998 (Paris)	71	127 (90th percentile)			RSP	G	8 h
73 workers in smoking workplaces	Phillips et al. 1998 (Kuala Lumpur)	52	88 (90th percentile)			RSP	G	8 h
72 workers in nonsmoking workplaces	Phillips et al. 1998 (Kuala Lumpur)			48	92 (90th percentile)	RSP	G	8 h
63 workers in smoking offices	Phillips et al. 1998 (Bremen)	46	105 (90th percentile)			RSP	G	8 h
53 workers in smoking workplaces	Phillips et al. 1996 (Sweden)	24	9.7–70			RSP	G	8 h

Field Studies—Indoor Particulate Matter Concentrations ($\mu g/m^3$)

Location/Condition	Reference	Smoking		Controls		Method		
		Mean	Range	Mean	Range	Measure	Type	Interval
57 workers in nonsmoking workplaces	Phillips et al. 1998 (Sydney)			32	41 (90th percentile)	RSP	G	3 8-h shifts >24 h
40 workers in nonsmoking offices	Phillips et al. 1998 (Bremen)			28	53 (90th percentile)	RSP	G	8 h
62 workers in smoking workplaces	Phillips et al. 1999 (Basil)	40	78 (90th percentile)			RSP	G	8 h
110 workers in smoking workplaces	Phillips et al. 1998 (Prague)	79	161 (90th percentile)			RSP	G	8 h
25 workers in smoking offices	Sterling et al. 1996	30	13.3–49.6			RSP	G	8 h
Offices								
12 offices (both smoking and nonsmoking)	Baek et al. 1997	99	12–392			RSP	G	4 h
31 offices/smoking	Carson & Erickson 1988	44	6–426			RSP	UV	59–84 min
1 office	Conner et al. 1990	65	12–147			RSP	UV	1 h
1 office	Conner et al. 1990	448	167–1088			RSP	G	1 h

Description	Reference							
30 offices/smoking	Crouse & Carson 1989	61	11–279			RSP	G	1 h
30 offices/smoking	Crouse & Carson 1989	47	11–84			RSP	UV	1 h
2 offices/smoking	Eatough et al. 1989	40	28–52			RSP	P,G	NS
194 offices	HBI 1990	28	6–180			RSP	G	1 h
6 office buildings — smoking restricted to rooms with local air filtration (22 smoking, 26 nonsmoking)	Hedge et al. 1994	157	–	37	–	RSP	G	3 h
4 office buildings — smoking restricted to separately ventilated areas (12 smoking, 20 nonsmoking)	Hedge et al. 1994	110	–	33.6	–	RSP	G	3 h
5 office buildings — smoking restricted to rooms with no additional air treatment (13 smoking, 23 nonsmoking)	Hedge et al. 1994	117	–	39	–	RSP	G	3 h
Offices/nonsmoking	Kirk et al. 1988			300		RSP	O	30 min

Field Studies—Indoor Particulate Matter Concentrations ($\mu g/m^3$)

Location/Condition	Reference	Smoking Mean	Smoking Range	Controls Mean	Controls Range	Measure	Method Type	Method Interval
Office/smoking	Kirk et al. 1988	105	100–110			RSP	P	10 h
Office/nonsmoking	Kirk et al. 1988			62	46–77	TSP	G	10 h
Offices/smoking	Kirk et al. 1988	720				RSP	O	30 min
Office/nonsmoking	Kirk et al. 1988			55	50–60	RSP	P	10 h
Office/smoking	Kirk et al. 1988	1118	117–120			TSP	G	10 h
11 offices/nonsmoking	Miesner et al. 1989			17.3	5.6–44.3	RSP	G	396–1106 min
3 designated smoking areas	Miesner et al. 1989	277	114–121			RSP	G	180–912 min
2 offices/smoking	Miesner et al. 1989	54	28–80			RSP	G	360–636 min
70 offices in 4 buildings	Oldaker et al. 1995	5[a]–34[a]	<7–74			RSP	G	7 h
131 offices	Oldaker et al. 1990	126	0–1088			RSP	G	1 h
7 offices/smoking	Proctor, Warren & Bevan 1989	103[b]	19–225			RSP	G	7 h
3 offices/ nonsmoking	Proctor et al. 1989			95[b]	27–208	RSP	O	1–8 h

Location	Reference					Measure	Cat.	Time
3 offices/smoking	Quant et al. 1982	51	37–89			RSP	P	30 min
3 offices/smokers	Repace 1987; Nelson et al. 1982	53	37–89			TSP	–	10-h avg
1 office	Ross et al. 1996	27	<12.5–40.2	22.8	20.8–24.8	RSP	G	8 h
4 offices—different buildings	Sterling et al. 1997	37	26–55	23	18–31	RSP	G	4 h
2 offices—different buildings	Sterling et al. 1997	53	46–66	25	22–30	RSP	G	4 h
Office area/nonsmoking	Sterling & Mueller 1988			6	4–11	RSP	O	1–8 h
2 offices—different buildings	Sterling et al. 1997	37	23–48	30	20–35	RSP	G	4 h
22 offices/smokers	Sterling & Sterling 1984	32				TSP		
Office area/nonsmoking	Sterling & Mueller 1988			7	6–8	RSP	O	1–8 h
9 offices/mixed ventilation	Turner & Binnie 1990	46	10–130			RSP	P	NS
17 offices/naturally ventilated	Turner & Binnie 1990	63	15–357			RSP	P	NS

Field Studies—Indoor Particulate Matter Concentrations ($\mu g/m^3$)

Location/Condition	Reference	Smoking		Controls		Measure	Method	
		Mean	Range	Mean	Range		Type	Interval
6 offices/mechanically ventilated	Turner & Binnie 1990	27	10–53			RSP	P	NS
44 offices/smoking	Weber & Fischer	133[b]				RSP	P	2 min
Public Buildings								
8 public buildings/smoking occupants	First 1984	260	40–660			TSP	P	NS
40 public buildings	Grimsrud et al. 1990		<50–308	24[c]	5–63	RSP	G	NS
45 public buildings	HBI 1990	36	6–284			RSP	G	1 h
7 public buildings/smoking occupants	Leaderer 1986	205	58–452			TSP	G	2–24 h
3 public buildings/nonsmoking	Leaderer 1986			18	9–32	TSP	G	4–21 h
4 clinic waiting rooms/nonsmoking	Miesner et al. 1989			11	6.7–17.3	RSP	G	726–749 min

2 laundromats/nonsmoking	Miesner et al. 1989			20.7	17.2–24.3	RSP	G	360–898 min
Grocery store/nonsmoking (2 locations)	Miesner et al. 1989			13	11.9–14.0	RSP	G	1002 min
Clinic/smoking room	Miesner et al. 1989	119	–			RSP	G	738 min
4 high school/classrooms	Miesner et al. 1989			41.9	19.0–109	RSP	G	402–510 min
2 museums/nonsmoking	Miesner et al. 1989			9.4	6.5–12.9	RSP	G	738–762 min
Hospital public areas/smoking	Miesner et al. 1989	36.3	20.1–52.5			RSP	G	828–846 min
Library/nonsmoking	Miesner et al. 1989			9.1	5.8–12.8	RSP	G	996–1014 min
Hospital public areas/nonsmoking	Miesner et al. 1989			17.4	14.5–23.3	RSP	G	432–942 min
Department store	Ogden & Maiolo 1989	55				RSP	G	4 h
27 public buildings/smoking occupants	Repace & Lowrey 1980	278	86–1140			RSP	P	2 min

Field Studies—Indoor Particulate Matter Concentrations ($\mu g/m^3$)

| Location/Condition | Reference | Smoking | | Controls | | Measure | Method | |
		Mean	Range	Mean	Range		Type	Interval
30 outside sites near 40 buildings	Turk et al. 1987			14[c]	<5–68	RSP	G	75–100 h
106 nonsmoking venues in 40 buildings	Turk et al. 1987			15[c]	5–63	RSP	G	75–100 h
70 smoking locations in 40 buildings	Turk et al. 1987	44[c]	<5–308			RSP	G	75–100 h
Restaurants								
5 restaurants nonsmoking	Brauer and 't Mannetje 1998			38	7–65	RSP	G	6 h
11 restaurants —restricted smoking	Brauer and 't Mannetje 1998	57	11–163			RSP	G	6 h
4 restaurants —unrestricted smoking	Brauer and 't Mannetje 1998	190	47–253			RSP	G	6 h
36 restaurants	Crouse et al. 1988	26.1[c]	15–168			RSP	UV	1 h
37 restaurants	Crouse et al. 1988	62.0[c]	16–231			RSP	G	1 h

Location	Reference	Mean	Range			Measure	Method	Duration
30 restaurants/smoking	Crouse & Carson 1989	31[c]	10–194			RSP	UV	1 h
30 restaurants/smoking	Crouse & Carson 1989	111[c]	16–366			RSP	G	1 h
62 restaurants	HBI 1990	53	10–228			RSP	G	1 h
Restaurant bar	Husgafvel-Pursiainen et al. 1986	720	260–1190			TSP	G	6–7 h
Restaurant discotheque	Husgafvel-Pursiainen et al. 1986	680	280–1370			TSP	G	6–7 h
82 waiters—personal exposure	Jenkins and Counts 1999	109	386 (95th percentile)			RSP	G	4–8 h
Restaurant/smoking	Kirk et al. 1988	690				RSP	O	30 min
Restaurant/nonsmoking	Kirk et al. 1988			400		RSP	O	30 min
7 restaurants	Lambert et al. 1993	53[a]	22–131	28[d]	21–69	RSP	G	24 h
2 restaurants	Miesner et al. 1989	85	363–133.1			RSP	G	708–882 min
Office building cafeteria/smoking	Miesner et al. 1989	26				RSP	G	692 min
41 restaurants	Ogden et al. 1990	54				RSP	FPM	1 h

Field Studies—Indoor Particulate Matter Concentrations ($\mu g/m^3$)

Location/Condition	Reference	Smoking		Controls		Measure	Method	
		Mean	Range	Mean	Range		Type	Interval
41 restaurants	Ogden et al. 1990	67				RSP	UV	1 h
41 restaurants	Ogden et al. 1990	106				RSP	G	1 h
83 restaurants	Oldaker et al. 1990	126	0–685			RSP	G	1 h
Barbecue/0.89[a]	Repace 1987	136		40[e]		RSP	P	2 min
Fast food/0.42*[a]	Repace 1987	109		24[e]		RSP	P	2 min
Shopping plaza restr a/0.18[a]	Repace 1987	163		36[e]		RSP	P	2 min
Sandwich a/0.13*[a]	Repace 1987	86		55		RSP	P	2 min
Sandwich a/nonsmoking section[a]	Repace 1987	51		–		RSP	P	2 min
Roadside/1.12*[a]	Repace 1987	107		30		RSP	P	2 min
Pizzeria/2.94*[a]	Repace 1987	414		40[e]		RSP	P	2 min
Dinner theater/0.14*[a]	Repace 1987	145		47		RSP	P	2 min
Restaurant & bar/0.40*	Repace 1987	93		55[e]		RSP	P	2 min

Location	Reference							
2 cafeterias/smoking	Sterling & Mueller 1988	70	23–129			RSP	O	1–8 h
2 cafeterias/nonsmoking	Sterling & Mueller 1988			32	15–57	RSP	O	1–8 h
Bars								
3 nightclubs (33 measurements)	Bergman et al. 1996	502	110–1714			TSP	G	7–8 h
10 neighborhood pubs	Collett et al. 1992	95	25–180			RSP	O	2 h
8 taverns	Collett et al. 1992	93	80–110			RSP	O	2 h
13 nightclubs	Collett et al. 1992	151	92–246			RSP	O	2 h
2 bars	HBI 1990	75	70–79			RSP	G	1 h
80 bartenders —personal exposure	Jenkins and Counts 1999	151	428 (95th percentile)			RSP	G	4–8 h
2 bars	Kirk et al. 1988	119	83–155			TSP	G	10 h
2 bars	Kirk et al. 1988	93	70–150			RSP	P	10 h
Pub/smoking	Kirk et al. 1988	1320				RSP	O	30 min
Pub/nonsmoking	Kirk et al. 1988			520		RSP	O	30 min
Taverns/smoking	Löfroth et al. 1989	430	390–470	ND	ND	TSP	G	3–4 h

Field Studies—Indoor Particulate Matter Concentrations ($\mu g/m^3$)

Location/Condition	Reference	Smoking		Controls			Method	
		Mean	Range	Mean	Range	Measure	Type	Interval
Taverns/smoking	Löfroth et al. 1989	370	320–420	45	40–50	RSP	P	3–4 h
3 bars	Miesner et al. 1989	85	30–140.9			RSP	G	372–786 min
1 tavern (26 measurements over 2 years prior to smoking ban)	Ott et al. 1996	83	25.3–182	26.1	0–67	RSP	P	0.4–2 h
Tavern[a]	Repace 1987	310	233–346	–		RSP	P	2 min
Cocktail lounge/3.24[*a]	Repace 1987	334		50[e]		RSP	P	2 min
Bar & grill/1.78[*a]	Repace 1987	589		63[e]		RSP	P	2 min
Hotel bar/0.59[*a]	Repace 1987	93		30		RSP	P	2 min
Functions								
2 discos	Eatough et al. 1989	788	774–801			RSP	P,G	NS
3 arenas/nonactivity day	Elliott & Rowe 1975			55	42–92	TSP	G	24 h

Description	Reference							
3 arenas/smoking occupants	Elliott & Rowe 1975	350	148–620			TSP	G	0.3 h
Sports arena/ smoking restrictions	Georghiou et al. 1989			303	187–426	TSP	G	2.5 h
Sports arena; no smoking restrictions	Georghiou et al. 1989	441	17–680			TSP	G	2.5 h
Leisure (108 samples) nonsmoking	Kirk et al. 1988			330	70–1240	RSP	O	30 min
Leisure (703 samples) smoking	Kirk et al. 1988	910	70–6220			RSP	O	30 min
Billiard parlor	Ogden & Maiolo 1989	355				RSP	G	2 h
5 betting shops/smoking	Proctor et al. 1989b	333	73–767					
Outdoors	Proctor et al. 1989b			313	108–624	RSP	G	1 h
1 betting shop/nonsmoking	Proctor et al. 1989b			48	33–63	RSP	G	1 h
Church bingo/0.47*	Repace 1987	279		30		RSP	P	2 min
Bowling alley/1.53*	Repace 1987	202		49		RSP	P	2 min
Reception hall/1.19*	Repace 1987	301		33[a]		RSP	P	2 min

Field Studies—Indoor Particulate Matter Concentrations ($\mu g/m^3$)

Location/Condition	Reference	Smoking Mean	Smoking Range	Controls Mean	Controls Range	Measure	Method Type	Method Interval
Lodge hall/1.25*	Repace 1987	697		60[e]		RSP	P	2 min
Outdoors	Repace 1987			55	—	RSP	P	2 min
Conference room/3.54*	Repace 1987	1947		55		RSP	P	2 min
Firehouse bingo/2.77*	Repace 1987	417	—	51		RSP	P	2 min
Sports arena/0.1*¹a	Repace 1987	94	—			RSP	P	2 min
Cocktail party/0.75*	Repace 1987	351		24		RSP	P	2 min
Bingo hall mech vent/0.93*¹a	Repace 1987	443		40[e]		RSP	P	2 min
Bingo hall nat vent/0.93*¹a	Repace 1987	1140		40[e]		RSP	P	2 min
Transportation								
Aircraft cabin	Conner et al. 1990	73	33–119	22	3–98	RSP	G	5 h
Aircraft cabin	Conner et al. 1990	67	20–106	19	13–30	RSP	UV	5 h
Automobiles/smoking	Kirk et al. 1988	300				RSP	O	30 min

Situation	Reference							Duration
Automobiles/nonsmoking	Kirk et al. 1988			300		RSP	O	30 min
Walking/smoking	Kirk et al. 1988	300				RSP	O	30 min
Bus/smoking	Kirk et al. 1988	1000				RSP	O	30 min
Walking/nonsmoking	Kirk et al. 1988			300		RSP	O	30 min
Bus/nonsmoking	Kirk et al. 1988			500		RSP	O	30 min
Travel (241 situations nonsmoking)	Kirk et al. 1988			420	70–1830	RSP	O	30 min
Trains/smoking	Kirk et al. 1988	900				RSP	O	30 min
Trains/nonsmoking	Kirk et al. 1988			400		RSP	O	30 min
Travel (297 situations smoking)	Kirk et al. 1988	790	0–4980			RSP	O	30 min
2 airport smoking lounges	Klepeis et al	114	65–177	13	5–23	RSP	P	1–3 h
2 subway stations/smoking	Miesner et al. 1989	93	55.1–157.3			RSP	G	232–295 min
Bus station/nonsmoking	Miesner et al. 1989			43.3		RSP	G	184 min
92 aircraft	Nagda et al. 1990	?–883	34.8			RSP	O,G	NS
Automobile	Ogden & Maiolo 1989	18				RSP	G	8 h

Field Studies—Indoor Particulate Matter Concentrations ($\mu g/m^3$)

Location/Condition	Reference	Smoking		Controls		Measure	Method	
		Mean	Range	Mean	Range		Type	Interval
29 Boeing 747s/smoking	Oldaker et al. 1990	39	3–185			RSP	G	3–12 h
28 Boeing 747s/nonsmoking	Oldaker et al. 1990			15	3–98	RSP	G	3–12 h
10 train compartments/smoking	Proctor 1989	216	70.8–325			RSP	G	1 h
Subway train/smoking compartment	Proctor 1987	630				RSP	O	–
10 train compartments/ nonsmoking	Proctor 1989			186	63.3–450	RSP	G	1 h
Subway train/nonsmoking compartment	Proctor 1987			180		RSP	O	–
Other Work Situations								
Hair salon	Eatough et al. 1989	29	–	–	–	RSP	P,G	NS
Common areas (4)	Eatough et al. 1990	52	31–94	–	–	RSP	P,G	NS
286 work situations	HBI 1990	36	6–519			RSP	G	1 h

Workshop/smoking	Kirk et al. 1988	500			RSP	O	30 min
Shop/smoking	Kirk et al. 1988	500			RSP	O	30 min
Common room/nonsmoking	Kirk et al. 1988		290		RSP	O	30 min
Shop/nonsmoking	Kirk et al. 1988		290		RSP	O	30 min
Workshop/nonsmoking	Kirk et al. 1988		400		RSP	O	30 min
Workshop (supposedly nonsmoking)	Kirk et al. 1988		65	50–80	RSP	P	12 h
Workshop (supposedly nonsmoking)	Kirk et al. 1988		125	110–140	TSP	G	12 h
Common room/smoking	Kirk et al. 1988	500			RSP	O	30 min
Work (224 samples)/smoking	Kirk et al. 1988	610	70–5780		RSP	O	30 min

Field Studies—Indoor Particulate Matter Concentrations (μg/m^3)

Location/Condition	Reference	Smoking		Controls		Method		
		Mean	Range	Mean	Range	Measure	Type	Interval
Work (480 samples)/nonsmoking	Kirk et al. 1988			310	0–2200	RSP	O	30 min

[a]Included in public buildings as well.
[b]Mean of means.
[c]Geometric mean.
[d]Median value. Mean not reported.
[e]Simultaneous outdoors.
[*]number of active smokers per 100 m^3.
FPM=Fluorescing Particulate Matter.
G=Gravimetric methods.
NS=Not Specified.
O=Optical.
P=Piezoelectric balance.
QCMI=Quartz crystal microbalance cascade impactor.
RSP=Respirable suspended particulates.
TSP=Total suspended particulates.
UV=Ultraviolet absorbing particulate matter.

Field Studies—Nicotine Concentrations, $\mu g/m^3$			
Locations/Conditions	**References**	**Mean**	**Range**
Residences—Area Sampling			
4 homes (several locations)	Eatough et al. 1989	5.5	0.02–13.6
10 homes—smoking	Mumford et al. 1989	3.4	1–6.3
13 homes—nonsmoking	Mumford et al. 1989	0.4	
Kitchens—smoking	Kirk et al. 1988	12.5	
Homes—smoking	Kirk et al. 1988	19	7–292
Homes—nonsmoking	Kirk et al. 1988	8	7–82
Workshop	Kirk et al. 1988	2.0[c]	
Flat	Kirk et al. 1988	6.0[c]	
Living rooms—smoking	Kirk et al. 1988	21	
Kitchens—nonsmoking	Kirk et al. 1988	8.5	
Living rooms—nonsmoking	Kirk et al. 1988	8.5	
47 homes	Leaderer and Hammond 1991	2.2	0.1–9.4
3 houses	Muramatsu, et al. 1984	11.1	7.6–14.6
7 households	Muramatsu et al. 1987	7	
48 homes, light-use area 120-h duration—smoking	Ogden et al. 1993	0.56[f]	
48 homes, heavy-use area 120-h duration—smoking	Ogden et al. 1993	2.45[f]	
2 homes	Ogden and Maiolo 1989	13.2	12.1–14.4
137 homes, 5-day duration: 1 adult smoker	Rickert 1995	1.7	
33 homes—5 day duration—all adults smoke	Rickert 1995	4.5	
21 nonsmoking homes	Rickert 1995	< 0.05	
20 homes—3-h duration—smoking	Scherer et al. 1995	7.33	0.35–35.2
20 homes—168-h duration—smoking	Scherer et al. 1995	5.09	0.20–34.6
42 homes—48-h duration—smoking	Williams et al. 1993	8.03	< 0.05–94.2

Field Studies—Nicotine Concentrations, $\mu g/m^3$			
Locations/Conditions	References	Mean	Range
Residences—personal exposure			
899 employed subjects away from work—nonsmoking homes 16-h duration (U.S.)	Jenkins et al. 1996	0.072	0.194 (95th percentile)
306 employed subjects away from work—smoking homes 16-h duration (U.S.)	Jenkins et al. 1996	2.71	7.93 (95th percentile)
44 housepersons—nonsmoking homes— 24-h duration (China)	Philips et al. 1998	0.38	0.72 (90th percentile)
66 employed subjects away from work—nonsmoking homes—16-h duration (Czech Republic)	Philips et al. 1997	0.43	0.42 (90th percentile)
48 housepersons smoking homes—24-h duration (France)	Philips et al. 1998	0.93	2.4 (90th percentile)
33 housepersons—nonsmoking homes— 24-h duration (Sweden)	Phillips et al. 1996	0.34	0.04–3.2
21 housepersons smoking homes 24-h duration (Germany)	Phillips et al. 1998	0.63	1.5 (90th percentile)
23 employed subjects away from work—smoking homes 16-h duration (Germany)	Phillips et al. 1998	0.65	1.7 (90th percentile)
59 housepersons —nonsmoking homes—24-h duration (Germany)	Phillips et al. 1998	0.10	0.22 (90th percentile)
82 employed subjects away from work—nonsmoking homes—16-h duration (Germany)	Phillips et al. 1998	0.18	0.33 (90th percentile)
30 employed subjects while in the home only—smoking homes-24-h duration (Australia)	Phillips et al. 1998	0.67	1.6 (90th percentile)
46 employed subjects while in the home only—24-h duration—nonsmoking homes (Australia)	Phillips et al. 1998	0.06	< 0.10 (90th percentile)
48 employed subjects away from work—smoking homes 16-h duration (Hong Kong)	Phillips et al. 1998	0.95	2.1 (90th percentile)

Field Studies—Nicotine Concentrations, $\mu g/m^3$			
Locations/Conditions	**References**	**Mean**	**Range**
56 employed subjects away from work—smoking homes 16-h duration (France)	Phillips et al. 1998	1.2	2.9 (90th percentile)
42 housepersons —nonsmoking homes—24-h duration (France)	Phillips et al. 1998	0.18	0.29 (90th percentile)
65 employed subjects away from work—nonsmoking homes—16-h duration (France)	Phillips et al. 1998	0.23	0.34 (90th percentile)
43 housepersons—smoking homes—24-h duration (Spain)	Phillips et al. 1997	1.4	2.8 (90th percentile)
28 employed subjects away from work—smoking homes 16-h duration (Spain)	Phillips et al. 1997	1.8	4.4 (90th percentile)
40 housepersons—nonsmoking homes—24-h duration (Spain)	Phillips et al. 1997	0.14	0.33 (90th percentile)
41 employed subjects away from work—nonsmoking homes—16-h duration (Spain)	Phillips et al. 1997	0.48	0.61 (90th percentile)
26 housepersons—smoking homes—24-h duration (Switzerland)	Phillips et al. 1999	0.70	1.5 (90th percentile)
37 employed subjects away from work—smoking homes 16-h duration (Switzerland)	Phillips et al. 1999	1.1	3.0 (90th percentile)
56 housepersons—nonsmoking homes— 24-h duration (Switzerland)	Phillips et al. 1999	0.23	0.31 (90th percentile)
71 employed subjects away from work—nonsmoking homes—16-h duration (Switzerland)	Phillips et al. 1999	0.32	0.69 (90th percentile)
36 housepersons —smoking homes—24-h duration (Italy)	Phillips et al. 1997	1.9	4.9 (90th percentile)
29 employed subjects away from work—smoking homes 16-h duration (Italy)	Phillips 1997	1.6	2.8 (90th percentile)
47 housepersons—nonsmoking homes— 24-h duration (Italy)	Phillips et al. 1997	0.32	0.60 (90th percentile)

Field Studies—Nicotine Concentrations, $\mu g/m^3$			
Locations/Conditions	References	Mean	Range
75 employed subjects away from work—nonsmoking homes—16-h duration (Italy)	Phillips et al. 1997	0.27	0.40 (90th percentile)
51 housepersons—smoking homes—24-h duration (Czech Republic)	Phillips et al. 1998	1.3	3.1 (90th percentile)
72 employed subjects away from work—smoking homes 16-h duration (Czech Republic)	Phillips et al. 1998	1.3	3.3 (90th percentile)
38 housepersons—nonsmoking homes— 24-h duration (Czech Republic)	Phillips et al. 1997	0.31	0.51 (90th percentile)
24 housepersons smoking homes—24-h duration (Portugal)	Phillips et al. 1998	0.59	1.2 (90th percentile)
53 housepersons—nonsmoking homes— 24-h duration (Portugal)	Phillips et al. 1998	0.14	0.31 (90th percentile)
10 employed subjects away from work—smoking homes—16-h duration (Sweden)	Phillips et al. 1996	0.30	0.07–1.6
119 employed subjects away from work—nonsmoking homes—16-h duration (Sweden)	Phillips et al. 1996	0.19	0.05–9.3
31 housepersons—smoking homes—24-h duration (Hong Kong)	Phillips et al. 1998	0.26	0.51(90th percentile)
87 employed subjects away from work—nonsmoking homes—16-h duration (Malaysia)	Phillips et al. 1998	0.22	0.24 (90th percentile)
33 housepersons—nonsmoking homes— 24-h duration (Hong Kong)	Phillips et al. 1998	0.10	0.27 (90th percentile)
68 employed subjects away from work—nonsmoking homes—16-h duration (Hong Kong)	Phillips et al. 1998	0.11	0.17 (90th percentile)
40 housepersons smoking homes—24-h duration (Malaysia)	Phillips et al. 1998	0.65	1.3 (90th percentile)

Field Studies—Nicotine Concentrations, $\mu g/m^3$			
Locations/Conditions	**References**	**Mean**	**Range**
59 employed subjects away from work—smoking homes 16-h duration (Malaysia)	Phillips et al. 1998	0.28	0.61 (90th percentile)
51 housepersons—nonsmoking homes— 24-h duration (Malaysia)	Phillips et al. 1998	0.24	0.24 (90th percentile)
54 housepersons—smoking homes 24-h duration (China)	Phillips et al. 1998	1.7	3.6 (90th percentile)
12 working women (smoking households)	Proctor et al. 1991	1.6	0–9.6
17 nonworking women (nonsmoking households)	Proctor et al. 1991	0.5	0–7.2
12 working women (nonsmoking households)	Proctor et al. 1991	0.8	0–2.6
11 nonworking women (smoking households)	Proctor et al. 1991	7.4	0–45.4
Offices—area sampling			
3 offices—nonsmoking	Bayer and Black 1987	0.8[b]	0.1–2.1
3 offices—smoking	Bayer and Black 1987	14.1	9.8–18.0
4 floors in 1 office building where smoking was permitted	Brickus et al. 1998	0.7	0.4–1.7
28 offices—smoking	Carson and Erickson 1988	7.2[b]	< 1.2–69.7
32 offices—smoking	Crouse and Carson 1989	3.8[b]	1.2–24.3
2 offices	Eatough, et al. 1989	6.0	4.1–7.8
35 open offices restricted smoking	Hammond et al. 1995	3.4g	9[g] (90th percentile)
4 office workers—nonsmokers	Hammond et al. 1987	14.6	3.1-28.2
1 office worker—smoker	Hammond et al. 1987	36.6	25.1-48.0
61 open offices with unrestricted smoking	Hammond et al. 1995	14.4[g]	34[g] (90th percentile)
29 open offices—smoking banned	Hammond et al. 1995	0.7[g]	1.7[g] (90th percentile)

Field Studies—Nicotine Concentrations, μg/m³			
Locations/Conditions	**References**	**Mean**	**Range**
194 offices	Healthy Buildings International 1990	3.5	< 1.6-71.5
6 office buildings smoking restricted to rooms with local air filtration; 22 locations	Hedge et al. 1994	44.2	-
5 office buildings smoking restricted to rooms with no additional air treatment; 23 locations	Hedge et al. 1994	0.3	
6 office buildings smoking restricted to rooms with local air filtration; 26 locations	Hedge et al. 1994	0.9	-
5 office buildings smoking restricted to rooms with no additional air treatment; 13 locations	Hedge et al. 1994	10.3	
4 office buildings— smoking restricted to separately ventilated areas; 12 locations	Hedge et al. 1994	8.2	-
Offices—nonsmoking	Kirk et al. 1988	9.5	
Offices—smoking	Kirk et al. 1988	14.5	
Office—nonsmoking	Kirk et al. 1988	1.1[c]	
Office—smoking	Kirk et al. 1988	6.0[c]	
Office—smoking (including cigars)	Klus at al 1985		43.3–199
4 offices	Miesner et al. 1989	1.8	0.4–4.3
Office	Muramatsu, et al. 1984	19.4	9.3–31.6
3 offices	Muramatsu et al. 1987	-	5.9–19.8
Office	Muramatsu, et al. 1984	22.1	14.6–26.1
156 offices	Oldaker et al. 1990	4.8[b]	0-69.7
3 offices—nonsmoking	Proctor et al. 1989a	1.6[c]	0.1–2.1
7 offices—smoking	Proctor et al. 1989a	5.8[c]	0.7–26
1 building, several locations where smoking permitted	Ross et al. 1996	2.4	0.2–5.6

Field Studies—Nicotine Concentrations, $\mu g/m^3$			
Locations/Conditions	**References**	**Mean**	**Range**
2 offices—different buildings negative smoking area pressurization, but no additional smoking area ventilation	Sterling et al. 1997	3.3	1.9–5.6
Offices—smoking prohibited	Sterling 1988	< 1.6	
2 office buildings where smoking unrestricted; 16 locations	Sterling et al. 1996	2.2	3.9 (95th percentile)
4 offices—different buildings workplace divided into smoking and nonsmoking offices	Sterling et al. 1997	6.1	2.2–16.7
2 offices—different buildings negative smoking area pressurization, with additional smoking area ventilation	Sterling et al. 1997	5.6	1.5–13.0
Offices—designated smoking	Sterling 1988	75	
Offices—smoking permitted	Sterling. 1988	4.8	
10 offices	Thompson et al. 1989	2.3[c]	0.3-6.7
10 offices	Thompson et al. 1989	2.6	0.3–6.7
9 offices-mixed ventilation	Turner and Binnie 1990	4.7	<DL-25.7
17 offices—naturally ventilated areas	Turner and Binnie 1990	10.0	<DL-41.9
7 offices-mechanically ventilated areas	Turner and Binnie 1990	3.4	<DL-6.6
10 offices—nonsmoking	Vaughan and Hammond 1990	2.1	0.9-5.3
6 offices—smoking	Vaughan and Hammond 1990	10.7	3.4-33.3
44 offices	Weber and Fischer 1980	1.1	0–16.0
Workplaces—personal exposure			
15 workers (U.S.)	Coultas et al. 1990	20.4	0–53.2
52 workers where smoking restricted to designated areas (U.S.)	Jenkins and Counts 1999	0.30	2.21 (95th percentile)
703 workers: smoking banned (U.S.)	Jenkins and Counts 1999	0.086	0.21 (95th percentile)

Field Studies—Nicotine Concentrations, $\mu g/m^3$			
Locations/Conditions	References	Mean	Range
134 workers in unrestricted smoking workplaces (U.S.)	Jenkins and Counts 1999	3.4	15.0 (95th percentile)
23 office workers in smoking environments (Germany)	Philips et al. 1998	0.65	1.7 (90th percentile)
74 workers in smoking environments (Malaysia)	Philips et al. 1998	1.9	2.2 (90th percentile)
57 workers: 24-h over ca 3 work shifts nonsmoking environments (Australia)	Phillips et al. 1998	0.15	0.19 (90th percentile)
102 workers in smoking environments (France)	Phillips et al. 1998	1.9	4.6 (90th percentile)
20 workers in nonsmoking environments (France)	Phillips et al. 1998	0.51	0.99 (90th percentile)
61 workers in smoking environments (Spain)	Phillips et al. 1997	4.1	9.0 (90th percentile)
65 workers in smoking environments (Switzerland)	Phillips et al. 1999	1.8	3.2 (90th percentile)
41 workers in nonsmoking environments (Switzerland)	Phillips et al. 1999	0.20	0.26 (90th percentile)
20 workers: 24-h over ca 3 work shifts in smoking environments (Australia)	Phillips et al. 1998	1.1	2.9 (90th percentile)
42 office workers nonsmoking environments (Germany)	Phillips et al. 1998	0.22	0.45 (90th percentile)
33 workers in nonsmoking environments (Italy)	Phillips et al. 1997	0.52	0.92 (90th percentile)
108 workers in smoking environments (Czech Republic)	Phillips et al. 1998	3.2	8.2 (90th percentile)
32 workers in nonsmoking environments (Czech Republic)	Phillips et al. 1998	0.90	1.3 (90th percentile)
54 workers in smoking environments (Sweden)	Phillips et al. 1996	0.48	0.11–3.1
80 workers in nonsmoking environments (Sweden)	Phillips et al. 1996	0.23	0.10–1.8
40 workers in smoking environments (Hong Kong)	Phillips et al. 1998	3.3	4.0 (90th percentile)

Field Studies—Nicotine Concentrations, $\mu g/m^3$

Locations/Conditions	References	Mean	Range
69 workers in nonsmoking environments (Hong Kong)	Phillips et al. 1998	0.28	0.32 (90th percentile)
72 workers in smoking environments (Italy)	Phillips et al. 1997	1.9	5.4 (90th percentile)
71 workers in nonsmoking environments (Malaysia)	Phillips et al. 1998	0.37	0.54 (90th percentile)
18 casino workers in smoking environments (U.S.)	Trout et al. 1998	9.44	4–14

Other work situations area sampling

Hair salon	Eatough, et al. 1989	0.6	
8 common areas	Eatough, et al. 1989	2.6	0.1–6.1
53 shop production areas and fire stations—smoking banned	Hammond et al. 1995	0.2^9	0.6^9 (90th percentile)
54 shop production areas and fire stations: restricted smoking	Hammond et al. 1995	2.2^9	4^9 (90th percentile)
114 shop production areas and fire stations: unrestricted smoking	Hammond et al. 1995	4.4^9	7.2^9 (90th percentile)
Freight car repair workers (smokers)	Hammond et al. 1987	11.7	0.9-41.1
Freight car repair workers (nonsmokers)	Hammond et al. 1987	0.1	0.0-0.3
282 reception areas, conf. rms.	HBI 1990	4.3	< 1.6–126.3
Shops—nonsmoking	Kirk et al. 1988	9.5	
Common room—smoking	Kirk et al. 1988	18	
Common room—nonsmoking	Kirk et al. 1988	8.5	
Workshops—nonsmoking	Kirk et al. 1988	9.5	
Working—nonsmoking (495 samples)	Kirk et al. 1988	9	7–99
Shops—smoking	Kirk et al. 1988	13.5	
Working—smoking (238 samples)	Kirk et al. 1988	14	7–167
Workshops—smoking	Kirk et al. 1988	12	

Field Studies—Nicotine Concentrations, $\mu g/m^3$			
Locations/Conditions	**References**	**Mean**	**Range**
3 work areas	Thompson et al. 1989	2.7[c]	1.0–3.8
6 common areas	Thompson et al. 1989	36.5[c]	12.6–60.3
3 common areas	Thompson et al. 1989	1.1[c]	0.8–1.7
Public buildings			
Hospital lobby/12-30 smoking occupants	Badre et al. 1978	37	
17 supermarkets	Crouse et al. 1989	0.7	0.3–3.3
1 public building	First 1984	5.5	
8 public buildings	First 1984	13.2	2.7–30.0
45 lobbies, customer areas, etc.	HBI 1990	5.9	< 1.6–90.8
Bus waiting room	Hinds and First 1975	1.0[a]	
Airline waiting room	Hinds and First 1975	3.1[a]	
8 bowling alleys/video parlors—area and personal monitors	Jenkins et al. 1991	13.0	4.6–24.5
2 airport gates—area and personal monitors	Jenkins et al. 1991	6.5	0.9–I5.1
Bus waiting area—area and personal monitors	Jenkins et al. 1991	1.9	1.0–3.0
3 laundromats —area and personal monitors	Jenkins et al. 1991	2.0	0.2–4.8
Hospital waiting room	Miesner et al. 1989	1.6	
Subway station	Miesner et al. 1989	1.0	
2 smoking areas in nonsmoking building	Miesner et al. 1989	22	17.1–26.5
4 hotel lobbies	Muramatsu 1984	11.2	5.5–18.1
Hospital lobby	Muramatsu 1984	3.0	1.9–5.0
3 bus, railway waiting rooms	Muramatsu 1984	19.1	10.1–36.4
Department store	Ogden and Maiolo 1989	0.6	
Billiard parlor	Ogden and Maiolo 1989	19.4	
Subway platform	Proctor 1987		0.20

Field Studies—Nicotine Concentrations, $\mu g/m^3$			
Locations/Conditions	**References**	**Mean**	**Range**
Restaurants—area sampling			
6 cafes	Badre et al. 1978		25–52
31 taverns and nightclubs (Canada)	Collett et al. 1992	38.6–58.0	
36 restaurants	Crouse and Carson 1989	4.1[b]	1.0–36
21 restaurants-area samples	Crouse and Oldaker 1990	4.3	0.3–24.0
62 cafeterias, lunch rooms	HBI 1990	7.5	<1.6–84.5
Restaurant	Hinds and First 1975	5.2[a]	
7 restaurants—area and personal monitors	Jenkins et al. 1991	3.4	0.0–16.1
32 restaurants	Jenkins and Counts 1999	5.8	36 (95th percentile)
Restaurants—smoking	Kirk et al. 1988	15	
Restaurants—nonsmoking	Kirk et al. 1988	10	
7 restaurants—nonsmoking sections	Lambert et al. 1993	1.0	0.2–2.8
7 restaurants—smoking sections	Lambert et al. 1993	2.8	1.5–3.8
2 restaurants/1 cafeteria	Miesner et al. 1989	3.4	2.0–6.2
3 cafeterias	Muramatsu, et al. 1984	26.4	11.6–42.2
5 restaurants	Muramatsu, et al. 1984	14.8	7.1–27.8
7 coffee shops	Muramatsu et al. 1987	34	
170 restaurants	Oldaker et al. 1990	5.1	0–23 .8
46 restaurants	Oldaker et al. 1991	8.4[b]	0.0–43.2
2 cafeterias— smoking areas	Sterling and Mueller 1988	14	<1.6–43.7
2 cafeterias—nonsmoking areas	Sterling and Mueller 1988	6.2	<1.6–10.9
Cafeteria	Thompson et al. 1989	-	2.3–4.4

Field Studies—Nicotine Concentrations, $\mu g/m^3$			
Locations/Conditions	References	Mean	Range
Restaurants—personal exposure			
21 restaurants-personal samples	Crouse and Oldaker 1990	6.3	0.3–24.8
82 restaurant servers—4- to 8-h duration	Jenkins and Counts 1999	5.9	28.9 (95th percentile)
34 restaurants	Thompson et al. 1989	3.5[b]	0.5–37.2
3 mall food courts	Thompson et al. 1989	2.3	1.6–3.1
Bars: area and personal sampling			
2 bars	Healthy Buildings International 1990	12.9	4.1–21.6
Cocktail lounge	Hinds and First 1975	10.3[a]	
80 bartenders, 5–9-h work shifts—personal sampling	Jenkins and Counts 1999	14.1	43.6 (95th percentile)
53 bars and bar areas in restaurants—area sampling	Jenkins and Counts 1999	14.4	49.6 (95th percentile)
8 cocktail lounges—area and personal sampling	Jenkins et al. 1991	17.6	1.8–90.6
Pubs—nonsmoking	Kirk et al. 1988	7.5	
2 bars	Kirk et al. 1988	8.4[c]	4.7–13
Pubs—smoking	Kirk et al. 1988	36	
Tavern (sampled twice)	Löfroth et al. 1989	65.5	60–71
5 bars	Miesner et al. 1989	7.4	2.0–13.1
3 pubs	Muramatsu et al. 1987	31	
Tavern	Oldaker and Conrad 1989	59.2[c]	6.1–108.6
Functions			
2 discos	Eatough et al. 1989	106	96–115
5 conference rooms	Muramatsu et al. 1984	38.7	16.5–53.0
1 betting shop—nonsmoking	Proctor et al. 1989a	1.2	0.4–2.0
5 betting shops—smoking	Proctor et al. 1989a	19.3	3–57

Field Studies—Nicotine Concentrations, $\mu g/m^3$			
Locations/Conditions	References	Mean	Range
Transportation			
Automobile/natural ventilation	Badre et al. 1978	65	
Automobile/vent closed	Badre et al. 1978	1010	
2 train compartments	Badre et al. 1978		36–50
Train/vent closed	Harmsden and Effenberger 1957		0.7–3.1
Train	Hinds and First 1975	4.9	
Bus	Hinds and First 1975	6.3	
Trains—smoking	Kirk et al. 1988	26.5	
Buses—nonsmoking	Kirk et al. 1988	9	
Automobiles—nonsmoking	Kirk et al. 1988	7	
Automobiles—smoking	Kirk et al. 1988	9	
Trains—nonsmoking	Kirk et al. 1988	8	
Walking—smoking	Kirk et al. 1988	9	
Walking—nonsmoking	Kirk et al. 1988	7	
Buses—smoking	Kirk et al. 1988	37.7	
Aircraft-48 te—smoking seats	Malmfors et al. 1989	32	15–98
Aircraft-48 b[d]—smoking seats	Malmfors et al. 1989	41	13–98
Aircraft-48 t[e] —nonsmoking seats	Malmfors et al. 1989	21	4.8–62.1
Aircraft-48 b[d] —nonsmoking seats	Malmfors et al. 1989	5.0	0.8–17
Airline flight attendants	Mattson et al. 1989	2.3	0.1–10.5
Airline passengers	Mattson et al. 1989	4.2	0.1–71.0
Airplanes/24 smoking seats	Muramatsu et al. 1987	13.5	?–28.8
Trains/20 nonsmoking seats	Muramatsu et al. 1987	1.3	
Trains/48 smoking seats	Muramatsu et al. 1987	16.7	?–48.6
7 airplanes	Muramatsu, et al. 1984	15.2	6.3–28.8
4 automobiles	Muramatsu, et al. 1984	47.7	7.7–83.1

Field Studies—Nicotine Concentrations, $\mu g/m^3$			
Locations/Conditions	**References**	**Mean**	**Range**
Airplanes/19 nonsmoking seats	Muramatsu et al. 1987	5.3	
8 trains	Muramatsu, et al. 1984	16.4	8.6–26.1
7 private automobiles	Muramatsu et al. 1987	43	
Aircraft seats—nonsmoking 69 flights	Nagda et al. 1992	0.04	
Aircraft seats—boundary between smoking and nonsmoking sections— 69 flights	Nagda et al. 1992	0.26	
69 aircraft—smoking seats	Nagda et al. 1992	13.4	
Airplanes-14 nonsmoking seats	Ogden et al. 1989b	6.8	<1.8–17.1
Airplanes-17 smoking seats	Ogden et al. 1989b	25.0	5.3–55.4
Automobile	Ogden and Maiolo 1989	0.4	
B-747 aircraft, nonsmoking	Oldaker et al. 1989	1.1[b]	0.1–12.4
B-747 aircraft, smoking seats	Oldaker et al. 1989	7.1[b]	1.8–42.7
Airplane/26 nonsmoking seats	Oldaker and Conrad 1987	5.5	≤ 0.08–40.2
Airplane/49 smoking seats	Oldaker and Conrad 1987	9.2	≤ 0.03–112.4
Subway train/smoking section	Proctor 1987	32	16–74
10 trains—nonsmoking sections	Proctor 1989	4.5	0.5–21.2
10 trains—smoking sections	Proctor 1989	15.3	0.6–49.3
Subway train/nonsmoking section	Proctor 1987	7	

[a]Background subtracted.
[b]Geometric mean.
[c]Mean of means.
[d]Business class.
[e]Tourist class.
[f] Median
[g] Due to unusual calculation procedure, value may be an overestimate by as much as a factor of 3.
HBI=Healthy Buildings International

Field Studies—Carbon Monoxide Concentrations

Location/conditions	Reference	Smoking Mean	Smoking Range	Controls Mean	Controls Range	Comment
		Residences				
12 residences	Baek et al. 1997	2.1	0.1–6.2	2.1	0.2–7.0	Korean summer and winter, outside controls
		Offices				
12 offices	Baek et al. 1997	2.4	0.4–6.8	2.2	0.5–6.1	Korean summer and winter, outside controls
28 offices	Carson and Erikson 1988	1.9	<0.1–8.7	1.9	<0.1–5.8	17 x 2- to 3-min samples
10 offices	Chappell and Parker 1977	2.5	1.5–4.5		1.5–4.5[a]	2- to 3-min sampling, 30 min after smoking
Office—30 min after smoking	Chappell and Parker 1977	1.0				
Office—72 m³, ~40 cigarettes per day	Harke 1974		<2.5–4.6			30-min samples
Office—78 m³, ~70 cigarettes per day	Harke 1974		<2.5–9.0			30-min samples
118 offices	Healthy Buildings Intl. 1990	3.6 ± 3.3[b]	1.4–8.7	3.3 ± 1.0[b]	2.0–6.6	76 controls
6 office bldgs.—NS	Hedge et al. 1994	0.2	0.1–0.3			26 locations; smoking restricted to rooms w/local air filtration
6 office bldgs.—smoking	Hedge et al. 1994	1.2	0.7–2.1			22 locations; smoking restricted to rooms w/local air filtration
5 office bldgs.—smoking	Hedge et al. 1994	1.0	0.4–2.2			13 locations; smoking restricted to rooms w/no additional air treatment
5 office bldgs.—NS	Hedge et al. 1994	0.4	0.1–2.1			23 locations; smoking restricted to rooms w/out additional air treatment
4 office bldgs.—NS	Hedge et al. 1994	0.3	0.2–0.7			20 locations; smoking restricted to separately ventilated areas
4 office bldgs.—smoking	Hedge et al. 1994	1.5	0.7–3.4			12 locations where smoking is restricted to separately ventilated areas

Field Studies—Carbon Monoxide Concentrations

Location/conditions	Reference	Carbon Monoxide, ppm				Comment
		Smoking		Controls		
		Mean	Range	Mean	Range	
Offices—NS	Kirk et al. 1988			2.4		
Offices—smoking	Kirk et al. 1988	2.0				30- to 760-m³ office space
7 offices—smoking	Proctor et al. 1989a	1.4[c]	0.9–2.3			30- to 80-m³ offices
3 offices—NS	Proctor et al. 1989a			1.2[c]	1.0–1.4	
66 offices—urban	Reynolds/Lorillard, 5-City	2.3 ± 2.0	0.1–10.5	2.5 ± 2.3[a,b]	NR–10.4[a]	Control is 57 outdoor measurements in Washington, D.C., and Ottawa
Offices—smoking	Sterling and Mueller 1988	1.8	1.5–2.4			Recirculated air
4 offices—different bldgs.	Sterling et al. 1997	1.6	1.3–2.3	1.5	1.2–2.1	Workplace divided into smoking and NS offices
2 offices—different bldgs.	Sterling et al. 1997	2.0	1.4–2.7	1.5	1.1–2.0	Negative smoking area pressurization, with add'l smoking area ventilation
2 offices—different bldgs.	Sterling et al. 1997	1.5	1.2–1.8	1.2	1.0–1.4	Negative smoking area pressurization, but no add'l smoking area ventilation
Offices—smoking	Sterling and Mueller 1988	1.35	1.3–1.4			Nonrecirculated air
25 offices	Szadkowski et al. 1976	2.8				Continuous
NS offices	Szadkowski et al. 1976	2.6				Continuous
Offices and workrooms	Weber and Fischer 1980	1.1[d]	0–6.5[d]			490 samples
Other Work Areas						
44 working rooms	Aviado 1984	2.8				Maximum value 8.9
Hair salon	Eatough et al. 1988 APCA	0.27				
142 workplaces	HBI 1990	3.2 ± 0.9[b]	1.7–6.7	3.0 ± 0.8[b]	2.0–6.4	140 controls
Workshops—smoking	Kirk et al. 1988	2.8				
Workshops—NS	Kirk et al. 1988			1.7		
Shops—smoking	Kirk et al. 1988	2.3				

Work—NS; 450 situations	Kirk et al. 1988	2.1		0-21.9	
Common rooms—smoking	Kirk et al. 1988		1.4		
Common rooms—NS	Kirk et al. 1988	2.0			
Shops—NS	Kirk et al. 1988	2.9			
Work—smoking; 221 situations	Kirk et al. 1988		2.2	0.0-31.9	UK-wide survey, 30-week study
Work areas—smoking prohibited	Sterling and Mueller 1988	2.1			
Work areas—designated smoking	Sterling and Mueller 1988		4.2		
Indoor smoking; 194 situations	Sterling et al. 1987		3.1	ND-242	Median value given
Designated smoking area	Sterling 1988		4.2		
Work areas—smoking permitted	Sterling and Mueller 1988		2.5		
Indoor NS; 15 situations	Sterling et al. 1987	3.4		ND-75	Median value given
15 bars and shops without smokers	Valerio et al. 1997		6.2		Street-level businesses

Field Studies—Carbon Monoxide Concentrations

Location/conditions	Reference	Carbon Monoxide, ppm				Comment
		Smoking		Controls		
		Mean	Range	Mean	Range	
23 bars and shops with smokers	Valerio et al. 1997	7.0				Street-level businesses
Public Buildings						
21 public bldgs.	HBI 1990	3.8 ± 1.3	2.2-6.4	3.3 1.2	2.0-7.9	
14 public places	Perry 1973	<10				1 grab sample, Drager Tube ± 25% accuracy
33 stores	Sebben et al. 1977	10.0		11.5 [a]		24 controls
Functions						
Disco, by stage	Eatough et al. 1988	20.5				
Arena 2/2000 occup	Elliott and Rowe 1975	25.0	-			MSA Monitaire Sampler, ± 25% accuracy
Arena 2/nonactivity day	Elliott and Rowe 1975	3.0	-			MSA Monitaire Sampler, ± 25% accuracy
Arena 1/nonactivity day	Elliott and Rowe 1975	3.0	-			MSA Monitaire Sampler, ± 25% accuracy
Arena 1/ 11,806 occup	Elliott and Rowe 1975	9.0	-			MSA Monitaire Sampler, ± 25% accuracy

	Reference						Notes
Theatre foyer	Perry 1973	3.4	1.4				
1 betting shop—NS	Proctor et al. 1989b	7	5-9	3.5	2.8-4.2		1-min samples acquired over 1 h
5 betting shops—smoking	Proctor et al. 1989b	5.1	2.4-9.4	3.9	2.2-9.0		12-30 smokers
Hospital lobby—smoking	Repace 1987	5					
9 nightclubs	Repace 1987	13.4	6.5-41.9				
Ice-skating rinks, apartments, offices w/attached or underground garages	Spengler et al. 1983		25->100				
Transportation							
Submarines	Aviado 1984	<40					
Automobile—3 smokers, natural vent	Badre et al. 1978	14		O[a]			20-min samples
Automobile—2 smokers, vent closed	Badre et al. 1978	20		O[a]			20-min samples
2 train compartments—2–3 smokers	Badre et al. 1978	-	4–5				
8 domestic planes—27–113 occupants	DOT 1971	≤2					1.25–2.5 h continuous
18 military planes/165-219 occup	DOT 1971		<2–5				6-7 h continuous
Intercity bus—3 cigarettes continuously	DOT 1973	18					
Intercity bus—23 cigarettes continuously	DOT 1973	33					

Field Studies—Carbon Monoxide Concentrations

Location/conditions	Reference	Carbon Monoxide, ppm				Comment
		Smoking		Controls		
		Mean	Range	Mean	Range	
Automobile—2 smokers, 4 cigarettes	Harke and Peters 1974		NR–42		NR–14	Peak concentration
Train—1–18 smoking occupants	Harmsden and Effenberger 1957		0–40			20-min samples
Walking—smoking	Kirk et al. 1988	2.3		2.7	0–13.1	30-min samples
Travel—NS, 235 situations	Kirk et al. 1988					
Automobiles—smoking	Kirk et al. 1988	5.5				
Automobiles—NS	Kirk et al. 1988			5.9		
Bus—NS	Kirk et al. 1988			2.5		
Trains—NS	Kirk et al. 1988			1.4		
Walking—NS	Kirk et al. 1988			1.6		30-min samples
Trains—smoking	Kirk et al. 1988	1.9				UK-wide survey, 30-week study
Travel—smoking, 283 situations	Kirk et al. 1988	2.9	0–13.1			
Bus—smoking	Kirk et al. 1988	3.3				
69 aircraft	Nagda et al. 1992	0.8				Commercial aircraft flights where smoking is permitted, remote area of nonsmoking section
23 aircraft	Nagda et al. 1992			0.6		NS commercial aircraft flights
69 aircraft	Nagda et al. 1992	1.4				Commercial aircraft flights where smoking is permitted, smoking section

Setting	Reference					Comments
Subway train—NS section	Proctor 1987			3.0	0.5–2.9	
10 train compartments—NS	Proctor 1989	1.6	1.0–2.2	1.3		Smoking/NS compartment separated by partition
10 train compartments—smoking	Proctor 1989					
Subway train—smoking section	Proctor 1987	3.5				
Ferryboat	Repace 1987	18.4		3.0		
Restaurants						
3 restaurants—multiple locations and observations	Akbar-Khanzadeh and Greco 1996	7.52	3.0–13.6	7.0	4.3–10.2	Dining, kitchen and bar areas
12 restaurants	Baek et al. 1997	11.3	0.7–89.9	2.4	0.4–6.6	Korean summer and winter; outside controls
15 restaurants	Chappell and Parker 1977	4.0	1.0–9.5	2.5[a]	1.0–5.0[a]	Just a few smokers
Small restaurants	First 1984		5–6.5			
Cafeteria	First 1984		0.5–2.0			
Restaurant—50–80 occupants in 470 m^3	Fischer et al. 1978	5.1	2.1–9.9	4.8[a]		27 x 30-min samples
49 restaurants	HBI 1990	3.4 ± 1.2[b]	2.0–7.9	3.0 ± 0.6[b]	2.0–4.1	13 controls
Restaurants—smoking	Kirk et al. 1988	2.7				
6 cafeterias	Repace 1987		2–23			
99 restaurants	Reynolds/Lorillard[c], 6-Cities	4.2 ± 2.7[b]	-1.5–42.3	2.5 ± 2.1[a,b]	-0.3–13.7	99 outdoor bkg. Hong Kong, Washington, D.C., and Winston-Salem, NC
14 restaurants	Sebben et al. 1977	9.9				Spot checks
45 restaurants	Sebben et al. 1977	8.2				Spot checks
Cafeteria—smoking section	Sterling and Mueller 1988	3.8	1.2–11.5	7.1[a]		0.5–1.67 cigarettes/h/10m^3
Cafeteria—NS section	Sterling and Mueller 1988			2.6	1.3–4.6	
Cafeteria—NS section	Weber et al. 1979			0.5	0.3–0.8	11 ACH; 24 x 30-min 3
Restaurant—6–100 occupants in 440 m^3	Weber et al. 1979	2.6	1.4–3.4	1.5[a]		29 x 30-min samples

Field Studies—Carbon Monoxide Concentrations

Location/conditions	Reference	Carbon Monoxide, ppm				Comment
		Smoking		Controls		
		Mean	Range	Mean	Range	
Cafeteria—80–150 occupants in 574 m³	Weber et al. 1979	1.2	0.7–1.7	3.0ᵃ	1.0–5.0ᵃ	11 ACH; 24 x 30-min samples
		Bars/Taverns				
14 nightclubs and taverns	Chappell and Parker 1977	13.0	3.0–29.0			19 x 2-3 min samples
Tavern	Chappell and Parker 1977	8.5		~12		
Tavern 2—post smoking	Cuddeback et al. 1976	-	-	2ᵃ	-	2-h post smoking, 1-2 ACH
Tavern 1—10–294 occupants	Cuddeback et al. 1976	11.5	10–12			8-h continuous, 6 ACH
Tavern 1—post smoking	Cuddeback et al. 1976	-	-	~1	-	8-h post smoking, 6 ACH
Tavern 2	Cuddeback et al. 1976	17	3–22	-	-	8-h continuous, 1-2 ACH
Taverns	First 1984		7–8			Sidestream smoke from taverns
2 taverns	HBI 1990	3.1	3.0–3.2	2		
Pub	Jarvis et al. 1983	13		3.0		
Pubs—NS	Kirk et al. 1988	3.0				3-h sampling
Pubs—smoking	Kirk et al. 1988	4.0		< 0.9	0.9–1,8	4-h sampling
Tavern 1	Löfroth et al. 1989	4.3	6.5–41.9			77 x 1-min samples
Tavern 2	Löfroth et al. 1989	13.4		9.2ᵃ	3.0–35.0ᵃ	28 x 30-min samples
9 nightclubs	Sebben et al. 1977	4.8	2.4–9.6	1.7ᵃ		
Bar—30–40 occupants per 50 m³	Weber et al. 1979					

ᵃOutdoors.
ᵇMean ± std. dev. Number of observations noted.
ᶜMean of means.

ᵈCorrected for CO concentration in unoccupied workrooms.
NR = Not reported.
ND = Not detected.

Field Studies—Nitrogen Oxides Concentrations

Location/Conditions	Reference	Nitrogen Oxides (ppb)						Comment
		Smoking				Controls[a]		
		Means		Ranges		Mean		
		NO	NO₂	NO	NO₂	NO	NO₂	
Office								
Office	Eatough et al. 1988	42						Heavy a.m. smoking
Office	Eatough et al. 1988	51						Heavy a.m. smoking
"Sick" office bldgs.	Nguyen and Martel 1987		Trace					3 bldgs.
5 office bldgs.	Turk et al. 1987	16 ± 5[b]				7–20	14 ± 6[a,b]	Restricted smoking
10 office bldgs.	Turk at al. 1987	24 ± 7[b]				11–32	27 ± 11[a,b]	Open smoking
Buildings and Workplaces								
Library	Eatough at al. 1988	44[d]						No smokers observed
Library	Eatough et al. 1988	27[d]						No smokers observed
Lunchroom	Eatough et al. 1988	53[d]						During breaks
Lunchroom	Eatough et al. 1988	42[d]						During breaks
Hair salon, break area	Eatough et al. 1988	7±1[d]						Infrequent smoking
5 Classrooms	Lee and Chang 1999					17–10 8	20–43	
"Worst" workplaces	Nguyen and Goyer 1988	500	50					Workplaces selected by workers
45 bldgs.	Sterling at al. 1987		NDe		ND–100			

Field Studies—Nitrogen Oxides Concentrations

Location/ Conditions	Reference	Nitrogen Oxides (ppb)						Comment
		Smoking				Controls[a]		
		Means		Ranges		Mean		
		NO	NO2	NO	NO2	NO	NO2	
17 outdoors	Sterling et al. 1987			ND–570		27.5[e]		
97 bldgs.	Sterling at al. 1987	ND[a]		ND–160				
7 outdoors	Sterling et al. 1987				ND–67		7[e]	
92 bldgs., smoking permitted	Sterling et al. 1987	ND		ND–160				
43 bldgs., smoking permitted	Sterling at al. 1987		2		ND–100			
6 bldgs., smoking restricted	Sterling at al. 1987	26.6		ND–70				
2 bldgs., smoking restricted	Sterling et al. 1987		ND		ND			
7 bldgs., smoking	Sterling and Sterling 1984	41.6						
32 bldgs., overall	Sterling and Sterling 1984	38.8[he]	33.9[e]					
4 bldgs., smoking	Sterling and Sterling 1984		33.9					
1 bldg., NS	Sterling and Sterling 1984					26		
44 workrooms, 492 measurements	Weber and Fischer 1980		60 ± 25[b]		Max 200			Uncorrected data
44 workrooms, 227 measurements	Weber and Fischer 1980	82	64					Window ventilation

Location	Reference						Notes
44 workrooms, 487 measurements	Weber and Fischer 1980	22 ± 48[b]	Max 320				Corrected for outdoor concentration
44 workrooms, 354 measurements	Weber and Fischer 1980		24 ± 22[b]	Max 115			Corrected for unoccup. data
44 workrooms, 348 measurements	Weber and Fischer 1980	32 ± 60[b]	Max 280				Corrected for unoccup. room AC
44 workrooms, 102 measurements	Weber and Fischer 1980	66	49				
44 workrooms, 492 measurements	Weber and Fischer 1980	84 ± 80[b]	Max 450				Uncorrected data
Eating establishments							
Bar	Triebig and Zober 1984	195	66–414	1–61	44	48	Natural ventilation
Restaurant	Triebig and Zober 1984	120	36–218	59–105	115	63	Natural ventilation
Cafeteria	Triebig and Zober 1984	5,9	2–38	15–103	5	27	AC
Restaurant	Weber 1984	80	14–121	24–99	11	50	Natural ventilation
Residences with smokers							
Dining room	Eatough at al. 1988	18[d]					Moderate smoking
Living room	Eatough et al. 1988	8±1[d]					Infrequent smoking
Dining room	Eatough et al. 1988	4[d]			5 ± 0		Moderate smoking NS control, fireplace
NS room	Weber et al. 1979				5	27	
Transportation							
Newark airport	Thurston 1987		150–330	218–350			Jetway doors to tarmac opened
Others							
Disco, by stage	Eatough at al. 1988	131 ± 24					20–100 smokers

Field Studies—Nitrogen Oxides Concentrations

Location/ Conditions	Reference	Nitrogen Oxides (ppb)							Comment
		Smoking				Controls[a]			
		Means		Ranges		Mean			
		NO	NO₂	NO	NO₂	NO	NO₂		
Smoke chamber	Klus et al. 1987			75–240	3–25				Increases in NO, NO₂, smokers present
Chamber, 13.6 m³, 3.55 ach	Löfroth et al. 1989	123 ± 9[c]		75 ± 3[c,d]					Smoking rate I cigt/30 min
Chamber, 13.6 m³, 3.55 ach	Löfroth at al. 1989	262 ± 9[c]		171 ± 6					Smoking rate I cigt/15 min
Chamber, 13.6 m³, 3.55 ACH	Löfroth et al. 1989	137 ± 7[c]		83 ± 6[c,d]					Smoking rate1 cigt/30 min

[a]Denotes outdoor background unless specifically indicated otherwise.
[b]Mean ± standard deviation, with number of samples noted.
[c]Eight determinations.
[d]Expressed as NO from total concentration of NO and NO2.
[e]Median values, not mean values.
ND=Not Detected.
AC=air conditioning.

Field Studies—Formaldehyde Concentrations

Location/condition	Reference	Smoking		Control		Comment
		Mean	Range	Mean	Range[a]	
Residences (qualifications noted)						
15 indoor environments	De Bortoli et al. 1985	27	8–52	6.6	2–14	4 apartments, 10 detached houses, 1 office
Living room, 4-person household	Klus et al. 1987	–	23–50	53	–	
7 rural homes	Lemus et al. 1998	260	153–470			Some with smokers, homes 10–34 years old
17 urban homes	Lemus et al. 1998	524	183–1314			Some with smokers, homes 10–34 years old
90 residential living rooms	Sega et al. 1992	32	2–100			Winter measurements. 56% of residences had smokers present
90 residential kitchens	Sega et al. 1992	37	6–109			Summer measurements. 56% of residences had smokers present
90 residential living rooms	Sega et al. 1992	40	10–112			Summer measurements. 56% of residences had smokers present
90 residential kitchens	Sega et al. 1992	29	4–87			Winter measurements. 56% of residences had smokers present
48 kitchens, 1-week exposures	Sexton et al. 1986	50	–	–	–	Not mobile homes
45 bedrooms, 1-week exposures	Sexton et al. 1986	44	–	–	–	Not mobile homes
51 bedrooms, 1-week exposures	Sexton et al. 1986	40	<12–110	–	–	Stationary homes. Smoking present in some
51 kitchens, 1-week exposures	Sexton et al. 1986	44	<12–110	–	–	Stationary homes. Smoking present in some

Field Studies—Formaldehyde Concentrations

Location/condition	Reference	Formaldehyde, $\mu g/m^3$				Comment
		Smoking		Control		
		Mean	Range	Mean	Range[a]	
51 whole homes, 1-week exposures	Sexton et al. 1986	43	16–105	–	–	Stationary homes. Smoking present in some
Energy-efficient non-UFFI residences	Sterling 1985	–	160–210	–	–	
Apartments	Sterling 1985	98	37–250	–	–	Part of study, 78 structures
Non-UFFI residence	Sterling 1985	–	74–98	–	–	
28 residences with UFFI structures	Sterling 1985	86	25–160	–	37–86	
UFFI residence	Sterling 1985	–	135–200	–	–	
UFFI and non-UFFI structures	Sterling 1985	61	37–250	–	–	Part of study, 78 structures
6 houses	Triebig and Zober 1984	–	2–280	–	–	Urea foam insulation present
Outside homes w/o woodstove	Zweidinger et al. 1988	–	5–6	–	–	Variation is weekday/ weekend, day/night
Background rural site	Zweidinger et al. 1988	–	0.9–1.4	–	–	Variation is weekday/ weekend, day/night, airport radio station
10 homes w/o woodstove	Zweidinger et al. 1988	–	19–21	–	–	Variation is weekday/ weekend, day/night
Woodsmoke-impacted site	Zweidinger et al. 1988	–	4–6	–	–	Variation is weekday/ weekend, day/night, Boise Fire Station site
10 homes w/wood stove	Zweidinger et al. 1988	–	5–31	–	–	Variation is weekday, weekend, day/night
Woodsmoke-impacted site	Zweidinger et al. 1988	–	4–6	–	–	Variation is weekday, weekend, day/night, Elm Grove site

Offices and the Workplace

Location	Reference					Comments
4 floors in an office building	Brickus et al. 1998	40	12.2–99.7	14.5	7.1–21	11 inside and 11 outside (control) samples
23 office NS locations	Hedge et al. 1994			63	48–79	Smoking restricted to rooms w/no add'l air treatment
13 office smoking locations	Hedge et al. 1994	75	49–99			Smoking restricted to rooms w/no add'l air treatment
26 office NS locations	Hedge et al. 1994			25	17.5–28	Smoking restricted to rooms w/local air filtration
22 office smoking locations	Hedge et al. 1994	50	38–61			Smoking restricted to rooms w/local air filtration
1 cafeteria	Hodgson et al. 1996	4.9	–			Cafeteria in office building
4 smoking lounges	Hodgson et al. 1996	27.5	11.7–42			California city and county office bldgs.
Public bldgs. and energy-efficient homes, ocuppied and unoccupied	Hollowell and Miksch 1981	–	0–280	–	–	
"Worst" workplace atmospheres	Nguyen and Goyer 1988	–	6–38	–	–	Situations chosen by the workers
Public bldgs., smoking	Sterling et al. 1987	20	ND–740	–	–	Median of 200 situations
Public bldgs.	Sterling et al. 1988	12	ND–2330	–	–	Median of 259 measurements

Field Studies—Formaldehyde Concentrations

Location/condition	meas.	Reference	Formaldehyde, $\mu g/m^3$					Comment
			Smoking		Control			
			Mean	Range	Mean	Range[a]		
Public bldgs., 207		Sterling et al. 1987	16	ND–740	1	ND–38		Medians shown. Controls are 24 outdoor measurements
Public bldgs., NS		Sterling et al. 1987	–	–	ND	ND–270		Median of 7 situations. Smoking restricted
Offices		Triebig and Zober 1984	50	40–58	–	25–26		
Working room		Triebig and Zober 1984	–	< 290	–	–		Formaldehyde source wood material
Dwellings, offices, schools		Triebig and Zober 1984	350	12–1100	–	–		Formaldehyde source is resins
Indoors		Triebig and Zober 1984	–	74–184	–	–		Realistic domestic environment, smokers not included
Dwellings, offices, schools		Triebig and Zober 1984	290	6–940	–	–		Formaldehyde source is resins
Office bldgs.		Triebig and Zober 1984	120	12–1300	–	–		Formaldehyde source is resins
			Miscellaneous					
4 Suburban office bldgs.		Turk et al. 1987	32 ± 9	< 2 5–60	26 ± 1	< 25–26		Open smoking policy
10 Suburban office bldgs.		Turk et al. 1987	33 ± 14	< 25–92	26 ± 1	< 25–26		Restricted smoking policy
3 urban office bldgs.		Turk et al. 1987	34 ± 16	< 25–236	38 ± 13	<25–58		Restricted smoking policy
22 urban office bldgs.		Turk et al. 1987	29 ± 6	< 25–93	38 ± 13	<25–58		Open smoking policy
Automobile		Klus et al. 1987	27	–	12	–		
Pub		Klus et al. 1987	–	7–38	–	–		
Railway compartment		Klus et al. 1987	35	–	12	–		

Sample	Reference	Mobile Homes				Comments
Restaurant	Klus et al. 1987	–	15–40	8	–	
2 taverns	Löfroth et al. 1989	89	–	ND	–	Outdoor background not determined
2 taverns	Löfroth et al. 1989	104	–	ND	–	Outdoor background not determined
Anatomy dissection lab	Wantke et al. 1996			155	74–274	Mean of 17 daily measurements
65 mobile homes	D. Sterling 1985	200	122–990	–	–	Median, not mean, value reported
164 mobile homes	D. Sterling 1985	180	<25–96	–	–	
65 mobile homes (complaint)	D. Sterling 1985	580	<12–4500	–	–	Median, not mean, value reported
Mobile homes	Hollowell and Miksch 1981	1100	28–5200	–	–	Mobile homes with complaints, Wisconsin
2 mobile homes	Hollowell and Miksch 1981	442	122–1000	–	–	Location Pittsburgh, PA
Mobile homes	Hollowell and Miksch 1981	500	0–3300	–	–	Mobile homes with complaints, Minnesota
391 mobile homes, 1981, summer	Sexton et al. 1986	100	16–570	–	–	Mobile homes. Smokers present in some
293 mobile homes, 1981, winter	Sexton et al. 1986	110	28–390	–	–	Mobile homes. Smokers present in some
266 mobile homes, 1981, summer	Sexton et al. 1986	75	<12–470	–	–	Mobile homes. Smokers present in some
523 mobile homes, winter	Sexton et al. 1986	96	21–390	–	–	Mobile homes. Smokers present in some
222 mobile homes, 1981, winter	Sexton et al. 1986	79	21–380	–	–	Mobile homes. Smokers present in some
663 mobile homes, summer	Sexton et al. 1986	88	<12–570	–	–	Mobile homes. Smokers present in some

[a]Strict smoking or NS conditions not noted.

Appendix 6　　　　Volatile Organic Compounds

Field Studies—Volatile Organic Compounds

Constituent	Reference	Smoking		Controls		Comment
		Mean	Range[a]	Mean	Range[a]	
Aliphatic Hydrocarbons						
Ethane (μg/m³)	Löfroth et al. 1989		68-180		8-9	2 studies of a tavern, 60 and 71 μg/m³ nicotine, outdoors as control
Ethene (μg/m³)	Löfroth et al. 1989		56-100		12-16	2 studies of a tavern, 60 and 71 μg/m³ nicotine, outdoors as control
	Persson et al. 1988		85-340		3-6	1 office, 40 min continuous smoking, 4 conditions including 0 ventilation
Propane (μg/m³)	Löfroth et al. 1989		33-70		6-7	2 studies of a tavern, 60 and 71 μg/m³ nicotine, outdoors as control
Propane (μg/m³)	Löfroth et al. 1989		40-70		6	2 studies of a tavern, 60 and 71 μg/m³ nicotine, outdoors as control
	Persson et al. 1988		60-230		3-6	1 office, 40 min continuous smkg, 4 conditions including 0 ventilation
1,3-Butadiene (μg/m³)	Brunnemann et al. 1989, 1990	3.5	2.7-4.5			1 bar, 3 sampling days, approx. 20 μg/m3 nicotine
	Heavner et al. 1996	1.29	0.0-3.86	0.60	0.0-1.63	Personal exposure (28 and 51 subjects) in smoking and nonsmoking workplaces

Field Studies—Volatile Organic Compounds

Constituent	Reference	Smoking		Controls		Comment
		Mean	Range[a]	Mean	Range[a]	
	Heavner et al. 1996	1.15	0.0–4.14	0.86	0.0–12.1	Personal exposure (29 and 60subjects) in smoking and nonsmoking homes.
	Löfroth et al. 1989		11-19		<1-1	2 studies of a tavern, 60 and 71 μg/m³) nicotine, outdoors as control
	Phillips, Howard, & Bentley 1998	4.3[g]	5.6[c]	1.7[g]	3.6[c]	Turin, Italy Indoors smoking vs. indoors nonsmoking
	Phillips, Howard, & Bentley 1998	19[b]	10[c]	10[b]	41[c]	Paris, France Indoors smoking vs. indoors nonsmoking
	Phillips, Howard, & Bentley 1998	0.43[b]	0.59[c]	1.1[b]	2.7[c]	Basel, Switzerland indoors smoking vs. indoors nonsmoking
	Phillips, Howard, & Bentley 1998	0.20[b]	0.71[c]	0.28[b]	0.59[c]	Beijing, PRC indoors smoking vs. indoors nonsmoking
Isoprene (μg/m³)	Brunnemann et al. 1989, 1990	97	80-106			1 bar, 3 sampling days, approx 20 μg/m³) nicotine
	Guerin 1991 & Higgins et al. 1990		83.3		2.3	Residence A, smkg room
	Guerin 1991 & Higgins et al. 1990		18.6		1.8	Residence A, remote from smkg room
	Guerin 1991 & Higgins et al. 1990	42.6	16.6-90			4 restaurants
	Guerin 1991 & Higgins et al. 1990		65.1			Bowling alley

Compound	Reference					Description
	Heavner et al. 1996	18.2	2.9–68.0	4.65	0.99–17.3	Personal exposure (29 and 60 subjects) in smkg and nonsmoking homes
	Heavner et al. 1996	22.8	1.02–61.2	5.29	0.0–19.7	Personal exposure (28 and 51 subjects) in smoking and nonsmoking workplaces
	Löfroth et al. 1989		85–150		< 1-2	2 studies of a tavern, 60 and 71 µg/m³ nicotine, outdoors as control
	Phillips, Howard, & Bentley 1998	5.8[b]	25 [c]	3.1[b]	5.5 [c]	Turin, Italy indoors smoking vs. indoors nonsmoking
	Phillips, Howard, & Bentley 1998	8.2[b]	31 [c]	3.2[b]	6.1 [c]	Paris, France Indoors smoking vs. indoors nonsmoking
	Phillips, Howard, & Bentley 1998	0.08[b]	0.23 [c]	0.25[b]	1.1 [c]	Basel, Switzerland Indoors smoking vs. indoors nonsmoking
	Phillips, Howard, & Bentley 1998	0.86[b]	3.7 [c]	0.78[b]	1.2 [c]	Beijing, PRC Indoors smoking vs. indoors nonsmoking
Cyclohexane (µg/m³)	Badre et al. 1978	257	20–640			6 cafes
	Badre et al. 1978	247	10–392			3 train spaces
n-Octane (µg/m³)	Proctor et al. 1991	7.8	1.7–22.2	12.4	2.4–77.4	11 nonworking women in smkg households vs. 17 nonworking women in nonsmoking households
	Proctor 1989	47	1–530	4	0.9–15	10 offices, smkg vs. nonsmoking, 5 smplg times per office
	Proctor 1989	1.6	0.2–2.9	1.4	0.8–2.0	Train compartments, 10 smkg and 10 nonsmoking journeys
	Wallace and Pellizzari 1986	4.7			3.1	TEAM[d] study, fall and winter
	Wallace and Pellizzari 1986	1.5			1.7	TEAM[d] study, spring and summer

Field Studies—Volatile Organic Compounds

Constituent	Reference	Smoking		Controls		Comment
		Mean	Range[a]	Mean	Range[a]	
n-Nonane	Heavner et al. 1995	3.84	.5–16.7	3.0	0.5–9.0	Personal exposure (25 & 21 subjects) in smoking and nonsmoking homes
	Heavner et al. 1996	2.79	0.0–29.9	4.4	0.0–75.0	Personal exposure (32 and 61 subjects) in smoking and nonsmoking homes
	Heavner et al. 1996	10.7	0.0–155	5.5	0.0–140	Personal exposure (29 and 51 subjects) in smoking and nonsmoking workplaces
n-Decane ($\mu g/m^3$)	Brickus et al. 1998	27	13–53	13	12.6–13.4	Area samples on 1st, 13th, and 25th floors of office building, outside acts as control
	Heavner et al. 1996	18.3	0.0–335	9.2	0.0–245	Personal exposure (29 and 51 subjects) in smoking and nonsmoking workplaces
	Heavner et al. 1995	5.0	0–46.0	5.07	1.5–16.9	Personal exposure (25 & 24 subjects) in smoking and nonsmoking homes
	Heavner et al. 1996	6.31	0.79–66.9	6.07	0.0–118	Personal exposure (32 and 61 subjects) in smoking and nonsmoking homes
	Proctor et al. 1991	21.8	2.5–62.2	27.6	1.8–80.3	11 nonworking women in smoking households vs. 17 nonworking women in nonsmoking households
	Proctor 1989	8	1–24	6	1–16	10 offices, smoking vs. nonsmoking, 5 smplg trips per office

	Reference					Description
	Proctor 1989	4.6	1.3-11.2	2.7	0.7-6.7	Train compartments, 10 smoking and 10 nonsmoking journeys
	Proctor 1989	4.6	1.3-11.2	2.7	0.7-6.7	Train compartments, 10 smoking and 10 nonsmoking journeys
n-Undecane ($\mu g/m^3$)	Heavner et al. 1996	4.48	0.43-36.3	3.95	0.0-58.6	Personal exposure (32 and 61 subjects) in smoking and nonsmoking homes
	Heavner et al. 1995	14.5	.6-250	3.9	0.14-17.2	Personal exposure (25 & 24 subjects) in smoking and nonsmoking homes
	Heavner et al. 1996	14.9	0.2-155	10.7	0.0-120.4	Personal exposure (29 and 51 subjects) in smoking and nonsmoking workplaces
	Proctor 1989	5	2-12	5	0.8-12	10 offices, smoking vs. nonsmoking, 5 smplg trips per office
	Proctor et al. 1991	9.9	2.2-29.9	10.5	2.3-35.4	11 nonworking women in smoking households vs. 17 nonworking women in nonsmoking households
	Proctor 1989	5.6	1.2-13.6	4.0	1.4-9.7	Train compartments, 10 smoking and 10 nonsmoking journeys
n-Dodecane ($\mu g/m^3$)	Heavner et al. 1996	1.60	0.0-7.21	1.99	0.0-28.7	Personal exposure (32 and 61 subjects) in smoking and nonsmoking homes
	Heavner et al. 1995	7.34	0.19-128	1.63	0.32-4.5	Personal exposure (25 & 24 subjects) in smoking and nonsmoking homes
	Heavner et al. 1996	6.02	0.0-17.7	9.64	0.0-85.4	Personal exposure (29 and 51 subjects) in smoking and nonsmoking workplaces

Field Studies—Volatile Organic Compounds

Constituent	Reference	Smoking		Controls		Comment
		Mean	Range[a]	Mean	Range[a]	
α-Pinene ($\mu g/m^3$)	Proctor et al. 1991	5.9	0.8-21.1	5.8	0.7-30.9	11 nonworking women in smoking households vs. 17 nonworking women in nonsmoking households
	Proctor 1989	4	1-11	4	2-8	10 offices, smoking vs. nonsmoking, 5 smplg trips per office
	Proctor 1989	3.3	0.6-6.3	3.7	0.7-18.1	Train compartments, 10 smoking and 10 nonsmoking journeys
	Proctor et al. 1991	21.4	2.4-116	11.8	2.1-76.2	11 nonworking women in smoking households vs. 17 nonworking women in nonsmoking households
Aromatic Hydrocarbons						
Limonene ($\mu g/m^3$)	Brickus et al. 1998	25	5.6-52	ND	ND	Area samples on 1st, 13th, and 2sth floors of office building, outside acts as control
	Guerin 1991 & Higgins et al. 1990		22.0		12.2	Residence A, smoking room
	Guerin 1991 & Higgins et al. 1990		11.6		8.9	Residence A, remote from smoking room
	Guerin 1991 & Higgins et al. 1990	104	52-239			4 restaurants
	Guerin 1991 & Higgins et al. 1990		21.2			Bowling alley
	Heavner et al. 1996	34.8	1.2-157	15.1	0.0-227	Personal exposure (29 and 51 subjects) in smoking and nonsmoking workplaces

Reference					Description
Heavner et al. 1996	33.2	0.0–629	13.3	0.0–124	Personal exposure (32 and 61 subjects) in smoking and nonsmoking homes
Proctor 1989	1.6	0.05–4.9	1.9	0.2–7.5	Train compartments, 10 smoking and 10 nonsmoking journeys
Heavner et al. 1995	23.2	5.7–57.4	17.8	3.45–42	Personal exposure (25 & 24 subjects) in smoking and nonsmoking homes
Hodgson et al. 1996	39	7–82			4 smoking lounges in office bldgs.
Hodgson et al. 1996	11.7				Cafeteria in office bldg.
Proctor 1989	6.8	0.1–19.4	2.5	0.05–8.9	Train compartments, 10 smoking and 10 nonsmoking journeys
Proctor et al. 1991	16.9	0.3–45.0	15.9	0.4–72.0	11 nonworking women in smoking households vs. 17 nonworking women in nonsmoking households
Proctor 1989	7	1–31	3	0.4–8	10 offices, smoking vs. nonsmoking, 5 smplg times per office

Benzene (μg/m³)

Reference					Description
Badre et al. 1978	68	20–100			3 train spaces
Badre et al. 1978	100	50–150			6 cafes
Badre et al. 1978	30	20,40			Automobile, 3 smokers, natural ventilation
Badre et al. 1978		150			Automobile, 2 smokers, no ventilation
Baek et al. 1997	12.6	0.1–109	8.2	0.1–49.9	Indoor and outside background area levels for 12 Korean offices
Baek et al. 1997	8.2	0.1–27.2	9.4	0.1–47.2	Indoor and outside background area levels for 12 Korean residences
Baek et al. 1997	12.0	0.1–51.7	8.0	0.3–31.1	Indoor and outside background area levels for 12 Korean restaurants

Field Studies—Volatile Organic Compounds

Constituent	Reference	Smoking		Controls		Comment
		Mean	Range[a]	Mean	Range[a]	
	Bayer and Black 1987		ND-18.3		ND-110.8	3 office complexes, smoker/nonsmoker, 4-h sample 0 and 71 $\mu g/m^3$
	Brickus et al. 1998	23	16–35	9	3.3–12.2	Area samples on 1st, 13th, and 25th floors of office building, outside acts as control
	Proctor 1989	2	0.6–4	2	0.2–11	10 offices, smoking vs. nonsmoking, 5 smplg trips per office
	Brown and Crump 1998	7.7		6.1		10 living rooms in smoking/nonsmoking homes
	Brown and Crump 1998	6.7		6.2		34 bedrooms in smoking and nonsmoking homes
	Brunnemann et al. 1989, 1990	31	31-36			1 bar, 3 sampling trips, approx 20 $\mu g/m^3$ nicotine
	Gilli et al. 1996	91		70		Personal exposure of students passive exposure to ETS vs. no exposure
	Guerin 1991 & Higgins et al. 1990	8.6	5.6-13.8			4 restaurants
	Guerin 1991 & Higgins et al. 1990		17.6		3.6	Residence A, smoking room
	Guerin 1991 & Higgins et al. 1990		10.2			Bowling alley
	Guerin 1991 & Higgins et al. 1990		6.9		2.8	Residence A, remote from smoking room
	Heavner et al. 1995	5.54	0.98-27.0	3.86	1.3–19.0	Personal exposure (25 & 24 subjects) in smoking and nonsmoking homes
	Heavner et al. 1996	4.25	0.93–9.93	4.12	0.55-33.4	Personal exposure (32 and 61 subjects) in smoking and nonsmoking homes

Reference					Description
Heavner et al. 1996	7.3	1.23–101.2	2.40	0.38–9.71	Personal exposure (29 and 51 subjects) in smoking and nonsmoking workplaces
Hodgson et al. 1996	3.5				Cafeteria in office bldg.
Hodgson et al. 1996	9.7	4.7–14.8			4 smoking lounges in office bldgs.
Jo and Moon 1999	16.9[b]		12.9[b]		69 indoor and 50 outdoor area samples in Korea
Jo and Park 1999			49.1	13.6–258	In-vehicle levels in Taegu, Korea
Lawyrk et al. 1995			13.1		In-vehicle level on suburban roads, N=52
Leung and Harrison 1998			15.0	2.3–283	12-h personal exposure of 28 subjects in urban environments
Löfroth et al. 1989		21–27		6-8	2 studies of a tavern, 6 nicotine, 3- and 4-h samples, outdoors as control
Phillips, Howard, & Bentley 1998	6.0[b]	10[c]	7.7[b]	9.8[c]	Turin, Italy indoors smoking vs. indoors nonsmoking
Phillips, Howard, & Bentley 1998	5.3[b]	12[c]	3.2[b]	8.8[c]	Paris, France Indoors smoking vs. indoors nonsmoking
Phillips, Howard, & Bentley 1998	1.4[b]	2.9[c]	3.6[b]	13[c]	Basel, Switzerland Indoors smoking vs. indoors nonsmoking
Phillips, Howard, & Bentley 1998	15[b]	34[c]	11[b]	32[c]	Beijing, PRC Indoors smoking vs. indoors nonsmoking
Proctor 1989	13	3–49	12	3–31	10 offices, smoking vs. nonsmoking, 5 smplg trips per office
Proctor et al. 1989[a]	17	9–30	9.5	9–10	5 smoking betting shops, 1 nonsmoking shop

Field Studies—Volatile Organic Compounds

Constituent	Reference	Smoking		Controls		Comment
		Mean	Range[a]	Mean	Range[a]	
	Proctor et al. 1991	21.6	5.2-103	13.2	0.2-32.1	11 nonworking women in smoking households vs. 17 nonworking women in nonsmoking households
	Proctor 1989	11.8	0.9-28.6	7.4	2.9-29.3	Train compartment, 10 smoking and 10 nonsmoking journeys
	Scherer et al. 1995	10.9	1.6-32.6	6.3	2.2-17.0	Personal exposures (23 and 24 subjects in suburban smoking and nonsmoking homes
	Scherer et al. 1995	16.1	4.5-44.1	20.6	5.4-129	Personal exposures (20 and 15 subjects) in urban smoking and nonsmoking homes
	Wallace et al. 1987	4.9		4.4		TEAM[b] Study, Antioch, CA, spring
	Wallace et al. 1987	4.8		4.5		TEAM[b] Study, L.A., CA, spring
	Wallace et al. 1987	16		8.4		TEAM b Study, New Jersey; fall
	Wallace et al. 1987	17		11		TEAM[b] Study, L.A., CA, winter
Toluene (μg/m^3)	Badre et al. 1978	1125	180-1870			4 train compartments
	Badre et al. 1978	545	40-1040			4 cafes
	Badre et al. 1978		500			Automobile, 2 smkrs, ventilation closed
	Badre et al. 1978	30	50,70			Automobile, 3 smkrs, natural ventilation
	Baek et al. 1997	52	0.1-234	49.5	1.3-213	Indoor and outside background area levels for 12 Korean restaurants

Reference					Description
Baek et al. 1997	80.4	0.8–341	50.4	0.5–271	Indoor and outside background area levels for 12 Korean offices
Baek et al. 1997	42.3	0.6–202	42.6	0.1–222	Indoor and outside background area levels for 12 Korean residences
Bayer and Black 1987		0.6–248		0.9–142	3 office complexes, smoker/nonsmoker, 4-h sample
Brickus et al. 1998	180	102–321	41	8.9–60.2	Area samples on 1st, 13th, and 25th floors of office building, outside acts as control
Brunneman et al. 1989, 1990	55	41–80			1 bar, 3 sampling trips, approx 20 $\mu g/m^3$ nicotine
Guerin 1991 & Higgins et al. 1990		51.2		20.9	Residence A, smoking room
Guerin 1991 & Higgins et al. 1990		25.0		16.0	Residence A, remote from smoking room
Guerin 1991 & Higgins et al. 1990	80.6	16.2–215			4 restaurants
Guerin 1991 & Higgins et al. 1990		40.2			Bowling alley
Heavner et al. 1996	26.0	9.7–74.2	23.3	5.9–102.1	Personal exposure (32 and 61 subjects) in smoking and nonsmoking homes
Heavner et al. 1995	27.8	4.4–118	19.3	3.0–47.4	Personal exposure (25 & 21 subjects) in smoking and nonsmoking homes
Heavner et al. 1996	59.1	4.6–570	17.9	2.4–110.3	Personal exposure (29 and 51 subjects) in smoking and nonsmoking workplaces
Hodgson et al. 1996	7.6				Cafeteria in office bldg.
Hodgson et al. 1996	32.2	12.6–49	115	33.9–394	4 smoking lounges in office bldgs.
Jo and Park 1999					In-vehicle levels in Taegu, Korea

Field Studies—Volatile Organic Compounds

Constituent	Reference	Smoking Mean	Smoking Range[a]	Controls Mean	Controls Range[a]	Comment
	Jo and Moon 1999	79.8[b]		51.0[b]		69 indoor and 50 outdoor area samples in Korea
	Lawyrk et al. 1995			60.2		In-vehicle level on suburban roads, N=52
	Leung and Harrison 1998			31.8	3.6-536	12-h personal exposure of 28 subjects in urban environments
	Phillips, Howard, & Bentley 1998	9.0[b]	17[c]	6.3[b]	12[c]	Turin, Italy; indoors smoking vs. indoors nonsmoking
	Phillips, Howard, & Bentley, 1998	12[g]	33 (90)[c]	6.9[g]	20 (90)[c]	Paris, France; indoors smoking vs. indoors nonsmoking
	Phillips, Howard, & Bentley, 1998	4.4[g]	97 (90)[c]	50[g]	119 (90)[c]	Basel, Switzerland; indoors smoking vs. indoors non-smoking
	Phillips, Howard, & Bentley 1998	44[b]	103[c]	25[b]	25[c]	Beijing, PRC; indoors smoking vs. indoors nonsmoking
	Proctor 1989a	59.6	28-120	37	35-39	5 smoking betting shops, 1 nonsmoking shop
	Proctor 1989	39	10-292	25	7-65	10 offices, smoking vs. nonsmoking, 5 smplg times per office
	Proctor 1989	31.7	3.3-88	27.3	7.8-105	Train compartments, 10 smoking and 10 nonsmoking journeys
	Proctor et al. 1991	112	22-208.2	144	0.2-1264	11 nonworking women in smoking households vs. 17 nonworking women in nonsmoking households
Ethylbenzene ($\mu g/m^3$)	Baek et al. 1997	8.2	0.1-33.6	7.4	1.0-87.7	Indoor and outside background area levels for 12 Korean restaurants

Reference					Description
Baek et al. 1997	7.6	0.6–44.8	5.5	0.6–28.1	Indoor and outside background area levels for 12 Korean offices
Baek et al. 1997	5.1	0.6–17	5.1	0.1–23.1	Indoor and outside background area levels for 12 Korean residences
Bayer and Black 1987		ND–0.04		ND–22	3 office complexes, smoker/nonsmoker, 4-h sample
Guerin 1991 & Higgins et al. 1990		2.5		3.1	Residence A, remote from smoking room
Guerin 1991 & Higgins et al. 1990	3.8	2.1–5.2			4 restaurants
Guerin 1991 & Higgins et al. 1990		8.0		3.9	Residence A, smkg room
Guerin 1991 & Higgins et al. 1990		22.2			Bowling alley
Heavner et al. 1996	3.63	0.42–10.5	4.79	1.1–25.9	Personal exposure (32 and 61 subjects) in smoking and nonsmoking homes
Heavner et al. 1995	3.07	0.82–19.5	3.35	0.86–25.4	Personal exposure (25 and 24 subjects) in smoking and nonsmoking homes
Heavner et al. 1996	6.19	0.75–23.4	5.87	0.25–66.3	Personal exposure (29 and 51 subjects) in smoking and nonsmoking workplaces
Hodgson et al. 1996	1.3				Cafeteria in office bldg.
Hodgson et al. 1996	4.73	2.0–8.7			4 smoking lounges in office bldgs.
Jo and Moon 1999	9.9[b]		7.3[b]		69 indoor and 50 outdoor area samples in Korea
Proctor et al. 1989[a]	13.9	9–32	5	4–6	5 smkg betting shops, 1 nonsmoking shop

Field Studies—Volatile Organic Compounds

Constituent	Reference	Smoking		Controls		Comment
		Mean	Range[a]	Mean	Range[a]	
	Proctor et al. 1991	10	4.0-14.6	11.1	0.3-52.1	11 non-working women in smkg households vs. 17 nonworking women in nonsmoking households
	Proctor 1989	12	2-122	5	1.4-13	10 offices, smkg vs. nonsmoking, 5 smplg trips per office
	Proctor 1989	4.7	0.3-9.4	4.7	1.4-M7	Train compartments, 10 smkg and 10 nonsmoking journeys
	Wallace et al. 1987	7.9		4.5		TEAM[b] Study, New Jersey, fall
	Wallace et al. 1987	2.8		3.1		TEAM[b] Study, L.A., CA, spring
	Wallace et al. 1987	5.6		7.5		TEAM[b] Study, New Jersey, winter
	Wallace et al. 1987	3.9		4.8		TEAM[b] Study, New Jersey, summer
	Wallace et al. 1987	9.1		5.6		TEAM[d] Study, L.A., CA, winter
	Wallace et al. 1987	1.8		2.1		TEAM[d] Study, Antioch, CA, spring
Styrene (μg/m³)	Baek et al. 1997	3.9	0.6-8.3	3.9	0.3–15.6	Indoor and outside background area levels for 12 Korean residences
	Baek et al. 1997	5.0	1.6-27.4	4.0	1.6-11.8	Indoor and outside background area levels for 12 Korean offices
	Baek et al. 1997	4.1	0.1-9.3	3.4	1.1-7.2	Indoor and outside background area levels for 12 Korean restaurants
	Guerin 1991 & Higgins et al. 1990		3.0		1.6	Residence A, remote from smoking room
	Guerin 1991 & Higgins et al. 1990	2.4	1.8-4.4			4 restaurants
	Guerin 1991 & Higgins et al. 1990		7.3		2.0	Residence A, smoking room

		185			Bowling alley
Guerin 1991 & Higgins et al. 1990					Bowling alley
Heavner et al. 1996	2.88	0.0–8.40	2.12	0.0–12.9	Personal exposure (29 and 51 subjects) in smoking and nonsmoking workplaces
Heavner et al. 1996	1.41	0.16–4.94	1.57	0.14–9.90	Personal exposure (32 and 61 subjects) in smoking and nonsmoking homes
Heavner et al. 1995	2.11	0.49–7.02	1.47	0.43–4.96	Personal exposure (25 and 24 subjects) in smoking and nonsmoking homes
Hodgson et al. 1996	4.1	1.2–5.6			4 smoking lounges in office bldgs.
Hodgson et al. 1996	1.0				Cafeteria in office bldg.
Proctor et al. 1991	3.0	1.0–9.7	2.9	0.8–12.4	11 nonworking women in smoking households vs. 17 nonworking women in nonsmoking households
Proctor 1989	14	2-59	17	4-79	10 offices, smoking vs. nonsmoking, 5 smpig trips per office
Proctor et al. 1989[a]	6.8	4-11	5	4-6	5 smoking betting shops, 1 nonsmoking shop
Proctor 1989	12.1	1.4–26.4	10.9	1.1-63.8	Train compartments, 10 smoking and 10 nonsmoking journeys
Wallace et al. 1987	1.4		1.2		Team[d] study, New Jersey, summer
Wallace et al. 1987	0.4		0.6		Team[d] study, L.A., CA, spring
Wallace et al. 1987	1.5		1.0		Team[d] study, New Jersey, winter
Wallace et al. 1987	3.4		1.7		Team[d] study, L.A., CA, winter
Wallace et al. 1987	1.9		1.0		Team[d] study, New Jersey, fall
Wallace et al. 1987	0.6		0.5		Team[d] study, Antioch, CA, spring
Wallace et al. 1987	6.3		3.8		Team[d] study, NewJersey, fall

o-Xylene
(μg/m^3)

Field Studies—Volatile Organic Compounds

Constituent	Reference	Smoking Mean	Smoking Range[a]	Controls Mean	Controls Range[a]	Comment
	Wallace et al. 1987	4.6		5.5		Team[d] study, New Jersey, summer
	Wallace et al. 1987	6.1		7.2		Team[d] study, New Jersey, winter
	Wallace et al. 1987	11		7.5		Team[d] study, L.A., CA, winter
	Wallace et al. 1987	2.8		2.7		Team[d] study, L.A., CA, spring
	Wallace et al. 1987	2.2		2.8		Team[d] study, Antioch, CA, spring
	Hodgson et al. 1996	5.2	2.4–10			4 smoking lounges in office bldgs.
	Hodgson et al. 1996	1.3				Cafeteria in office bldg.
	Jo and Moon 1999	10.3[b]		9.3[b]		69 indoor and 50 outdoor area samples in Korea
	Jo and Park 1999			13.2	3.3–58.8	In-vehicle levels in Taegu, Korea
	Baek et al. 1997	13	0.1–36.6	11.2	3.1–50.2	Indoor and outside background area levels for 12 Korean restaurants
	Jo and Moon 1999	10.3[b]		9.3[b]		69 indoor and 50 outdoor area samples in Korea
	Jo and Park 1999			13.2	3.3–58.8	In-vehicle levels in Taegu, Korea
	Baek et al. 1997	13	0.1–36.6	11.2	3.1–50.2	Indoor and outside background area levels for 12 Korean restaurants
	Baek et al. 1997	14.5	1.1–75.2	9.0	1.1–48.7	Indoor and outside background area levels for 12 Korean offices
	Baek et al. 1997	9.1	1.3–29.1	8.6	0.7–36.7	Indoor and outside background area levels for 12 Korean residences
	Heavner et al. 1995	4.2	1.2–25.4	4.2	1.1–34.2	Personal exposure (25 and 24 subjects) in smoking and nonsmoking homes

m + p-Xylenes (μg/m³)

Reference					
Heavner et al. 1996	2.71	0.0–6.6	3.91	0.90–34.6	Personal exposure (32 and 61 subjects) in smoking and nonsmoking homes
Heavner et al. 1996	3.90	0.65–17.6	4.04	0.35–93.7	Personal exposure (29 and 51 subjects) in smoking and nonsmoking workplaces
Guerin 1991 & Higgins et al. 1990		15.2			Bowling alley
Guerin 1991 & Higgins et al. 1990		7.1	5.4		Residence A, smoking room
Guerin 1991 & Higgins et al. 1990		4.9	3.7		Residence A, remote from smoking room
Guerin 1991 & Higgins et al. 1990	4.1	1.5–6.7			4 restaurants
Proctor 1989	14	3–68	12	5–27	10 offices, smoking vs. nonsmoking, 5 smplg trips per office
Proctor 1989	8.6	0.4–18.4	10.7	1.3–59.7	Train compartments, 10 smoking and 10 nonsmoking journeys
Proctor et al. 1989[a]	12.4	7–25	11	11–11	5 smoking betting shops, 1 nonsmoking shop
Proctor et al. 1991	10.2	3.8–16.4	9.5	1.0–17.0	11 nonworking women in smoking households vs. 17 nonworking women in nonsmoking households
Baek et al. 1997	23.4	2.0–137	16.2	2.2–96.4	Indoor and outside background area levels for 12 Korean offices
Baek et al. 1997	22.4	0.2–71.2	20.8	4.4–99.9	Indoor and outside background area levels for 12 Korean restaurants
Baek et al. 1997	14.4	2.0–57.4	14.8	0.8–59.9	Indoor and outside background area levels for 12 Korean residences

Field Studies—Volatile Organic Compounds

Constituent	Reference	Smoking		Controls		Comment
		Mean	Range[a]	Mean	Range[a]	
	Guerin 1991 & Higgins et al. 1990		22.4		11.3	Residence A, smoking room
	Guerin 1991 & Higgins et al. 1990		12.0		7.7	Residence A, remote from smoking room
	Guerin 1991 & Higgins et al. 1990	13.2	5.4-20.6			4 restaurants
	Guerin 1991 & Higgins et al. 1990		659			Bowling alley
	Heavner et al. 1996	11.8	1.97-57.8	10.2	0.68-204	Personal exposure (29 and 51 subjects) in smoking and nonsmoking workplaces
	Heavner et al. 1995	11.2	0.96-58	10.9	2-79	Personal exposure (25 and 24 subjects) in smoking and nonsmoking homes
	Heavner et al. 1996	7.8	0.5-21.1	10.4	0.6-93.8	Personal exposure (32 and 61 subjects) in smoking and nonsmoking homes
	Hodgson et al. 1996	3.9				Cafeteria in office bldg.
	Hodgson et al. 1996	17.1	7.9 = 32			4 smoking lounges in office bldgs.
	Jo and Moon 1999	22.9[b]		15.2[b]		69 indoor and 50 outdoor area samples in Korea
	Proctor et al. 1989[a]	35.4	16-77	27	27-27	5 smoking betting shops, 1 Nonsmoking shop
	Proctor et al. 1991	30.6	4.7-102	34.0	3.4-166	11 nonworking women in smoking households vs. 17 nonworking women in nonsmoking households

	Reference					Description
	Proctor 1989	73	14-328	69	23-170	10 offices, smoking vs. nonsmoking, 5 smplg trips per office
	Proctor 1989	44.4	4.1-99.6	50.9	2.1-300	Train compartments, 10 smoking and 10 nonsmoking journeys
	Wallace et al. 1987	11		13		TEAM[d] Study, New Jersey, summer
	Wallace et al. 1987	17		21		TEAM[d] Study, New Jersey, winter
	Wallace et al. 1987	25		17		TEAM[d] Study, L.A., CA, winter
	Wallace et al. 1987	6.3		7.2		TEAM[d] Study, Antioch CA, spring
	Wallace et al. 1987	19		11		TEAM[d] Study, New Jersey, fall
	Wallace et al. 1987	11		11		TEAM[d] Study, L.A., CA, spring
Carbonyls						
Formaldehyde[e]	Löfroth et al. 1989		89-104		ND	2 studies of a tavern, 60 and 71 µg/m³ nicotine, outdoors as control
Acetaldehyde (µg/m³)	Badre et al. 1978	462	170-630			5 cafes
	Badre et al. 1978	546	65-1040			4 train areas
	Badre et al. 1978	370	260,480			Automobile, 3 smokers, natural ventilation
	Badre et al. 1978		183-204	1080		Automobile, ND ventilation, 2 smokers
	Löfroth et al. 1989				ND	2 studies of a tavern, 60 and 71 µg/m³ nicotine, 3 and 4 h sampling, outdoors as control
Propionaldehyde (µg/m³)	Badre et al. 1978	110	20-270			6 cafes
	Badre et al. 1978	83	40-120			4 train spaces
	Badre et al. 1978	25	10,40			Automobile, 3 smkrs, natural ventilation
	Badre et al. 1978		1000			Automobile, 2 smkrs, ventilation closed
Aldehydes (µg/m³)	Just et al. 1972		10-20			Public bldgs.

Field Studies—Volatile Organic Compounds

Constituent ($\mu g/m^3$)	Reference	Smoking		Controls		Comment
		Mean	Range[a]	Mean	Range[a]	
Acrolein ($\mu g/m^3$)						
	Badre et al. 1978		20			Hospital lobby
	Badre et al. 1978	68	30-100			5 cafes
	Badre et al. 1978	25	20,30			Automobile, 3 smkrs, natural ventilation
	Badre et al. 1978	70	300			Automobile, 2 smkrs, ventilation closed
	Badre et al. 1978		20-120			4 train spaces
	Fischer et al. 1978 and Weber et al. 1979	18				Restaurant B, 440 m^3
	Fischer et al. 1978 and Weber et al. 1979	16				Restaurant A, 470 m^3
	Fischer et al. 1978 and Weber et al. 1979	23				Bar, 50 M3
	Fischer et al. 1978 and Weber et al. 1979	14				Cafeteria, 574 m^3
	Löfroth et al. 1989		21-24		ND	2 studies of a tavern, 5-25 smokers, 3- and 4-h sampling, outdoors as control
Acetone ($\mu g/m^3$)	Badre et al. 1978	1130	910-1400			5 cafes
	Badre et al. 1978		1200			Automobile, 2 smokers, ventilation closed
	Badre et al. 1978		1160			Hospital lobby
	Badre et al. 1978	360	320,400			Automobile, 3 smokers, natural ventilation
	Badre et al. 1978	593	360-750			3 train spaces
	Heavner et al. 1996	71.2	19.7-665	50.1	2.8-390	Personal exposure (29 and 60 subjects) in smoking and nonsmoking homes

	Reference					Location / Comments
	Heavner et al. 1996	953	8.3–21,084	59.8	5.5–414	Personal exposure (28 and 51 subjects) in smoking and nonsmoking workplaces
2-Butanone (μg/m³)	Hodgson et al. 1996	21	16–111			Cafeteria in office bldg.
	Hodgson et al. 1996	54.5				4 smoking lounges in office bldgs.
	Badre et al. 1978	2373	880–5310			6 cafes
	Badre et al. 1978	320	300,340			Automobile, 3 smokers, natural ventilation
	Badre et al. 1978	903	170–1440			3 train spaces
	Guerin 1991 & Higgins et al. 1990	12.8	0.1–22.4			4 restaurants
	Guerin 1991 & Higgins et al. 1990		12.4	6.4		Residence A, remote from smoking room
	Guerin 1991 & Higgins et al. 1990		18.5	9.5		Residence A, smoking room
	Guerin 1991 & Higgins et al. 1990		17.7			Bowling alley
	Hodgson et al. 1996	5.75	<1–10.1			4 smoking lounges in office bldgs.
	Hodgson et al. 1996	2.1				Cafeteria in office bldg.
Ethyl acetate (μg/m³)	Badre et al. 1978	2000	1050–3560			6 cafes
	Badre et al. 1978	1675	210–4640			4 train spaces
	Badre et al. 1978	285	140,430			Automobile, 3 smokers, natural ventilation
	Badre et al. 1978		1000			Automobile, 2 smokers, no ventilation
Other Oxygenated Ethanol (μg/m³)	Badre et al. 1978	477	300–690			6 cafes
	Badre et al. 1978	1236	240–2460			4 train spaces
	Badre et al. 1978	210	200,220			Automobile, 3 smokers, natural ventilation
	Badre et al. 1978		300			Automobile, 2 smokers, No ventilation
2-Propanol (μg/m³)	Badre et al. 1978	182	150–210			5 cafes
	Badre et al. 1978	145	90–190			4 train spaces

Field Studies—Volatile Organic Compounds

Constituent	Reference	Smoking		Controls		Comment
		Mean	Range[a]	Mean	Range[a]	
	Badre et al. 1978	30	30,30			Automobile, 3 smokers, natural ventilation
	Badre et al. 1978		500			Automobile, 2 smokers, No ventilation
2-Methylfuran (μg/m³)	Guerin 1991 & Higgins et al. 1990		6.5			Bowling alley
	Guerin 1991 & Higgins et al. 1990		12.1		1.4	Residence A, smoking room
	Guerin 1991 & Higgins et al. 1990		3.8		ND	Residence A, remote from smoking room
	Guerin 1991 & Higgins et al. 1990	6.4	3.2-11.5			4 restaurants
2,5-Dimethylfuran (μg/m³)	Guerin 1991 & Higgins et al. 1990		2.1			Bowling alley
	Guerin 1991 & Higgins et al. 1990		4.9		ND	Residence A, smoking room
	Guerin 1991 & Higgins et al. 1990		1.3		ND	Residence A, ND smkq room
	Guerin 1991 & Higgins et al. 1990	1.9	1.2-13.3			4 restaurants
Nitriles						
Acetonitrile (μg/m³)	Guerin 1991 & Higgins et al. 1990		15.9			Bowling alley
	Guerin 1991 & Higgins et al. 1990		17.3		ND	Residence A, smoking room
	Guerin 1991 & Higgins et al. 1990		3.4		ND	Residence A, remote from smoking room

Compound	Reference			Location
Acrylonitrile (μg/m³)	Guerin 1991 & Higgins et al. 1990	17.5	2.4-48.9	4 restaurants
	Guerin 1991 & Higgins et al. 1990		1.8	Bowling alley
	Guerin 1991 & Higgins et al. 1990	ND	0.8	Residence A, smoking room
	Guerin 1991 & Higgins et al. 1990	ND	0.6	Residence A, remote from smoking room
Propionitrile (μg/m³)	Guerin 1991 & Higgins et al. 1990	0.6	0.1-1.9	4 restaurants
	Guerin 1991 & Higgins et al. 1990		2.7	Bowling alley
	Guerin 1991 & Higgins et al. 1990	ND	3.6	Residence A, smoking room
	Guerin 1991 & Higgins et al. 1990	ND	0.9	Residence A, remote from smoking room
Butyronitrile (μg/m³)	Guerin 1991 & Higgins et al. 1990	2.6	0.3-7.0	4 restaurants
	Guerin 1991 & Higgins et al. 1990		0.7	Bowling alley
	Guerin 1991 & Higgins et al. 1990	ND	1.1	Residence A, smoking room
	Guerin 1991 & Higgins et al. 1990	ND	0.3	Residence A, remote from smoking room
Isobutyronitrile (μg/m³)	Guerin 1991 & Higgins et al. 1990	0.5	<0.1-1.3	4 restaurants
	Guerin 1991 & Higgins et al. 1990		0.6	Bowling alley
	Guerin 1991 & Higgins et al. 1990	ND	1.2	Residence A, smoking room
	Guerin 1991 & Higgins et al. 1990	ND	0.3	Residence A, remote from smoking room

Field Studies—Volatile Organic Compounds

Constituent	Reference	Smoking		Controls		Comment
		Mean	Range[a]	Mean	Range[a]	
Valeronitrile (μg/m³)	Guerin 1991 & Higgins et al. 1990	0.6	0.1-1.5			4 restaurants
	Guerin 1991 & Higgins et al. 1990		0.5			Bowling alley
	Guerin 1991 & Higgins et al. 1990		1.2		ND	Residence A, smoking room
	Guerin 1991 & Higgins et al. 1990		0.5		ND	Residence A, remote from smoking room
	Guerin 1991 & Higgins et al. 1990	0.5	ND-1.4			4 restaurants
Nitrogen Heterocyclics						
Pyridine (μg/m³)	Guerin 1991 & Higgins et al. 1990		6.5		0.2	Residence A, smoking room
	Guerin 1991 & Higgins et al. 1990		0.6		ND	Residence A, remote from smoking room
	Guerin 1991 & Higgins et al. 1990		3.8			Bowling alley
	Guerin 1991 & Higgins et al. 1990	5.0	0.8-15.7			4 restaurants
	Heavner et al. 1996	1.34	0.0-4.86	0.43	0.0-4.13	Personal exposure (32 and 61 subjects) in smoking and nonsmoking homes
	Heavner et al. 1996	1.68	0.0-5.34	0.89	0.0-7.56	Personal exposure (29 and 51 subjects) in smoking and nonsmoking workplaces
	Heavner et al. 1995	2.34	0.0-8.59	0.67	0.01-1.86	Personal exposure (25 & 24 subjects) in smoking and nonsmoking homes

Reference					Description
Hodgson et al. 1996	3.7	<1.5–7.1			4 smoking lounges in office bldgs.
Hodgson et al. 1996	<1.5				Cafeteria
Proctor et al. 1991	1.6	0.4–8.9	0.5	0.2–2.2	11 nonworking women in smoking household vs. 17 nonworking women in nonsmoking households

3-Ethenylpyridine ($\mu g/m^3$)

Reference					Description
Eatough et al. 1989		0.6–0.9			2 restaurants
Guerin 1991 & Higgins et al. 1990[i]		ND		ND	Residence A, remote from smoking room
Guerin 1991 & Higgins et al. 1990[i]	3.2	0.2–8.4			4 restaurants
Guerin 1991 & Higgins et al. 1990[i]		3.6			Bowling alley
Guerin 1991 & Higgins et al. 1990[i]		8.4		ND	Residence A, smoking room
Heavner et al. 1996	0.79	0.0–3.62	0.0	0.0–0.09	Personal exposure (32 and 61 subjects) in smoking and nonsmoking homes
Heavner et al. 1995	1.28	0.0–5.58	0.08	0.0–0.57	Personal exposure (25 & 24 subjects) in smoking and nonsmoking homes
Heavner et al. 1996	0.68	0.0–3.50	0.02	0.0–0.59	Personal exposure (29 and 51 subjects) in smoking and nonsmoking workplaces
Hodgson et al. 1996	2.1				Cafeteria in office bldg.
Hodgson et al. 1996	7.89	<1.5–13.3			4 smoking lounges in office bldgs.
Jenkins et al. 1996	1.15	3.41	0.065	0.20	Personal exposure (306 and 899 subjects) away from work in smoking and nonsmoking homes—16-h samples

Field Studies—Volatile Organic Compounds

Constituent	Reference	Smoking		Controls		Comment
		Mean	Range[a]	Mean	Range[a]	
	Jenkins and Counts 1999	3.5	10.4[g]			Area samples in 53 bar areas
	Jenkins et al. 1996	0.34	3.84[g]	0.09	0.28[g]	Personal exposure (331 and 867 subjects) in smoking and nonsmoking workplaces—8-h samples
	Jenkins and Counts 1999	3.3	10.3[g]			Personal exposures of 80 bartenders in ~40 facilities
	Jenkins and Counts 1999	1.8	6.7[g]			Personal exposures of 82 wait staff in ~30 facilities
	Jenkins and Counts 1999	1.4	5.4[g]			Area samples in 32 restaurants
	Ogden 1995	0.95[b]				5-day personal breathing zone levels of 48 male smokers
	Ogden 1995			0.02[b]		5-day personal breathing zone levels of 48 females married to nonsmokers
	Ogden 1995	0.44[b]				5-day personal breathing zone levels of 47 females married to smokers
	Philips et al. 1997	0.96	2.0[c]	0.14	0.48[c]	Personal exposure (28 and 41 subjects) away from work in smoking and nonsmoking homes (Italy)
	Phillips et al. 1999	0.53	1.1[c]	0.17	0.25[c]	Personal exposure (65 and 41 subjects) in smoking and nonsmoking workplaces (Switzerland)

Reference					Description
Phillips et al. 1999	0.53	1.5[c]	0.15	0.32[c]	Personal exposure (65 and 41 subjects) away from work in smoking and nonsmoking homes (Switzerland)
Phillips et al. 1998	1.4	3.0[c]	0.37	0.77[c]	Personal exposure (109 and 32 subjects) in smoking and nonsmoking workplaces (Czech Republic)
Phillips et al. 1998	0.69	1.4[c]	0.20	0.26[c]	Personal exposure (71 and 66 subjects) away from work in smoking and nonsmoking homes (Czech Republic)
Phillips et al. 1997	0.83	2.1[c]	0.24	0.44[c]	Personal exposure (72 and 33 subjects) in smoking and nonsmoking workplaces (Italy)
Phillips et al. 1997	0.90	2.0[c]	0.13	0.25[c]	Personal exposure (29 and 75 subjects) away from work in smoking and nonsmoking homes (Italy)
Phillips et al. 1997	1.5	2.8[c]	0.63	1.7[c]	Personal exposure (61 and 8 subjects) in smoking and nonsmoking workplaces (Spain)
Phillips et al. 1998	1.1	1.6[c]	0.21	0.35[c]	Personal exposure (40 and 69 subjects) in smoking and nonsmoking workplaces (Hong Kong)
Phillips et al. 1998	0.87	2.1[c]	0.27	0.68[c]	Personal exposure (102 and 20 subjects) in smoking and nonsmoking workplaces (France)

Field Studies—Volatile Organic Compounds

Constituent	Reference	Smoking		Controls		Comment
		Mean	Range[a]	Mean	Range[a]	
	Phillips et al. 1998	0.61	1.4[c]	0.10	0.18[c]	Personal exposure (56 and 65 subjects) away from work in smoking and nonsmoking homes (Italy)
	Phillips et al. 1998	0.41	0.86[c]	0.15	0.19[c]	Personal exposure (20 and 57 subjects) in smoking and nonsmoking workplaces (Australia)
	Phillips et al. 1998	0.39	0.92[c]	0.07	0.11[c]	Personal exposure (30 and 46 subjects) away from work in smoking and nonsmoking homes (Australia)
	Phillips et al. 1998	0.63	1.7[c]	0.20	0.33[c]	Personal exposure (63 and 42 subjects) in smoking and nonsmoking workplaces (Germany)
	Phillips et al. 1998	0.51	1.2[c]	0.14	0.32[c]	Personal exposure (23 and 82 subjects) away from work in smoking and nonsmoking homes (Germany)
	Phillips et al. 1998	0.43	0.96[c]	0.25	0.50[c]	Personal exposure (74 and 71 subjects) in smoking and nonsmoking workplaces (Malaysia)
	Phillips et al. 1998	0.13	0.21[c]	0.13	0.19[c]	Personal exposure (59 and 87 subjects) away from work in smoking and nonsmoking homes (Malaysia)

Reference					Description
Phillips et al. 1998	0.19	0.39[c]	0.09	0.20[c]	Personal exposure (24 and 53 housepersons) in smoking and nonsmoking homes (Portugal)
Phillips et al. 1998	0.53	1.00[c]	0.13	0.24[c]	Personal exposure (54 and 44 housepersons) in smoking and nonsmoking homes (China)
Phillips et al. 1998	0.35	0.78[c]	0.09	0.13[c]	Personal exposure (48 and 68 subjects) away from work in smoking and nonsmoking homes (Hong Kong)
Proctor et al. 1989[h]	1	0.4–3	3	0.6–7	2 30 m³ offices, 1 occupied by smoker 1 by nonsmoker
Proctor et al. 1989[h]	1	0.1–4	1	0.1–2	2 60 m³ offices, 1 occupied by smoker 1 by nonsmoker
Sterling et al. 1996	0.91	0.7–1.3			Area measurements in 2 bldgs. (16 locations) where smoking was unrestricted
Sterling et al. 1996	0.92	0.5–1.3			Personal exposures of 25 subjects in 2 bldgs. where smoking was unrestricted
Picoline (µg/m³)					
Guerin 1991 & Higgins et al. 1990[i]		4.2			Bowling alley
Guerin 1991 & Higgins et al. 1990[i]		4.0		ND	Residence A, smoking room
Guerin 1991 & Higgins et al. 1990[i]		ND		ND	Residence A, remote from smoking room
Guerin 1991 & Higgins et al. 1990[i]	1.3	ND-3.5			4 restaurants
Heavner et al. 1995 (2-Picoline)	0.45	0.0–1.55	0.07	0.0–0.67	Personal exposure (25 & 24 subjects) in smoking and nonsmoking homes
Heavner et al. 1995 (3-Picoline)	0.68	0.0–2.40	0.14	0.0–0.51	Personal exposure (25 and 24 subjects) in smoking and nonsmoking homes

Field Studies—Volatile Organic Compounds

Constituent	Reference	Smoking		Controls		Comment
		Mean	Range[a]	Mean	Range[a]	
	Heavner et al. 1995 (4-Picoline)	0.16	0.0–0.94	0.09	0.0–0.69	Personal exposure (25 and 24 subjects) in smoking and nonsmoking homes
	Heavner et al. 1996 (2-Picoline)	0.27	0.0–1.92	0.0	0.0–0.16	Personal exposure (32 and 61 subjects) in smoking and nonsmoking homes
	Heavner et al. 1996 (4-Picoline)	0.05	0.0–0.45	0.0	0.0–0.18	Personal exposure (32 and 61 subjects) in smoking and nonsmoking homes
	Heavner et al. 1996 (2-Picoline)	0.26	0.0–1.51	0.0	0.0–0.09	Personal exposure (29 and 51 subjects) in smoking and nonsmoking workplaces
	Heavner et al. 1996 (3-Picoline)	0.30	0.0–1.70	0.01	0.0–0.60	Personal exposure (29 and 51 subjects) in smoking and nonsmoking workplaces
	Heavner et al. 1996 (4-Picoline)	0.05	0.0–0.77	0.0	0.0–0.0	Personal exposure (29 and 51 subjects) in smoking and nonsmoking workplaces
Pyrrole ($\mu g/m^3$)	Guerin 1991 & Higgins et al. 1990		5.4			Bowling alley
	Guerin 1991 & Higgins et al. 1990		10.6		ND	Residence A, smoking room
	Guerin 1991 & Higgins et al. 1990		0.6		ND	Residence A, remote from smoking room
	Guerin 1991 & Higgins et al. 1990		<0.1–12.1			4 restaurants
	Hodgson et al. 1996	4.95	< 1–9.4			4 smoking lounges in office bldgs
	Hodgson et al. 1996	2.1				Cafeteria in office bldg.

[a] A single result indicates a single observation.
[b] Median.
[c] 90 Percentile.
[d] Also see Chapter 10 and Appendix 5.
[e] Given as 3-vinylpyridine plus 4-vinylpyridine.
[f] 9th Percentile.
[g] Given as 2-vinylpyridine.
[h] 3- plus 4-picoline.
ND: Not Detected. Single result indicates a single observation.
[b] Smoking versus Nonsmoking residences, 6 p.m. –6 a.m., weighted geometric means, personal sampling.
[c] Also see Chap. 10 and Appendix 5.
[d] Given as 3-vinylpyridine plus 4-vinylpyridine.
[e] Given as 2-vinylpyridine.
[f] 3- plus 4-picoline.
[g] Median.
ND: Not Detected.

Field Studies—Polycyclic Aromatic Hydrocarbons

		Concentration, ng/m^3				
		Smoking		Controls		
Constituent	Reference	Mean	Range	Mean	Range	Comment
Naphthalene						
	Chuang et al. 1999			2190	334–9700	Indoor air of 24 low-income homes
	Chuang et al. 1999			433	57–1820	Outdoor air of 24 low-income homes
	Mumford et al. 1990			2300		Kerosene heater off, XAD trap sampler
	Mumford et al. 1990			950		Kerosene heater on, XAD trap sampler
	Wilson et al. 1989				860 ± 240	Vented room, N=10
	Wilson et al. 1989				1160 ± 360	Nonvented room, N=10
Acenaphthylene						
	Chuang et al. 1999			44	2–533	Indoor air of 24 lo- income homes
	Chuang et al. 1999			7.6	0.6–34	Outdoor air of 24 low-income homes
	Mumford et al. 1990			19		Kerosene heater off, XAD trap sampler
	Mumford et al. 1990			68		Kerosene heater on, XAD trap sampler
	Wilson et al. 1989				17 ± 3	Nonvented room, N=10
	Wilson et al. 1989				15 ± 7	Vented room, N=10
Acenaphthylene, dihydro						
	Wilson et al. 1989				120 ± 50	Vented room, N=10
	Wilson et al. 1989				120 ± 20	Nonvented room,N=10
Phenanthrene						
	Chuang et al. 1999			66	10–184	Indoor air of 24 low-income homes
	Chuang et al. 1999			30	4.8–147	Outdoor air of 24 low-income homes
	Guenther et al. 1988			191		Outdoor air, Fairbanks, AK
	Husgafvel–Pursiainen et al. 1986	4	2.26–6			3 restaurants, 2 sites per restaurant
	Mumford et al. 1990			43		Kerosene heater off, XAD trap sampler
	Mumford et al. 1990			33		Kerosene heater on, XAD trap sampler
	Wilson et al. 1989				210 ± 60	Vented room, N=10

Field Studies—Polycyclic Aromatic Hydrocarbons

Constituent	Reference	Concentration, ng/m³				Comment
		Smoking		Controls		
		Mean	Range	Mean	Range	
	Wilson et al. 1990	87			11–210	Smoking, gas heat and stove. Range is for all homes
	Wilson et al. 1989			31		Nonsmoking, gas heat and stove
	Wilson et al. 1989			19		Nonsmoking, electric heat and stove
	Wilson et al. 1989				240 ± 90	Nonvented room, N=10
Anthracene	Chuang et al. 1999			6.3	0.82–17	Indoor air of 24 low-income homes
	Chuang et al. 1999			1.94	0.22–11	Outdoor air of 24 low-income homes
	Guenther et al. 1988			24.6		Outdoor air, Fairbanks, AK
	Wilson et al. 1989				9.6 ± 2.6	Nonvented room, N=10
	Wilson et al. 1989				9.6 ± 4.2	Vented room, N=10
Fluoranthene	Chuang et al. 1999			8.6	1.2–19.3	Indoor air of 24 low-income homes
	Chuang et al. 1999			5.8	1.1–23.1	Outdoor air of 24 low-income homes
	Grimmer et al. 1977	99	80–116	50		In sidestream smoke, 36 m³ room, 5 cigt/hr
	Guenther et al. 1988			72.3		Outdoor air, Fairbanks, AK
	Gundel et al. 1990	0.222		<0.037		Background is outdoors
	Gundel et al. 1990	0.855		0.116		Background is outdoors
	Husgafvel–Pursiainen et al. 1986	8	2.9–13.3			3 restaurants, 2 sites per restaurant
	Mumford et al. 1990			14		Kerosene heater off, XAD trap sampler
	Mumford et al. 1990			9.5		Kerosene heater on, XAD trap sampler
	Wilson et al. 1989				22 ± 10	Vented room, N=10
	Wilson et al. 1990	8.6			2.4–23	Smoking, gas heat and stove. Range for all homes
	Wilson et al. 1989				20 ± 2	Nonvented room, N=10
	Wilson et al. 1989			3.7		Nonsmoking, gas heat and stove

Pyrene					
Wilson et al. 1989			2.99		Nonsmoking, electric heat and stove
Chuang et al. 1999	66		2.95	0.54–10.7	Outdoor air of 24 low-income homes
Grimmer, et al. 1977		53–84	26		5 cigt/hr, 36 m³ room, 1 air change. Sidestream smoke
Guenther et al. 1988			72.3		Outdoor air, Fairbanks, AK
Gundel et al. 1990	0.82		<0.11		Background is outdoors
Husgafvel–Pursiainen et al. 1986	7	2.7–11.8			3 restaurants, 2 sites per restaurant
Just et al. 1972; Gold et al. 1990		4.1–9.4		0.1–1.7	Coffee houses; background is outdoors
Matsushita et al. 1990	0.92	0.52–1.94	0.75	0.36–1.01	7 living rooms in Kawasaki, Japan. Outdoors is background
Matsushita et al. 1990	1.64	0.53–4.29	1.58	0.57–3.69	10 kitchens, Manila, Philippines. Outdoors is background
Matsushita et al. 1990	1.67	0.43–5–0 4	1.58	0.57–3.69	10 living rooms, Manila, Philippines. Outdoors is background
Matsushita et al. 1990	0.91	0.52–1.85	0.75	0.36–1.01	7 kitchens, Kawasaki, Japan. Outdoors is background
Mumford et al. 1990			8.3		Kerosene heater off, XAD trap sampler
Mumford et al. 1990			12		Kerosene heater on, XAD trap sampler
Wilson et al. 1989			2.8		Nonsmoking, gas heat and stove
Wilson et al. 1990	5.2			1.5–18.1	Smoking, gas heat and stove. Range for all homes
Wilson et al. 1989				11 ± 5	Vented room
Wilson et al. 1989				9.1 ± 1.2	Nonvented room
Wilson et al. 1989			2.13		Nonsmoking, electric heat and stove
Benz[a]anthracene					
Chuang et al. 1999			0.59	0.11–3.13	Indoor air of 24 low-income homes
Chuang et al. 1999			0.44	0.07–2.18	Outdoor air of 24 low-income homes
Grimmer et al. 1977	100	90–107	31		Sidestream smoke. Includes chrysene
Guenther et al. 1988			14.4		Outdoor air, Fairbanks, AK
Gundel et al. 1990	0.127		0.055		Background is outdoor air
Gundel et al. 1990	0.516		<0.043		Background is outdoor air

Field Studies—Polycyclic Aromatic Hydrocarbons

Constituent	Reference	Concentration, ng/m³				Comment
		Smoking		Controls		
		Mean	Range	Mean	Range	
	Chuang et al. 1999			6.73	0.80–29.4	Indoor air of 24 low-income homes
	Husgafvel–Pursiainen et al. 1986	5	1.8–9.3			3 restaurants, 2 sites per restaurant
	Matsushita et al. 1990	0.48	0.22–1.58	0.48	0.21–0.63	7 living rooms in Kawasaki, Japan. Background is outdoors
	Matsushita et al. 1990	0.54	0.17–1.62	0.48	0.21–0.63	7 kitchens, Kawasaki, Japan. Background is outdoors
	Matsushita et al. 1990	1.48	0.34–4.45	1.36	0.43–2.49	10 kitchens, Manila, Philippines. Background is outdoors
	Matsushita et al. 1990	1.47	0.25–4.35	1.36	0.43–2.49	10 living rooms, Manila, Philippines. Background is outdoors
	Mumford et al. 1990			0.72		Kerosene heater off, includes particle– and vapor-phase PAH
	Mumford et al. 1990			2.8		Kerosene heater on, includes particle– and vapor-phase PAH
	Wilson et al. 1989				0.22 ± 0.10	Vented room
	Wilson et al. 1990	1.4			0.085–3.4	Smoking, gas heat and stove. Range is for all homes
	Wilson et al. 1989				0.24 ± 0.12	Nonvented room
	Wilson et al. 1989			0.46		Nonsmoking, gas heat and stove
	Wilson et al. 1989			0.2		Nonsmoking, electric heat and stove
Chrysene	Guenther et al. 1988	3.58		18.7		Outdoor air, Fairbanks, AK
	Gundel et al. 1990	5.54		0.043		Background is outdoors
	Gundel et al. 1990			0.094		Background is outdoors

Reference					Comments
Matsushita et al. 1990	2.13	0.42–6.19	1.88	0.67–3.5	10 living rooms, Manila, Philippines. Background is outdoors
Husgafvel–Pursiainen et al. 1986	16	6.9–29.1			3 restaurants, 2 sites per restaurant. Includes triphenylene
Mumford et al. 1990			1.49		Kerosene heater, particle- and vapor-phase, on
Mumford et al. 1990			3.1		Kerosene heater, particle- and vapor-phase, off
Turk et al. 1987			0.53 ± 94[a]	0.00–4.23	Outdoor concentrations, 20 measurements
Turk et al. 1987			1.23 ± 2.14[a]	0–10.01	Indoor concentrations, 31 observations
Wilson et al. 1989			0.67		Nonsmoking, gas heat and stove
Wilson et al. 1989			0.37		Nonsmoking, electric heat and stove
Wilson et al. 1990	2.6			0.18–7.2	Smoking, gas heat and stove. Range is for all homes
Wilson et al. 1989				0.45 ± 0.18	Nonvented room

Benzo[b/j/k]fluoranthenes

Reference					Comments
Grimmer et al. 1977	35	29–39	5		Sidestream smoke, all 3 isomers. 36 m³ room, 5 cigt/h
Guenther et al. 1988			19.9		Outdoor air, Fairbanks, AK. All 3 isomers
Gundel et al. 1990	0.148		0.021		Background is outdoors. Benzo[k]fluoranthene
Gundel et al. 1990	0.423		0.097		Background is outdoors. Benzo[b]fluoranthene
Gundel et al. 1990	0.578		0.098		Background is outdoors. Benzo[b]fluoranthene
Gundel et al. 1990	0.132		0.042		Background is outdoors. Benzo[k]fluoranthene
Husgafvel–Pursiainen et al. 1986	2	ND–4.9			3 restaurants, 2 sites per restaurant
Husgafvel–Pursiainen et al. 1986					Benzo(k)f luoranthene

Field Studies—Polycyclic Aromatic Hydrocarbons

Constituent	Reference	Concentration, ng/m³				Comment
		Smoking		Controls		
		Mean	Range	Mean	Range	
Benzo[b/j]fluoranthenes						
	Husgafvel-Pursiainen et al. 1986	9	3.6–19.6			3 restaurants, 2 sites per restaurant
	Matsushita et al. 1990	1.95	0.67–4.65	1.88	0.67–3.5	10 kitchens, Manila, Philippines. Background is outdoors
	Matsushita et al. 1990	1.18	0.42–2.52	1.19	0.44–1.84	7 kitchens, Kawasaki, Japan. Background is outdoors
	Matsushita et al. 1990	1.13	0.4–2.43	1.19	0.44–1.84	7 living rooms in Kawasaki, Japan. Background is outdoors
	Mumford et al. 1990			0.45		Kerosene heater off, particulate phase. Total isomers
	Mumford et al. 1990			5.5		Kerosene heater on, particulate phase, Total isomers
	Turk et al. 1987			0.64 ± 1.49[a]	0.00–8.29	Indoor air, 31 observations. Benzo(k)fluoranthene
	Turk et al. 1987			0.86 ± 0.95[a]	0.00–3.72	Outdoor air, 20 measurements. Benzo(b)fluoranthene
	Turk et al. 1987			1.54 ± 3.35[a]	0.00–18.76	Indoor air, 31 observations. Benzo(b)fluoranthene
	Turk et al. 1987			0.37 ± 0.44[a]	0.00–1.83	Outdoor air, 20 measurements. Benzo(k)fluoranthene
	Wilson et al. 1990	2.5			0.18–5.83	Smoking, gas heat and stove. All 3 isomers
	Wilson et al. 1989			0.92		Nonsmoking, gas heat and stove. All 3 isomers
	Wilson et al. 1989'			0.47		Nonsmoking, electric heat and stove. All 3 isomers

Benzo[e]pyrene

Reference					Comments
Chuang et al. 1999			1.05	0.20–5.28	Indoor air of 24 low-income homes
Chuang et al. 1999			0.73	0.17–2.31	Outdoor air of 24 low-income homes
Chuang et al. 1999			0.73	0.17–2.31	Outdoor air of 24 low-income homes
Grimmer et al. 1977	18	16–21	<2		Sidestream smoke, 36 m³ room, 5 cigt/h
Guenther et al. 1988			7.2		Outdoor air, Fairbanks, AK
Gundel et al. 1990	0.39		<0.21		Background is outdoors
Gundel et al. 1990	0.93		0.19		Background is outdoors
Husgafvel–Pursiainen et al. 1986	3	1.3–6			3 restaurants, 2 sites per restaurant
Just et al. 1972; Gold et al. 1990		3.3–23.4		3.0–5.1	Coffee houses. Background is outdoors
Triebig and Zober 1984	18	<2–21			Working room
Triebig and Zober 1984					Cafes
Triebig and Zober 1984	9.9	3.3–23.4			Arena
Wilson et al. 1989				0.20± 0.11	Vented room, N=10
Wilson et al. 1989				0.27 ± 0.12	Nonvented room, N=10

Benzo[a]pyrene

Reference					Comments
Adlkofer et al. in Bieva et al. 1989		3–25			Room air
Chuang et al. 1999			0.46	0.04–4.33	Outdoor air of 24 low-income homes
Elliott and Rowe 1975	9.9		0.69		Arena, 12000–13000 attend
Elliott and Rowe 1975	21.7		0.69		Arena, 13000–14000 attend
Elliott and Rowe 1975	12.5		0.69		Arena, 12000 attend
Elliott and Rowe 1975	7.1		0.69		Arena, 8700–11000 attend
Galuskinova 1964		28–144		2.8–46	Smoky social rooms. Control is Prague atmosphere
Gold et al. 1990		2.8–760		4.0–9.3	Restaurant, public places. Background is outdoors
Grimmer et al. 1977	22	16–28	<3		5 cigts/hr, 36 m³ room, 1 air change. Sidestream smoke
Grimmer and Naujack 1987		88–214		0–68	Smoker's room vs. ventilated room. 30 cigts
Chuang et al. 1999			0.73	0.17–2.31	Outdoor air of 24 low-income homes
Grimmer et al. 1977	18	16–21	<2		Sidestream smoke, 36 m³ room, 5 cigt/h
Guenther et al. 1988			7.2		Outdoor air, Fairbanks, AK

Field Studies—Polycyclic Aromatic Hydrocarbons

Constituent	Reference	Smoking		Controls		Comment
		Mean	Range	Mean	Range	
	Gundel et al. 1990	0.39		<0.21		Background is outdoors
	Gundel et al. 1990	0.93		0.19		Background is outdoors
	Husgafvel–Pursiainen et al. 1986	3	1.3–6			3 restaurants, 2 sites per restaurant
	Just et al. 1972; Gold et al. 1990		3.3–23.4		3.0–5.1	Coffee houses. Background is outdoors
	Triebig and Zober 1984	18	<2–21			Working room
	Guenther et al. 1988			11.3		Outdoor air, Fairbanks, AK
	Guerin 1988		0.2–10			Common smoking conditions
	Guerin 1988		10–20			Heavy smoking conditions
	Gundel et al. 1990	0.38		0.03		Control is outdoors
	Gundel et al. 1990	0.605		0.094		Control is outdoors
	Husgafvel–Pursiainen et al. 1986	7	2.2–13.3			3 restaurants, 2 stations per restaurant
	Mahajan and Huber 1990		3.3–23.4			Restaurants and public places
	Matsushita et al. 1990	2.2	0.42–4.79	1.84	0.56–2.64	7 living rooms in Kawasaki, Japan. Background is outdoors
	Matsushita et al. 1990	2.28	0.66–5.05	1.84	0.56–2.64	7 kitchens, Kawasaki, Japan. Background is outdoors
	Matsushita et al. 1990	3.7	0.96–10.14	2.54	0.81–4.35	10 kitchens, Manila, Philippines. Background is outdoors
	Matsushita et al. 1990	3.34	0.57–9.64	2.54	0.81–4.35	10 living rooms, Manila, Philippines. Background is outdoors
	Mumford et al. 1990			0.24		Kerosene heater, off, particulate phase
	Mumford et al. 1990			2		Kerosene heater, on, particulate phase
	Triebig and Zober 1984	6.2	28–44			Restaurant

Reference					Description
Triebig and Zober 1984	7.1	28–44			Arena
Triebig and Zober 1984		0.25–10.1			Cafes
Triebig and Zober 1984	22	<3–25			Working room
Turk et al. 1987	1.07	ND–3.60	0.39±	ND–1.35	18 smoking/nonsmoking sites
Turk et al. 1987			0.54±	0.04–0.11	Outdoors, 20 measurements
Turk et al. 1987			0.54±	0.04–0.11	Outdoors, 20 measurements
Wilson et al. 1990	1.2		0.69[a]	0–3.33	Smoking, gas heat and stove. Range is for all homes
Wilson et al. 1989			0.63		Nonsmoking home, gas heat and stove
Wilson et al. 1989			0.28		Nonsmoking home, electric heat and stove
Wilson et al. 1989				0.25 ±0.17	Nonvented room, 10 samples
Wilson et al. 1989				0.16 ±0.12	Vented room, 10 samples
Yocom 1982	12		<1		Indoors, fireplace. Control is outdoors
Yocom 1982	4.7		<1		Indoors, woodstove. Control is outdoors

Dibenz[a,h]anthracene

Reference					Description
Chuang et al. 1999			0.54	0.02–2.75	Indoor air of 24 low-income homes
Chuang et al. 1999			0.17	0.02–0.39	Outdoor air of 24 low-income homes
Grimmer et al. 1977	13	10–14	3		Sidestream smoke. Also includes indeno[1,2,3–cd]pyrene
Gundel et al. 1990	<0.11		<0.11		Control is outdoors
Gundel et al. 1990	0.18		<0.11		Control is outdoors
Turk et al. 1987			0.20 ±0.18[a]	0.00–0.68	Indoor air, 31 measurements
Turk et al. 1987			0.20 ±0.17[a]	0.00–0.60	Outdoor air, 20 measurements
Wilson et al. 1989			0.12 ± 0.09		Vented room, 10 samples
Wilson et al. 1989			0.17 ± 0.11		Nonvented room, 10 samples

Field Studies—Polycyclic Aromatic Hydrocarbons

Constituent	Reference	Smoking (Concentration, ng/m³)		Controls		Comment
		Mean	Range	Mean	Range	
Benzo(ghi)perylene						
	Grimmer et al. 1977	17	15–18	3		Sidestream smoke, 36 m³ room, 5 cigt/h
	Guenther et al. 1988			13.7		Outside air, Fairbanks, AK
	Gundel et al. 1990	0.44		0.22		Control is outdoors
	Gundel et al. 1990	0.88		0.28		Control is outdoors
	Just et al. 1972; Gold et al. 1990		5–25			Restaurant, public places. No control
	Matsushita et al. 1990	8.14	1.44–24.96	6.69	1.94–13.18	10 living rooms, Manila, Philippines. Background is outdoors
	Matsushita et al. 1990	3.89	1.58–6.07	4.02	1.86–5.61	7 kitchens, Kawasaki, Japan. Background is outdoors
	Matsushita et al. 1990	3.85	1.71–6.03	4.02	1.86–5.61	7 living rooms in Kawasaki, Japan. Background is outdoors
	Mumford et al. 1990			0.22		Kerosene heater, off, particulate phase
	Mumford et al. 1990			3.7		Kerosene heater, on, particulate phase
	Turk et al. 1987			1.05 ± 1.14[a]	0.00–4.26	Outdoor air, 20 measurements
	Turk et al. 1987			0.96 ± 0.82[a]	0.00–2.94	Indoor air, 31 measurements
	Wilson et al. 1989			0.88		Nonsmoking home, gas heat and stove
	Wilson et al. 1989				0.49 ± 0.26	Nonvented room, 10 samples
	Wilson et al. 1989				0.31 ± 0.24	Vented room, 10 samples
	Wilson et al. 1989			0.53		Nonsmoking home, electric heat and stove
	Wilson et al. 1990	1.4			0–4.5	Smoking, gas heat and stove. Range is for all homes
	Grimmer et al. 1977	2	< 0–3	0		Sidestream smoke, 36 m³ room, 5 cigt/h

Compound	Reference					Comments
Anthanthrene	Guenther et al. 1968			5.4		Outside air, Fairbanks, AK
	Just et al. 1972; Gold et al. 1990		0.5–9.4		2.8–7.0	Coffee houses. Background is outdoors
	Triebig and Zober 1984	21.7				Arena
	Triebig and Zober 1984	3	0.5–1.9			Cafes
	Triebig and Zober 1984		< 3–3			Working room
Indeno[1,2,3-cd]pyrene	Guenther et al. 1988			9.5		Outdoor air, Fairbanks, AK
	Gundel et al. 1990			0.19		Control is outdoors.
	Gundel et al. 1990	0.55		0.2		Control is outdoors
	Mumford et al. 1990			0.15		Kerosene heater, off, particulate phase
	Mumford et al. 1990			1.3		Kerosene heater, on, particulate phase
	Turk et al. 1987			0.84 ± 0.76[a]	0.00–3.28	Indoor air, 31 measurements
	Turk et al. 1987			0.97 ± 1.27[a]	0.00–5.28	Outdoor air, 20 measurements
	Wilson et al. 1990	0.97			0–2.6	Smoking, gas heat and stove. Range is for all homes
	Wilson et al. 1990			0.82		Nonsmoking home, gas heat and stove
	Wilson et al. 1990			0.32		Nonsmoking home, electric heat and stove
Coronene	Guenther et al. 1988			9.8		Outdoor air, Fairbanks, AK
	Gundel et al. 1990	0.54		0.29		Control is outdoors
	Gundel et al. 1990	< 0.01		0.12		Control is outdoors
	Just et al. 1972, Gold et al. 1990		0.6–1.2		1.0–2.8	Coffee houses. Background is outdoors
	Mumford et al. 1990			2.3		Kerosene heater, on, particulate phase
	Mumford et al. 1990			0.069		Kerosene heater, off, particulate phase
	Wilson et al. 1990	0.6	0–3.7			Smoking, gas heat and stove. Range is for all homes
	Wilson et al. 1990			1.3		Nonsmoking home, gas heat and stove
	Wilson et al. 1990			0.52		Nonsmoking home, electric heat and stove

Field Studies—Polycyclic Aromatic Hydrocarbons

Constituent	Reference	Concentration, ng/m³				Comment
		Smoking		Controls		
		Mean	Range	Mean	Range	
	Wilson et al. 1989				0.19 ± 0.18	Vented room, 10 samples
	Wilson et al. 1989				0.39 ± 0.21	Nonvented room, 10 samples
Perylene	Just et al. 1972; Gold et al. 1990		0.7–1.3		0.1–1.7	Coffee houses. Background is outdoors

[a]Represents mean standard deviation. ND: Not Detected

Appendix References

Adlkofer, F. X., Scherer, G., Von Meyerinck, L., Von Maltzan, CH., & Jarczyk, L. (1989) Exposure to ETS and its biological effects: a review. In: Bieva, C. J., Courtious, Y., & Govaerts, M., eds. *Present and Future of Indoor Air Quality*, Elsevier Science Publishers, pp. 183-196.

Akbar-Khanzadeh, F. & Greco T. M. (1996) Health and social concerns of restaurant/bar workers exposed to environmental tobacco smoke. *Med. Lav.*, 87, 122-132.

Aviado, D. M. (1984) Carbon monoxide as an index of environmental tobacco exposure. In: Rylander, R., Peterson, Y, & Snella, M.-C., eds., *ETS–Environmental Tobacco Smoke. Report from a Workshop on Effects on Exposure Levels*, Switzerland, University of Geneva, pp. 47-60. Published simultaneously as *Eur. J. of Resp. Dis. Supp.* (1984), No. 133, 65.

Badre, R., Guillerm, R., Abran, N., Bourdin, M., & Dumas, C. (1978) Atmospheric pollution by smoking. *Annales Pharmaceutiques Francaises*, 36(9-10), 443-452.

Baek, S. O., Kim, U. S., & Perry, R. (1997) Indoor Air quality in homes, offices, and restaurants in Korean urban areas - indoor/outdoor relationships. *Atmos. Environ.*, 31, 529-544.

Bayer, C. W. & Black, M. S. (1987) Thermal desorption/gas chromatographic/mass spectrometric analysis of volatile organic compounds in the offices of smokers and nonsmokers. *Biomed. Environ. Mass Spectrom.*, 14, 363-367.

Benton, G., Miller, D. P., Reimold, M., & Sisson, R. (1981) A study of occupant exposure to particulates and gases from woodstoves in homes. *Proceedings, 1981 International Conference on Residential Solid Fuels-Environmental Impacts and Solutions*, Cooper, J. A. & Malek, D., eds., Portland, OR, June 1-4, 1981.

Bergman, T. A., Johnson, D. L., Boatright, D. T., Smallwood, K. G., & Rando, R. J. (1996) "Occupational Exposure of Nonsmoking Nightclub Musicians to Environmental Tobacco Smoke." *J. Am. Ind. Hyg. Assn.*, 57: 746–752

Brauer, M. & 't Mannetje, A. (1998) Restaurant smoking restrictions and environmental tobacco smoke exposure. *Am. J. Public Health*, 88, 1834-1836.

Brickus, L. S., Cardosa, J. N., & De Aquino Neto, F. R. (1998) Distributions of indoor and outdoor air pollutants in Rio de Janeiro, Brazil, implications to indoor air quality in bayside offices. *Environ. Sci. Technol.*, 32, 3485-3490.

Brown, V. M & Crump, D. R. (1998) The use of diffusive samplers for the measurement of volatile organic compounds in the indoor air of 44 homes in Southhampton. *Indoor and Built Environ.*, 7, 245-253.

Brunekreef, B. & Boleij, J. S. M. (1982) Long-term average suspended particulate concentrations in smokers' homes. *Int. Arch. Occup. Environ. Health*, 50, 299-302.

Brunnemann, K. D., Kagan, M. R., Cox, J. E., & Hoffmann, D. (1990) Analysis of 1,3-butadiene and other selected gas-phase components in cigarette mainstream and sidestream smoke by gas chromatography–mass selective detection. *Carcinogenesis*, 11(10), 1863-1868.

Brunnemann, K. D., Haley, N. J., Adams, J. D., Sepkovic, D. W., & Hoffmann, D. (1989) Model studies on the uptake of nicotine after exposure to environmental tobacco smoke (ETS). *41st Tobacco Chemists' Research Conference*, Greensboro, NC.

Burton, R. M., Scila, R. A., Wilson, W. E., Pahl, D. A., Mumford, J. L. & Koutrakis, P. (1990) Characterization of kerosene heater emissions inside two mobile homes. *Proceedings of the 5th International Conference on Indoor Air Quality and Climate, Vol. 2*, Toronto, Canada, pp. 337-342.

Carson, J. R. & Erickson, C. A. (1988) Results from survey of environmental tobacco smoke in offices in Ottawa, Ontario. *Environ. Technol. Letts.*, 9, 501-508.

Chappel, S. B. & Parker, R. J. (1977) Smoking and carbon monoxide levels in enclosed public places in New Brunswick. *Can. J. Public Health*, 68, 159-161.

Chuang, J.C., Callahan, P.J., Lyu, C.W., & Wilson, N.K. (1999) Polycyclic aromatic hydrocarbon exposures of children in low-income families. *J. Expo. Anal. Environ. Epidemiol.* 9, 85–98.

Collett, C. W., Ross, J. A., Levine, K. B.(1992) "Nicotine, RSP, and CO_2 Levels in Bars and Nightclubs," Environ. Int.18: 347-352.

Colome, S. D., Kado, N. Y., Jacques, P., & Kleinman, M. (1990) Indoor-outdoor relationship of particles less than 10 μm in aerodynamic diameter (PM10) in homes of asthmatics, *Proceeding of the 5th International Conference on Indoor Air Quality and Climate, Vol. 2*, Toronto, Canada, pp. 275-280.

Colome, S. D. & Spengler, J. D. (1981) Residential indoor and matched outdoor pollutant measurements with special consideration of wood-burning homes. *Proceedings, 1981 International Conference on Residential Solid Fuels-Environmental Impacts and Solutions*, Cooper, J. A. & Malek, D., eds., Portland, OR, pp. 435-455.

Conner, J. M., Oldaker, G. B., III, & Murphy, J. J. (1990) Method for assessing the contribution of environmental tobacco smoke to respirable suspended particles in indoor environments. *Environ. Technol.*, 11, 189-196.

Coultas, D. B., Samet, J. M., McCarthy, J. F., & Spengler, J. D. (1990) A personal monitoring study to assess workplace exposure to environmental tobacco smoke. *Am. J. Public Health*, 80, 988-990.

Crouse, W. E., Ireland, M. S., Johnson, J. M., Striegel, R. M., Jr., Williard, C. S., DePinto, R. M., Oldaker, G. B., III, & McBride, R. L. (1988) Results from a survey of environmental tobacco smoke (ETS) in restaurants. *Transactions of an International Specialty Conference Combustion Processes and the Quality of the Indoor Environment*, Harper, J. P., ed., Niagara Falls, NY, pp. 214-222.

Crouse, W. E. & Oldaker, G. B. (1990) Comparison of area and personal sampling methods for determining nicotine in environmental tobacco smoke. *1990 EPA/AWMA Conference on Toxic and Related Air Pollutants*, Raleigh, NC.

Crouse, W. E. & Carson, J. R. (1989) Surveys of environmental tobacco smoke (ETS) in Washington, D.C. offices and restaurants. *43rd Tobacco Chemists' Research Conference*, Richmond, VA.

Crouse, W. E., Ireland, M. S., Striegel, R. M., Jr., & Williard, C. S. (1989) Results from a survey of nicotine in supermarkets. *EPA/AWMA International Symposium Measurement of Toxic and Related Air Pollutants*, Raleigh, NC, pp. 583-589.

Cuddeback, J. E., Donovan, J. R., & Burg, W. R. (1976) Occupational aspects of passive smoking. *Am. Ind. Hyg. Assoc. J.*, 263-267.

De Bortoli, M., Knoppel, H., Pecchio, E., Peil, A., Rogora, L., Schauenburg, H., Schlitt, H., & Vissers, H. (1985) Measurements of indoor air quality and comparison with ambient air. A study on 15 homes in Northern Italy. EUR 9656 EN, Commission of the European Communities.

Department of Transportation (1971) *Health Aspects of Smoking in Transport Aircraft*, Washington, D.C., US National Technical Information Service.

Department of Transportation (1973) *Carbon monoxide as an indicator of cigarette-caused pollution levels in intercity buses.*

Eatough, D. J., Benner, C. L., Tang, H., Landon, V., Richards, G., Caka, F. M., Crawford, J., Lewis, E. A., Hansen, L. D., & Eatough, N. L. (1989a) The chemical composition of environmental tobacco smoke III. Identification of conservative tracers of environmental tobacco smoke. *Environ. Int.*, 15, pp. 19-28.

Eatough, D. J., Hansen, L. D. & Lewis, E. A. (1988) Assessing exposure to environmental tobacco smoke. In: Perry, R. & Kirk, P. W. W., eds., *Indoor and Ambient Air Quality*, London, Selper Ltd., pp. 131-140.

Elliott, L. P. & Rowe, D. R. (1975) Air quality during public gatherings. *JAPCA*, 25(6), 635-636.

First, M. W. (1984) 1.1 Environmental Levels: Environmental tobacco smoke measurements: retrospect and prospect. In: Rylander, R., Peterson, Y., & Snella, M.-C., eds., *ETS–Environmental Tobacco Smoke. Report from a Workshop on Effects and Exposure Levels*, pp. 9-16, *Eur. J. of Resp. Dis. Supp.*, No. 133, Vol. 65.

First, M. W. (1983) Air sampling and analysis for contaminants in workplaces. In: Lioy, P. J. & Lioy, M. J. Y., eds., *Air Sampling Instruments for Evaluation of Atmospheric Contaminants, 6th Edition*, Cincinnati, OH, pp. A2-A13.

Fischer, T., Weber, A., & Grandjean, E. (1978) Air pollution due to tobacco smoke in restaurants. *Int. Arch. Occup. Environ. Health*, 41, 267-280.

Galuskinova, V. (1964) 3,4-Benzopyrene determination in the smoking atmosphere of social meeting rooms and restaurants. A contribution to the problem of the noxiousness of so-called passive smoking. *Neoplasma*, 11(5), 465-468.

Georghiou, P. E., Blagden, P. A., Snow, D. A., Winsor, L., & Williams, D. T. (1989) Air levels and mutagenicity of PM-10 in an indoor ice arena. *JAPCA*, 39(12), 1583-1585.

Gilli, G., Scursatone, E., & Bono, R. (1996) Geographical distribution of benzene in air in Northwestern Italy and personal exposure. *Environ. Health Perspec.* 104(Supplement 6), 1137-1140.

Gold, K. W., Naugle, D. F., and Berry, M. A. (1990) *Indoor Concentrations of Environmental Carcinogens*, Research Triangle Institute, Report Number 4479/07-F, P. O. Box 12194, Research Triangle Park, NC 27709-2194, April 1990, 40 pp.

Griest, W. H., Jenkins, R. A., Tomkins, B. A., Moneyhun, J. H., Ilgner, R. H., Gayle, T. M., Higgins, C. E., & Guerin, M. R. (1988) *Sampling and Analysis of Diesel Engine Exhaust and the Motor Pool Workplace Atmosphere, Final Report*. ORNL/TM-10689.

Grimmer, G. & Naujack, K.-W. (1987) Gas chromatographic determination of polycyclic aromatic hydrocarbons in sidestream and mainstream smoke and the air of enclosed environments. In: O'Neill, I. K., Brunnemann, K. D., Dodet, B., & Hoffmanna, D., eds., *Enviornmental Carcinogens–Methods of Analysis and Exposure Measurements, Vol. 9–Passive Smoking* (IARC Scientific Publications No. 810, International Agency for Research on Cancer, Lyon, pp. 249-268.

Grimmer, G., Böhnke, H., & Harke, H.-P. (1977) Passive smoking: measuring of concentrations of polycyclic aromatic hydrocarbons in rooms after machine smoking of cigarettes. *Int. Arch. Occup. Environ. Health*, 40, 83-92.

Grimsrud, D. T., Turk, B. H., Prill, R. J., & Geisling-Sobotka, K. L. (1990) Pollutant concentrations in commercial buildings in the U.S. pacific northwest. *Proceedings of the 5ᵗʰ International Conference on Indoor Air Quality and Climate, Vol. 2* Toronto, Canada, pp. 483-488.

Guenther, F. R., Chesler, S. N., Gordon, G. E. & Zoller, W. H. (1988) Residential wood combustion: a source of atmospheric polycyclic aromatic hydrocarbons. *JHRCC&C*, 761-766.

Guerin, M. R. (1988) Formation and general characteristics of environmental tobacco smoke. *APCA Specialty Conference on Combustion Processes and the Quality of the Indoor Environment*, Niagara Falls, New York.

Guerin, M. R. (1991) Environmental tobacco smoke. In: Hansen, L. D. & Eatough, D. J., eds., *Organic Chemistry of the Atmosphere*, CRC Press, Boca Raton, FL, pp. 79-119.

Gundel, L. A., Daisey, J. M., & Offermann, F. J. (1990) Development of an indoor sampling and analysis method for particulate polycyclic aromatic hydrocarbons. *Proceedings of the 5ᵗʰ International Conference on Indoor Air Quality and Climate, Vol. 2*, Toronto, Canada, pp. 299-304.

Hammond, S. K., Coghlin, J., & Leaderer, B. P. (1987) Field study of passive smoking exposure with passive sampler. *Procs., 4ᵗʰ International Conference Indoor Air Quality and Climate, Vol. 2*, West Berlin, pp. 131-136.

Hammond, S. K. & Leaderer, B. P. (1987) A diffusion monitor to measure exposure to passive smoking. *Environ. Sci. Technol.*, 21, 494-497.

Hammond, S. K., Leaderer, B. P., Roche, A. C., & Schenker, M. (1987) Collection and analysis of nicotine as a marker for environmental tobacco smoke. *Atmos. Environ.*, 21(2), 457-462.

Hammond S. K., Sorenson G., Youngstrom, R., Ockene, J. K. Occupational Exposure to Environmental Tobacco Smoke. *JAMA* 1995;274, 956-960.

Harke, H. P. & Peters, H. (1974) The problem of passive smoking. III. The influence of smoking on the CO concentration in driving automobiles. *Int. Arch. Arbeitsmed*, 33, 221-229.

Harke, H. P. (1974) The problem of passive smoking. I. The influence of smoking on the CO concentration of office rooms. *Int. Arch. Arbeitsmed.*, 33, 199-206.

Harmsden, H. & Effenberger, E. (1957) Tabakrauch in verkehrsmitteln, wohn-und arbeitsraumen [Tobacco smoke in transportation vehicle, living, and working rooms]. *Archiv fur Hygiene und Bakteriologie* 14, 383-400.

Hawthorne, A. R., Gammage, R. B., Dudney, C. S., Hingerty, B. E., Schuresko, D. D., Parzyck, D. C., Womack, D. R., Morris, S. A., Westley, R. R., White, D. A., & Schrimsher, J. M. (1984) *An indoor air quality study of forty east Tennessee homes*. ORNL-5965.

Healthy Buildings International (1990) *Measurement of Environmental Tobacco Smoke in General Office Areas*.

Heavner, D. L., Morgan, W. T., & Ogden, M. W. (1995) Determination of volatile organic compounds and ETS apportionment in 49 homes. *Environ. Int.*, 21, 3-21.

Heavner, D. L., Morgan, W. T., & Ogden, M. W. (1996) Determination of volatile organic compounds and respirable suspended particulate matter in New Jersey and Pennsylvania homes and workplaces. *Environ. Int.*, 22, 159-183.

Hedge, A., Erickson, W. A., & Rubin, G. (1994) The effects of alternative smoking policies on indoor air quality in 27 smoking office buildings. *Ann. Occup. Hyg.* , 38, 265-278.

Henderson, F. W., Reid, H. F., Morris, R., Wang, O.-L., Hu, P. C., Helms, R. W., Forehand, L., Mumford, J., Lewtas, J., Haley, N. J., & Hammond, S. K. (1989) Home air nicotine levels and urinary cotinine excretion in preschool children. *Am. Rev. Respir. Dis.*, 140, 197-201.

Herning, R. I., Jones, R. T., Bachmann, J., & and Mines, A. H. (1981) Puff volume increases when low-nicotine cigarettes are smoked. *Brit. J. Med.*, 283, 187-189.

Heyder, J., Gebhard, J. & Stahlhofen, W. (1979) Inhalation of aerosols: particle deposition and retention. In: Willeke, K., ed., *Generation of Aerosols and Facilities for Exposure Experiments*, Ann Arbor Science, Publishers, pp. 65-103.

Higgins, C. E., Thompson, C. V., Ilgner, R. H., Jenkins, R. A., & Guerin, M. R. (1990) Determination of vapor phase hydrocarbons and nitrogen-containing constituents in environmental tobacco smoke. Internal Progress Report, Analytical Chemistry Division, Oak Ridge National Laboratory, Oak Ridge, TN 37831-6120.

Higgins, C. E., Jenkins, R. A., & Guerin, M. R. (1987) Organic vapor phase composition of sidestream and environmental tobacco smoke from cigarettes. *Proceedings of the 1987 EPA/APCA Symposium on Measurement of Toxic and Related Air Pollutants*, pp. 140-151.

Highsmith, V. R., & Rodes, C. E. (1988) Indoor particle concentrations associated with use of tap water in portable humidifiers. *Environ. Sci. Technol.*, 22, 1109-1112.

Hinds, W. C. & First, M. W. (1975) Concentrations of nicotine and tobacco smoke in public places. *New England J. Med.*, 292(16), 844-845.

Hodgson, A. T., Daisey, J. M., Mahanama, K. R. R., Brinke, J. T., & Alevantis, L. E. (1996) Use of volatile tracers to determine the contribution of environmental tobacco smoke to concentrations of volatile organic compounds in smoking environments. *Environ. Int.*, 22, 295-307.

Hollowell, C. D., & Miksch, R. R. (1981) Sources and concentrations of organic compounds in indoor environments. *Bull. N.Y. Acad. Med.*, 57(10), 962-977.

Husgafvel-Pursiainen, K., Sorsa, M., Moller, M., & Benestad, C. (1986) Genotoxicity and polynuclear aromatic hydrocarbon analysis of environmental tobacco smoke samples from restaurants. *Mutagenesis*, 1, 287-291.

Jarvis, M. J., Russell, M. A. H., Feyerabend, C. (1983) Absorption of nicotine and carbon monoxide from passive smoking under natural conditions of exposure. *Thorax*, 38, 829-833.

Jenkins, R. A. (1986) Occurrence of selected metals in cigarette tobaccos and smoke. In: O'Neill, I. K., Schuller, P. & Fishbein, L., eds., *Environmental Carcinogens Selected Methods of Analysis, Vol. 8–Some Metals: As, Be, Cd, Cr, Ni, Pb, Se, Zn (IARC Scientific Publications No. 71)*,, International Agency for Research on Cancer, Lyon, pp. 129-138.

Jenkins, R. A., Moody, R. L., Higgins, C. E., & Moneyhun, J. H. (1991) Nicotine in environmental tobacco smoke (ETS): comparison of mobile personal and stationary area sampling. *Proceedings of the EPA/AWMA Conference on Measurement of Toxic and Related Air Pollutants*, Durham, NC.

Jenkins, R. A., Holmberg, R. W., Wike, J. S., Moneyhun, J. H., & Brazell, R. S. (1984) *Chemical Characterization and Toxicologic Evaluation of Airborne Mixtures–Chemical and Physical Characterization of Diesel Fuel Smoke*. ORNL/TM-9196, June, 1984.

Jenkins, R. A., Palausky, A., Counts, R. W., Bayne, C. K., Dindal, A. B., & Guerin, M. R. (1996) Exposure to environmental tobacco smoke in sixteen cities in the United States as determined by personal breathing zone air sampling. *J. Expo. Anal. Environ. Epidemiol.*, 6, 473-502.

Jenkins, R. A. & Counts, R. W. (1999) Occupational exposure to environmental tobacco smoke, results of two personal exposure studies. *Environ. Health Perspec.*, 107(Supplement 2), 341-348.

Jo, W. K. & Moon, K. C. (1999) Housewives' exposure to volatile organic compounds relative to proximity to roadside service stations. *Atmos. Environ.*, 33, 2921-2928.

Jo, W. K. & Park, K. H. (1999) Concentrations of volatile organic compounds in the passenger side and the back seat of automobiles. *J. Exp. Anal. Environ. Epidemiol.*, 9, 217-227.

Just, J., Borkowska, M., & Maziarka, S. (1972) Zanieczyszczenie dymem tytoniowym powietrza kawiarn warszawskich (Tobacco smoke in the air of Warsaw coffee rooms.) *Rocz. Panestw. Zakh. Hyg.*, 23, 135-192.

Kirk, P. W. W., Hunter, M., Baek, S. O., Lester, J. N., & Perry, R. (1988) Environmental tobacco smoke in indoor air. In: Perry, R. & Kirk, P. W. W., eds., *Indoor and Ambient Air Quality*, London, Selper, Ltd., pp. 99-112.

Klepeis, N. E., Ott, W. R., & Switzer, P. (1996) A multiple-smoker model for predicting indoor air quality in public lounges. *Environ. Sci Technol.*, 30, 2813–2820.

Klus, H., Begutter, H., Ball, M., & Intorp, M. (1987a) Environmental tobacco smoke in real life situations. In: Siefert, B. et al., eds, *Poster Handout from the 4th International Conference on Indoor Air Quality and Climate, Vol. 2*, Berlin (West), pp. 137-141.

Klus, H., Begutter, H., Ball, M., & Intorp, M. (1987) Environmental tobacco smoke in real life situations. *Proceedings of the 4th International Conference on Indoor Air Quality and Climate, Vol. 2*, Berlin (West), pp. 137-141.

Klus, H., Begutter, H., Nowak, A., Pinterits, G., Ultsch, I., & Wihlidal, H. (1985) Indoor air pollution due to tobacco smoke under real conditions. Preliminary results. *Tokai J. Exp. Clin. Med.*, 10(4), 331-340.

Lambert, W. E., Sarnet, J. M., & Spengler, J. D. (1993) Environmental tobacco smoke concentrations in no-smoking and smoking sections of restaurants. *Am. J. of Public Health,* 83(9), 1339-1341.

Lawryk, N. J. Lioy, P. J., & Weisel, C. P. (1995) Exposure to volatile organic compounds in the passenger compartment of automobiles during periods of normal and malfunctioning operation. *J. Expo. Anal. & Environ. Epidem.*, 5, 511-531.

Leaderer, B. P. (1986) As cited in *Environmental Tobacco Smoke, Measuring Exposures and Assessing Health Effects*, National Research Council, National Academy Press, Washington, D.C., p. 72.

Leaderer, B. P. & Hammond, S. K. (1991) Evaluation of vapor-phase nicotine and respirable suspended particle mass as markers for environmental tobacco smoke. *Environ. Sci. Technol.*, 25, 770-777.

Lebret, E., Boleij, J., & Brunekreef, B. (1990) Environmental tobacco smoke in Dutch homes. *Proceedings of the 5th International Conference on Indoor Air Quality and Climate, Vol. 2*, Toronto, Canada, pp. 263-268.

Lebret, E., van de Wiel, H. J., Bos, H. P., Noij, D., & Boleij, J. S. M. (1984) Volatile hydrocarbons in Dutch homes. *3rd International Conference on Indoor Air Quality and Climate*, pp. 169-174.

Lee, S. C. & Chang, M. (1999) Indoor air quality investigations at five classrooms. *Indoor Air*, 9, 134-138.

Lemus, R., Abdelghani, A. A., Akers, T. G., & Horner, W. E. (1998) Potential health risks from exposure to indoor formaldehyde. *Rev. Environ. Health*, 13, 91-98.

Leung, P. L. & Harrison, R. M. (1998) Evaluation of personal exposure to monoaromatic hydrocarbons. *Occup. Environ. Med.*, 55, 249-257.

Lioy, P. J., Wallace, L., & Pellizzari, E. (1990) Indoor/outdoor and personal monitor relationships for selected volatile organic compounds measured at three homes during New Jersey teams–1987. *EPA/AWMA Symposium on Total Exposure Assessment Methodology*, Las Vegas, NV, pp. 17-37.

Löfroth, G., Burton, R. M., Forehand, L., Hammond, K. S., Seila, R. L., Zweidinger, R. B., & Lewtas, J. (1989). Characterization of environmental tobacco smoke. *Environ. Sci. Technol.*, 23, 610-614.

Mahajan, V. K., & Huber, G. L. (1990) Health effects of involuntary smoking: impact on tobacco use, smoking cessation, and public policies. *Seminars in Res. Med.*, 11(1), 87-114.

Malmfors, T., Thorburn, D., & Westlin, A. (1989) Air quality in passenger cabins of DC-9 and MD-80 aircraft. *Environ. Toxicol. Letts.*, 10, 613-628.

Matsushita, H. & Mori, T. (1984) Nitrogen dioxide and nitrosamine levels in indoor air and sidestream smoke of cigarettes. In: Berglund, B. et al. eds., *Indoor Air Vol. 2, Radon, Passive Smoking, Particulates and Housing Epidemiology*, Swedish Council for Building Research, Stockholm, pp. 335-340.

Matsushita, H., Tanabe, K., Koyano, M., Laquindanum, J., & Lim-Sylianco, C. Y. (1990) Automatic analysis for polynuclear aromatic hydrocarbons indoors and its application to human exposure assessment. *Proceedings of the 5th International Conference on Indoor Air Quality and Climate, Vol. 2*, Toronto, Canada, pp. 287-292.

Mattson, M. E., Boyd, G., Byar, D., Brown, C., Callahan, J. F., Corte, D. Cullen, J. W., Greenblatt, J., Haley, N. J., Hammond, S. K., Lewtas, J., & Reeves, W. (1989) Passive smoking on commercial airline flights. *JAMA*, 261(6), 867-871.

Miesner, E. A., Rudnick, S. N., Hu, F.-C., Spengler, J. D., Preller, L., Özkaynak, H., & Nelson, W. (1989) Particulate and nicotine sampling in public facilities and offices. *JAPCA*, 39(12), 1577-1582.

Mumford, J. L., Lewtas, J., Burton, R. M., Henderson, F. W., Forehand, L., Allison, J. C., & Hammond, S. K. (1989) Assessing environmental tobacco smoke exposure of pre-school children in homes by monitoring air particles, mutagenicity, and nicotine. *1989 EPA/AWMA International Symposium on Measurement of Toxic and Related Air Pollutants*, Raleigh, NC, pp. 606-616.

Mumford, J. L., Lewtas, J., Burton, R. M., Svendsgaard, D. V., Houk, V. S., Williams, R. W., Walsh, D. B., & Chuang, J. C. (1990) Unvented kerosene heater emissions in mobile homes: studies on indoor air particles, semivolatile organics, carbon monoxide, and mutagenicity. *Proceedings of the 5th International Conference on Indoor Air Quality and Climate, Vol. 2*, Toronto, Canada, pp. 257-262.

Muramatsu, M., Umemura, S., Okada, T., & Tomita, H. (1984) Estimation of personal exposure to tobacco smoke with a newly developed nicotine personal monitor. *Environ. Res.*, 35, 218-227.

Muramatsu, M., Umemura, S., Fukui, J., Arai, T., & Kira, S. (1987) Estimation of personal exposure to ambient nicotine in daily environment. *Int. Arch. Occup. Environ. Health*, 59, 545-550.

Nagda, N. L., Koontz, M. D., Konheim, A. G., & Hammond, S. K. (1992) Measurement of cabin air quality aboard commercial airlines. *Atmos. Environ.*, 26A, 2203-2310.

Nagda, N., Fortmann, R., Koontz, M. & Konheim, A. (1990) Investigation of cabin air quality aboard commercial airliners. *Proceedings of the 5th International Conference on Indoor Air Quality and Climate, Vol. 2*, Toronto, Canada, pp. 245-250.

Neas, L. M., Dockery, D. W., Ware, J. H., Spengler, J. D., Ferris, B. G., Jr., & Speizer, F. E. (1994) Concentration of indoor particulate matter as a determinant of respiratory health in children. *Am. J. Epidemiol.*, 139, 1088-1099.

Nelson, W. D., Wallace, L. A., Thomas, K. W., & Pellizzari, E. D. (1989) New Jersey team study: impact of attached garage use on aromatic levels in indoor air, personal exposure, and breath. *Proceedings of the 1989 EPA/A&WMA International Symposium Measurement of Toxic and Related Air Pollutants*, Raleigh, NC, pp. 419-427.

Nguyen, V. H. & Martel, J.-G. (1987) A field study in three office towers in Quebec, Canada. In: Seifert, B., Esdorn, H., Fischer, M., Ruden, H., & Wegner, J., eds., *Proceedings of the 4th International Conference on Indoor Air Quality and Climate, Vol. 2*, Berlin, pp. 512-514.

Nguyen, V. H. & Goyer, N. (1988) A global approach to investigate indoor air quality and performance of the ventilation system. In: Perry, R. & Kirk P. W. W., eds., *Indoor and Ambient Air Quality*, London, Selper, Ltd., pp. 327-332.

Nitschke, I. A., Clark, W. A., Clarkin, M. E., Traynor, G. W., & Wadach, J. B. (1985) *Indoor Air Quality, Infiltration, and Ventilation in Residential Buildings*, NYSERDA 85-10, Albany New York State Energy Research and Development Authority.

Ogden, M. W. Maiolo, K. C., Oldaker, G. B., III, & Conrad, F. W., Jr. (1990) Evaluation of methods for estimating the contribution of ETS to respirable suspended particles. *Proceedings of the 5th International Conference on Indoor Air Quality and Climate, Vol. 2*, Toronto, Canada, pp. 415-420.

Ogden, M. W., Nystrom, C. W., Oldaker, G. B., III, & Conrad, F. W., Jr. (1989) Evaluation of a personal passive sampling device for determining exposure to nicotine in environmental tobacco smoke. *EPA/AWMA International Symposium on Measurement of Toxic and Related Air Pollutants*, Raleigh, NC.

Ogden M. (1996) Environmental Tobacco Smoke Exposure of Smokers Relative to Nonsmokers. *Anal. Commun.* 33, 197-198.

Ogden, M. W. & Maiolo, K. C. (1989) Collection and determination of solanesol as a tracer of environmental tobacco smoke in indoor air. *Environ. Sci. Technol.*, 23(9), 1148-1154.

Ogden, M. W., Eudy, L. W., Heavner, D. L., Conrad, F. W., Jr., & Green, C. R. (1989) Improved gas chromatographic determination of nicotine in environmental tobacco smoke. *Analyst*, 114, 1005-1008.

Oldaker, G. B., III & Conrad, F. W., Jr. (1989) Results from measurements of nicotine in a tavern. *EPA/AWMA International Symposium on Measurement of Toxic and Related Air Pollutants*, Raleigh, NC, pp. 577-582.

Oldaker, G. B., III & Conrad, F. W., Jr. (1987) Estimation of effect of environmental tobacco smoke on air quality within passenger cabins of commercial aircraft. *Environ. Sci. Technol.*, 21, 994-999.

Oldaker, G. B., III & McBride, R. L. (1988) Portable air sampling system for surveying levels of environmental tobacco smoke in public places. *Symposium on Environment and Heritage, World Environment Day Hong Kong, 1988,* Hong Kong University, Hong Kong, June 6, 1988.

Oldaker, G. B., III, Stancill, Conrad, F. W., Jr., Morgan, W. T., Collie, B. B., Fenner, R. A., Lephardt, J. O., Baker, P. G., Lyons-Hart, J., & Parrish, M. E. (1991) Results from a survey of environmental tobacco smoke in Hong Kong restaurants. *Unpublished data.* Personal communication.

Oldaker, G. B., III, Ogden, M. W., Maiolo, K. C., Conner, J. M., Conrad, F. W., & DeLuca, P. O. (1990) Results from surveys of environmental tobacco smoke in restaurants in Winston-Salem, North Carolina. *Proceedings of the 5th International Conference on Indoor Air Quality and Climate, Vol. 2*, Toronto, Canada, pp. 281-285.

Oldaker, G. B., Perfetti, P. F., Conrad, F. C., Jr., Conner, J. M., & McBride, R. L. (1990) Results of surveys of environmental tobacco smoke in offices and restaurants. *Int. Arch. Occup. Environ. Health*, 99-104.

Oldaker, G.B. III, Ogden M.W., Maiolo K.C. Conner J.M., Conrad F.W., Jr., Stancill M.W., & DeLuca P. (1989) Results from surveys of environmental tobacco smoke in restaurants in Winston-Salem, NC. *43rd Tobacco Chemists' Research Conference*, Richmond VA.

Oldaker GB, Taylor WD, Parrish KB. (1995) Investigations of Ventilation Rate, Smoking Activity and Indoor Air Quality at Four Large Office Buildings. *Environ. Technol.* 16, 173-180.

Oldaker, G. B., III, Stancill, M. W., Conrad, F. W., Jr., Colie, B. B., Fenner, R. A., Lephardt, J. O., Baker, P. G., Lyons-Hart, J., & Parrish, M. E. (1990) Estimation of effect of environmental tobacco smoke on air quality within passenger cabins of commercial aircrafts. II. In: Lunau, F. & Reynolds, G. L., eds., *Indoor Air Quality and Ventilation*, Selper Ltd., London, pp. 447-454.

Ott, W. R., Switzer, P., & Robinson, J. (1996) Particle concentrations inside a tavern before and after prohibition of smoking: evaluating the performance of an indoor air quality model. *J. Air & Waste Manage. Assoc.*, 46, 1120–1134.

Parker, G. B., Wilfert, G. L., & Dennis, G. W. (1984) Indoor air quality and infiltration in multifamily naval housing. *Annual PNWIS/APCA Conference, November 12-14,1984*, Portland, OR, pp. 1-14.

Patrianakos, C. & Hoffmann, D. (1979) Chemical studies on tobacco smoke. LXIV: on the analysis of aromatic amines in cigarette smoke. *J. Anal. Toxicol.*, 3, 150-154.

Perritt, R. L., Hartwell, T. D., Sheldon, L. S., Cox, B. G., Smith, M. L., & Rizzuto, J. E. (1990) Distribution of NO_2, CO, and respirable suspended particulates in New York state homes. *Proceedings of the 5th International Conference on Indoor Air Quality and Climate, Vol. 2*, Toronto, Canada, pp. 251-256.

Perry, J. (1973) Fasten your seat belts; no smoking. *Br. Columbia Med. J.*, 15, 304-305.

Persson, K.-A., Berg, S., Tornqvist, M., Scalia-Tomba, G.-P., & Ehrenberg, L. (1988) Note on ethene and other low-molecular weight hydrocarbons in environmental tobacco smoke. *Acta Chemica Scandinavica*, B42, 690-696.

Phillips, K., Howard, D. A., Browne, D., & Lewsley, J. M. (1994) Assessment of personal exposures to environmental tobacco smoke in British nonsmokers. *Environ. Int.*, 20, 693-712.

Phillips, K., Howard, D. A., Bentley, M. C., & Alvan, G. (1998) Assessment by personal monitoring of respirable suspended particles and environmental tobacco smoke exposure for non-smokers in Sydney, Australia. *Indoor Built Environ.*, 7, 188-203.

Phillips, K., Howard, D. A., Bentley, M., & Alvan, G. (1998) Measured exposures by personal monitoring for respirable suspended particles and environmental tobacco smoke of housewives and office workers resident in Bremen, Germany. *Int. Arch. Occup. Environ. Health*, 71, 201-212.

Phillips, K., Bentley, M. C., Howard, D. A., & Alvan, G. (1998) Assessment of air quality in Paris by personal monitoring of nonsmokers for respirable suspended particles and environmental tobacco smoke. *Environ. Int.*, 24, 05-425.

Phillips, K, Bentley, M. C., Howard, D., & Alvan, G. (1998) Assessment of environmental tobacco smoke and respirable suspended particle exposures for nonsmokers in Kuala Lumpur using personal monitoring. *J. Expo. Anal. Environ. Epidemiol.*, 8, 519-541.

Phillips, K., Howard, D. A., Bentley, M. C., & Alvan, G. (1998) Assessment of environmental tobacco smoke and respirable suspended particle exposures for nonsmokers in Hong Kong using personal monitoring. *Environ. Int.*, 24, 851-870.

Phillips, K., Bentley, M. C., Howard, D. A., Alvan, G., & Huici, A. (1997) Assessment of air quality in Barcelona by personal monitoring of nonsmokers for respirable suspended particles and environmental tobacco smoke. *Environ. Int.*, 23, 173-196.

Phillips, K., Bentley, M., Howard, D., & Alvan, G. (1998) Assessment of environmental tobacco smoke and respirable suspended particle exposures for nonsmokers in Prague using personal monitoring. *Int. Arch. Occup. Environ. Health*, 71, 379-390.

Phillips, K., Howard, D. A., Bentley, M. C., & Alvan, G. (1998) Assessment of environmental tobacco smoke and respirable suspended particle exposure of nonsmokers in Lisbon by personal monitoring. *Environ. Int.*, 24, 301-324.

Phillips, K, Howard, D. A., & Bentley, M. C.. (1997) Assessment of air quality in Turin by personal monitoring of nonsmokers for respirable suspended particles and environmental tobacco smoke. *Environ. Int.*, 23, 851-871.

Phillips, K., Howard, D. A., & Bentley, M. C. (1998) Exposure to tobacco smoke in Sydney, Kuala Lumpur, European, and Chinese cities. *Proceedings of the 14ʰ International Clean Air & Environment Conference, Melbourne, Australia, Oct. 1998*, pp. 347-352.

Phillips, K., Bentley, M. C., Howard, D. A., & Alvan, G. (1996) Assessment of air quality in Stockholm by personal monitoring of nonsmokers for respirable suspended particles and environmental tobacco smoke. *Scan. J. Work. Environ. Health*, 22, 1-24.

Phillips, K., Howard, D. A., Bentley, M., & Alvan, G. (1999) Assessment of environmental tobacco smoke and respirable suspended particle exposures for nonsmokers in Basel by personal monitoring. *Atmos. Env.*, 33, 1889-1904.

Phillips, K., Howard, D. A., Bentley, M. C., and Alvan, G. (1998). Environmental tobacco smoke and respirable suspended particle exposures for non-smokers in Beijing. *Indoor Built Environ.*, 7, 254–269.

Proctor, C. J. (1989) A comparison of the volatile organic compounds present in the air of real-world environments with and without environmental tobacco smoke. In: *Session 80–Evaluating Indoor Sources of Gas Phase Organic Compounds*, AWMA 82ⁿᵈ Annual Meeting, June 25-30, Anaheim, CA.

Proctor, C. J., Warren, N. D., Bevan, M. A. J., & Baker-Rogers, J. (1991) A comparison of methods of assessing exposure to environmental tobacco smoke in nonsmoking British women. *Environ. Int.*, 17, 287-297.

Proctor, C. J. (1985) A study of the atmosphere in London underground trains before and after the ban on smoking. *Toxicol. Letts.*, 35, 131-134.

Proctor, C. J., Warren, N. D., & Bevan, M. A. J. (1989) Measurements of environmental tobacco smoke in an air-conditioned office building. *Procs., Conference on the Present and Future of Indoor Air Quality*, Brussels. Poster P-1

Proctor, C. J., Warren, N. D., & Bevan, M. A. J. (1989) An investigation of the contribution of environmental tobacco smoke to the air in betting shops. *Environ. Technol. Letts.*, 10, 333-338.

Quackenboss, J. J., Lebowitz, M. D., & Crutchfield, C. D. (1989) Indoor-outdoor relationships for particulate matter: exposure classifications and health effects. *Environ. Int.*, 15, 353-360.

Quant, F. R., Nelson, P. A., & Sem, G. J. (1982) Experimental measurements of aerosol concentrations in offices. *Environ. Int.*, 8, 223-227.

Ramakrishna, J., Durgaprasad, M. B., & Smith, K. R. (1989) Cooking in India: the impact of improved stoves on indoor air quality. *Environ. Int.*, 15, 341-352.

Repace, J. L. (1982) Problem of passive smoking. *Bull. N. Y. Acad. Med.*, 57(10), 936-946.

Repace, J. L. (1987) Indoor concentrations of environmental tobacco smoke: field surveys. In: O'Neill, I. K., Brunnemann, K. D., Dodet, B., & Hoffmann, D., eds., *Environmental Carcinogens Methods of Analysis and Exposure Measurement (IARC Scientific Publications No. 81)*, International Agency for Research on Cancer, Lyon, pp. 141-162.

Repace, J. L. & Lowrey, A. H. (1980) Indoor air pollution, tobacco smoke, and public health. *Science*, 208, 464-472.

Repace, J. L. (1987) Indoor concentrations of environmental tobacco smoke: models dealing with effects of ventilation and room size. *Environmental Carcinogens: Selected Methods of Analysis, Vol. 9 (IARC Scientific Publications No. 81)*, International Agency for Research on Cancer, Lyon, pp. 25-41.

Repace, J. L. & Lowrey, A. H. (1982) Tobacco smoke, ventilation and indoor air quality. *ASHRAE Trans.*, 88, 894-914.

Rickert, W. S. (1995) Levels of ETS particulates, nicotine, solanesol, and carbonyls in a 'random' selection of homes in a midsize Canadian city. *Presented at the 49th Tobacco Chemists' Research Conference*, September 24-27, 1995, Lexington, KY USA

Ross, J. A., Sterling, E., Collett, C., & Kjono, N. E. (1996) Controlling environmental tobacco smoke in offices. *Heating/Piping/Air Conditioning,* May, 76-83.

Scherer, G., Ruppert, T., Daube, H., Kossien, I., Riedel, K., Tricker, A. R., et al. (1995) Contribution of tobacco smoke to environmental benzene exposure in Germany. *Environ Int.*, 21, 779-789.

Sebben, J., Pimm, P., & Shephard, R. J. (1977) Cigarette smoke in enclosed public facilities. *Arch. Environ. Health*, 32 (2) 53-58.

Sega, K., Fugas, M., & Kalinic, N. (1992) Indoor concentration levels of selected pollutants and household characteristics. *J. Exp. Anal. Environ. Epidem.*, 2, 477-485.

Sexton, K., Liu, K.-S., & Petreas, M. X. (1986) Formaldehyde concentrations inside private residences: a mail-out approach to indoor air monitoring. *JAPCA*, 36(6), 698-704.

Spengler, J. D. & Sexton, K. (1983). Indoor air pollution: A public health perspective. *Science*, 221(4605), 9-17.

Spengler, J. D., Dockery, D. W., Turner, W. A., Wolfson, J. M., & Ferris, B. J., Jr. (1981) Long-term measurements of respirable sulfates and particles inside and outside homes. *Atmos. Environ.*, 15, 23-30.

Spengler, J. D., Treltman, R. D., Tosteson, T. D., Mage, D. T., & Soczek, M. L. (1985) Personal exposures to respirable particulates and implications for air pollution epidemiology. *Environ. Sci. Technol.*, 19(8), 700-706.

Sterling, T. D., Collett, C. W., & Sterling, E. M. (1987) Environmental tobacco smoke and indoor air quality in modern office work environments. *J. Occup. Med.*, 29(1), 57-62.

Sterling, E. M., Collett, C. W., Kleven, S., & Arundel, A. (1988) Typical pollutant concentrations in public buildings. In: Perry, R. & Kirk, P. W. W., eds., *Indoor and Ambient Air Quality*, London, Selper, Ltd., pp. 399-404.

Sterling, T. (1988) ETS concentrations under different conditions of ventilation and smoking regulations. *Proceedings, Indoor Ambient Air Quality Conference*, London, pp. 89-98.

Sterling, T. D. & Sterling, E. M. (1984) Environmental tobacco smoke. 1.2. Investigations on the effect of regulating smoking on levels of indoor pollution and on the perception of health and comfort on office workers. *Eur. J. of Resp. Dis.*, 65(Supplement 133), 17-32.

Sterling, D. (1985) Volatile organic compounds in indoor air: an overview of sources, concentrations, and health effects. In: Gammage, R. B. & Kaye, S. V., eds., *Indoor Air and Human Health*, Lewis Publishers, Inc., Chelsea, MI, pp. 387-402.

Sterling, E. M., Collett, C. W., & Ross, J. A. (1997) The effectiveness of designated smoking areas in controlling non-smokers exposure to environmental tobacco smoke. *Indoor Built Environ.*, 6, 29-44.

Sterling, E. M., Collett, C. W., & Ross, J. A. (1996) Assessment of non-smokers' exposure to environmental tobacco smoke using personal-exposure and fixed location monitoring. *Indoor Built Environ.*, 5, 112-125.

Sterling, T. D. & Mueller, B. (1988) Concentrations of nicotine, RSP, CO and CO_2 in nonsmoking areas of offices ventilated by air recirculated from smoking designated areas. *Am. Ind. Hyg. Assoc. J.*, 49(9), 423-426.

Szadkowski, D., Harke, H. P., & Angerer, J. (1976) Burden of carbon monoxide from passive smoking in offices. *Int. Med.*, 310-313.

Thompson, C. V., Jenkins, R. A., & Higgins, C. E. (1989) A thermal desorption method for the determination of nicotine in indoor environments. *Environ. Sci. Technol.*, 23, 429-435.

Thurston, G. C. (1987) A field study of nitrogen oxide levels inside and outside an international airport terminal. In: Seifert, B., Esdorn, H., Fischer, M., Ruden, H., & Wegner, J., eds., *Proceedings of the 4th International Conference on Indoor Air Quality and Climate, Vol. 1*, pp. 451-455.

Triebig, G. & Zober, M. A. (1984). Indoor air pollution by smoke constituents–A survey. *Prev. Med.*, 13, 570-581.

Trout, D., Decker, J., Mueller, C., Bernert, J.T., and Pirkle, J. (1998) Exposure of casino employees to environmental tobacco smoke. *J. Occup. Environ. Med.*, 40(3) 270-276.

Turk, B. H., Brown, J. T., Geisling-Sobotka, K., Froehlich, D. A., Grimsrud, D. T., Harrison, J., Koonce, J. F., Prill, R. J., & Rezvan, K. L. (1987) *Indoor Air Quality and Ventilation Measurements in 38 Pacific Northwest Commercial Buildings*. Final Report to the Bonneville Power Administration. Indoor Environment Program, Applied Science Division, Lawrence Berkeley Laboratory, Berkeley, CA. LBL 22315.

Turk, B. H., Brown, J. T., Geisling-Sobotka, K., Froehlich, D. A., Grimsrud, D. T., Harrison, J., Koonce, J. F., Prill, R. J., & Revzan, K. L. (1987) *Indoor Air Quality and Ventilation Measurements in 38 Pacific Northwest Commercial Buildings, Vol. I–Measurement Results and Interpretation (LBL 22315 ½), Vol. II–Appendices (LBL 22315 2/2)*, Lawrence Berkeley Laboratory, Berkeley, CA 94720.

Turner, S., Cyr, L., & Gross, A. J. (1992) The measurement of environmental tobacco smoke in 585 office environments. *Environ. Int.*, 18, 19-28.

Turner, S. & Binnie, P. W. H. (1990) An indoor air quality survey of twenty-six Swiss office buildings. *Proceedings of the 5th International Conference on Indoor Air Quality and Climate, Vol. 4*, Toronto, Canada, pp. 27-32.

Valerio, F., Pala, M., Lazzarotto, A., & Balducci, D. (1997) Preliminary evaluation, using passive tubes, of carbon monoxide concentrations in outdoor and indoor air at street level shops in Genoa (Italy). *Atmos. Environ.*, 31, 2871-2876.

Van der Wal, J. F., Moons, A. M. M. & Cornelissen, H. J. M. (1990) The indoor air quality in renovated Dutch homes. *Proceedings of the 5th International Conference on Indoor Air Quality and Climate, Vol. 2*, Toronto, Canada, pp. 441-446.

Vaughan, W. M. & Hammond, S. K. (1990) Impact of "designated smoking area" policy on nicotine vapor and particle concentrations in a modern office building. *J. Air Waste Manage. Assoc.*, 40, 1012-1017.

Wallace, L. A. & Pellizzari, E. D. (1986) Personal air exposures and breath concentrations of benzene and other volatile hydrocarbons for smokers and nonsmokers. *Toxicol. Letts.*, 35, 113-116.

Wallace, L., Pellizzari, E., Hartwell, T. D., Perritt, K., & Ziegenfus, R. (1987) Exposures to benzene and other volatile compounds from active and passive smoking. *Arch. Environ. Health*, 42(5), 272-279.

Wallace, L., Pellizzari, E. D., & Wendel, C. (1990) Total organic concentrations in 2500 personal, indoor, and outdoor air samples collected in the USEPA TEAM studies. *Proceedings of the 5th International Conference on Indoor Air Quality and Climate, Vol. 2*, Toronto, Canada, pp. 639-644.

Wallace, L. A., Pellizzari, E. D, Hartwell, T. d., Sparacino, C., Whitmore, R., Sheldon, L., Zelon, H., & Perritt, R. (1987) The TEAM study: personal exposures to toxic substances in air, drinking water, and breath of 400 residents of New Jersey, North Carolina, and North Dakota. *Environ. Res.*, 43, 290-307.

Wallace, L. A., Pellizzari, E. D., Hartwell, T. D., Whitmore, R., Zelon, H., Perritt, R., & Sheldon, L. (1988) The California TEAM study: breath concentrations and personal exposures to 26 volatile compounds in air and drinking water of 188 residents of Los Angeles, Antioch, and Pittsburg, CA. *Atmos. Environ.*, 22(10), 2141-2163.

Wantke, F., Focke, M., Hemmer, W., Tschabitscher, M., Gann, M., Tappler, P., Gotz, M., & Jarisch, R. (1996) Formaldehyde and phenol exposure during an anatomy dissection course, a possible source of IgE-mediated sensitization? *Allergy*, 51, 837-841.

Weber, A. & Fischer, T. (1980). Passive smoking at work. *Int. Arch. Occup. Environ. Health*, 47, 209-221.

Weber, A., Fischer, T., & Grandjean, E. (1979) Passive smoking in experimental and field conditions. *Environ. Res.*, 20(1), 205-216.

Weber, A., Fischer, T. & Grandjean, E. (1979) Passive smoking: Irritating effects of the total smoke and the gas phase. *Int. Arch. Occup. Environ. Health*, 43, 183-193.

Weber, A., (1984) Annoyance and irritation by passive smoking. *Prev. Med.*, 13, 618-625.

Williams, R., Collier, A., & Lewtas, J. (1993) Environmental tobacco smoke exposure of young children as assessed using a passive diffusion device for nicotine. *Indoor Environ.*, 2, 98–104.

Wilson, N. K., Chuang, J. C., Kuhlman, M. R., & Mack, G. A. (1989) Measurement of polycyclic aromatic hydrocarbons and other semivolatile organic compounds in indoor air. *EPA/AWMA Symposium on Total Exposure Assessment Methodology, Las Vegas, 11/89*, Pittsburgh, PA.

Wilson, N. K., Chuang, J. C., & Kuhlman, M. R. (1990) Sampling semivolatile organic compounds in indoor air. *Proceedings of the 5th International Conference on Indoor Air Quality and Climate, Vol. 2*, Toronto, Canada, pp. 645-650.

Yocom, J. E. (1982) Indoor-outdoor air quality relationships: a critical review. *JAPCA*, 32(5), 500-520.

Zhao, B. C. & YuFeng, L. (1990) Characterization and situation of indoor coal-smoke pollution in endemic fluorosis areas. *Proceeding of the 5th International Conference on Indoor Air Quality and Climate, Vol. 1*, Toronto, Canada, pp. 98-102.

Zweidinger, R., Tejada, S., Highsmith, R., Westburg, H., & Gage, L. (1988) Distribution of volatile organic hydrocarbons and aldehydes during the IACP Boise, Idaho residential study. *Proceedings, 1988 EPA/APCA International Symposium on Measurement of Toxic and Related Air Pollutants*, Research Triangle Park, NC, pp. 814-819.

Index